Environmental Stresses in Soybean Production

Environmental Stresses in Soybean Production

Soybean Production
Volume 2

Edited by

Mohammad Miransari
AbtinBerkeh Scientific Ltd. Company, Isfahan, Iran

AMSTERDAM • BOSTON • HEIDELBERG • LONDON
NEW YORK • OXFORD • PARIS • SAN DIEGO
SAN FRANCISCO • SINGAPORE • SYDNEY • TOKYO
Academic Press is an imprint of Elsevier

Academic Press is an imprint of Elsevier
125 London Wall, London EC2Y 5AS, United Kingdom
525 B Street, Suite 1800, San Diego, CA 92101-4495, United States
50 Hampshire Street, 5th Floor, Cambridge, MA 02139, United States
The Boulevard, Langford Lane, Kidlington, Oxford OX5 1GB, United Kingdom

Notices

Knowledge and best practice in this field are constantly changing. As new research and experience broaden our understanding, changes in research methods, professional practices, or medical treatment may become necessary.

Practitioners and researchers must always rely on their own experience and knowledge in evaluating and using any information, methods, compounds, or experiments described herein. In using such information or methods they should be mindful of their own safety and the safety of others, including parties for whom they have a professional responsibility.

To the fullest extent of the law, neither the Publisher nor the authors, contributors, or editors, assume any liability for any injury and/or damage to persons or property as a matter of products liability, negligence or otherwise, or from any use or operation of any methods, products, instructions, or ideas contained in the material herein.

British Library Cataloguing-in-Publication Data
A catalogue record for this book is available from the British Library

Library of Congress Cataloging-in-Publication Data
A catalog record for this book is available from the Library of Congress

ISBN: 978-0-12-801535-3

For information on all Academic Press publications
visit our website at https://www.elsevier.com/

Working together
to grow libraries in
developing countries

www.elsevier.com • www.bookaid.org

Publisher: Nikki Levy
Acquisition Editor: Nancy Maragioglio
Editorial Project Manager: Billie Jean Fernandez
Production Project Manager: Julie-Ann Stansfield
Designer: Maria Inês Cruz

Typeset by TNQ Books and Journals

Dedication

This book is dedicated to my parents, my wife,
and my two children, who have always supported me.

Contents

List of Contributors

N. Ahmad National Institute for Biotechnology and Genetic Engineering, Pakistan Atomic Energy Commission, Faisalabad, Pakistan

N.K. Arora Babasaheb Bhimrao Ambedkar University, Lucknow, Uttar Pradesh, India

M. Fujita Kagawa University, Miki-cho, Japan

M. Hasanuzzaman Sher-e-Bangla Agricultural University, Dhaka, Bangladesh

M.S. Hossain Kagawa University, Miki-cho, Japan

M. Hussain National Institute for Biotechnology and Genetic Engineering, Pakistan Atomic Energy Commission, Faisalabad, Pakistan

J.A. Mahmud Sher-e-Bangla Agricultural University, Dhaka, Bangladesh; Kagawa University, Miki-cho, Japan

M. Miransari AbtinBerkeh Scientific Ltd. Company, Isfahan, Iran

K. Nahar Sher-e-Bangla Agricultural University, Dhaka, Bangladesh; Kagawa University, Miki-cho, Japan

A. Rahman Sher-e-Bangla Agricultural University, Dhaka, Bangladesh; Kagawa University, Miki-cho, Japan

M. Rahman National Institute for Biotechnology and Genetic Engineering, Pakistan Atomic Energy Commission, Faisalabad, Pakistan

G. Raza National Institute for Biotechnology and Genetic Engineering, Pakistan Atomic Energy Commission, Faisalabad, Pakistan

M. Rezvani Islamic Azad University (Qaemshahr Branch), Qaemshahr, Iran

S. Tewari Babasaheb Bhimrao Ambedkar University, Lucknow, Uttar Pradesh, India

F. Zaefarian Faculty of Crop Sciences, Sari Agricultural Sciences and Natural Resources University, Sari, Iran

Y. Zafar International Atomic Energy Agency, Vienna, Austria

Foreword

The important role of academicians and researchers is to feed the world's increasing publications. However, such contributions must be directed using suitable and useful resources. My decision to write this two-volume book, as well as other books and contributions, including research articles, has been mainly due to the duties I feel toward the people of the world. Hence I have tried to prepare references that can be of use at different levels of science. I have spent a significant part of my research life working on the important legume crop soybean (*Glycine max* (L.) Merr.) at McGill University, Canada. Before the other important research I conducted, with the help of my supervisor, Professor A.F. Mackenzie, was related to the dynamics of nitrogen in the soil and in the plants, including wheat (*Triticum aestivum* L.) and corn (*Zea mays* L.), across the great province of Quebec. These experiments resulted in a large set of data, with some interesting and applicable results. Such experiments were also greatly useful for my experiments on the responses of soybeans under stress. I conducted some useful, great, and interesting research with the help of my supervisor, Professor Donald Smith, on new techniques and strategies that can be used for soybean production under stress, both under field and greenhouse conditions. When I came to Iran, I continued my research on stress for wheat and corn plants at Tarbiat Modares University, with the help of my supervisors, Dr. H. Bahrami and Professor M.J. Malakouti, and my great friend Dr. F. Rejali from the Soil and Water Research Institute, Karaj, Iran, using some new, great, and applicable techniques and strategies. Such efforts have so far resulted in 60 international articles, 18 authored and edited textbooks, and 38 book chapters, published by some of the most prestigious world publishers, including Elsevier and the Academic Press. I hope that this two-volume contribution can be used by academicians and researchers across the globe. I would be happy to have your comments and opinions about this volume and Volume 1.

Dr. Mohammad Miransari
AbtinBerkeh Scientific Ltd. Company, Isfahan, Iran

Preface

The word "stress" refers to a deviation from natural conditions. A significant part of the world is subjected to stresses such as flooding, acidity, compaction, nutrient deficiency, etc. The important role of researchers and academicians is to find techniques, methods, and strategies that can alleviate such adverse effects on the growth of plants. The soybean [*Glycine max* (L.) Merr.] is an important legume crop that feeds a large number of people as a source of protein and oil. The soybean and its symbiotic bacteria, *Bradyrhizobium japonicum*, are not tolerant under stress. However, it is possible to use some techniques, methods, and strategies that may result in the enhanced tolerance of soybeans and *B. japonicum* under stress. Some of the most recent and related details have been presented in this volume.

In Chapter 1, with respect to the importance of the soybean as the most important legume crop, Miransari has presented the latest developments related to the use of biotechnological techniques for the production of tolerant soybean genotypes and rhizobium strains under stress.

In Chapter 2, due to the high rate of rainfall worldwide, Arora et al. have analyzed soybean response under flooding stress. They have also presented how it is possible to alleviate the stress, especially by using Plant Growth-Promoting Rhizobacteria (PGPR).

In Chapter 3, the use of tillage as an interesting method for the alleviation of stresses such as soil compaction, drought, acidity, and suboptimal root zone temperature has been presented by Miransari. The alleviating effects of tillage on stress improve the properties of soil, such as soil moisture, organic matter, biological activities, etc.

In Chapter 4, Hasanuzzaman et al. have reviewed different stresses including salinity, drought, high temperature, chilling, waterlogging, metal toxicity, pollutants, and ultraviolet radiation, decreasing soybean production worldwide. They accordingly presented the related techniques, which may be used for the alleviation of such stresses, increasing soybean yield production.

In Chapter 5, Rezvani has analyzed the effects of organic farming on the production of soybeans with an emphasis on managing weeds and producing healthy food.

In Chapter 6, the effects of plant hormones on the growth and yield of soybeans have been analyzed by Miransari. Plant hormones can affect soybean growth, including the process of nodulation under different conditions including stress. If the production of plant hormones in crop plants, including soybeans, is regulated, it is possible to produce tolerant soybean genotypes under stress and increase soybean yield production.

In Chapter 7, with respect to the importance of soybeans as a source of oil and protein, Miransari has presented the related details affecting soybean oil and protein under different conditions including stress. The other important aspect related to the role of proteins in soybean growth and yield production is the effect of protein signaling on the alleviation of stress in crop plants and has also been analyzed.

In Chapter 8, Miransari has presented the important effects of PGPR on the growth and yield of soybeans under different conditions including stress. The most recent advancements related to the mechanisms used by PGPR to enhance soybean growth and yield production have been reviewed and analyzed.

In Chapter 9, Rahman et al. have presented the use of genetics and genomics in the alleviation of stresses, including salinization, frequent drought periods, flooding, unusual fluctuations in temperature, and rainfall pattern and its frequency, resulting from climate change. The authors have accordingly indicated the molecular methods including the use of quantitative trait locus and genes, which can be used for the enhanced growth of soybeans under stress by modifying the related traits.

In Chapter 10, the effects of acidity on soybean growth and rhizobium activity, including the process of nodulation, have been presented by Miransari. The related alleviating methods, including the use of liming, tolerant soybean genotypes and rhizobium strains, and the use of the signal molecule genistein, have also been reviewed and analyzed.

In Chapter 11, Miransari has presented the adverse effects of compaction stress on the growth and yield of soybeans. The compaction of soil is a result of using agricultural machinery, especially at high-field moisture. The use of different techniques that can alleviate the stress, including reduced or nontillage, subsoiler, organic matter, and soil microbes, have been reviewed and analyzed.

In Chapter 12, Miransari has presented the effects of different nutrients on soybean production under different conditions including stress. The parameters affecting nutrient uptake by soybeans, including the properties of soil, soybean genotypes, and climatic conditions, have also been reviewed and analyzed.

Dr. Mohammad Miransari
AbtinBerkeh Scientific Ltd. Company, Isfahan, Iran

Acknowledgments

I would like to appreciate all of the authors for their contributions and wish them all the best for their future research and academic activities. My sincere appreciation and acknowledgments are also conveyed to the great editorial and production team at Elsevier, including Ms. Billie Jean Fernandez, the editorial project manager, Ms. Nancy Maragioglio, the senior acquisition editor, Ms. Julie-Ann Stansfield, the production project manager, Ms. Maria Inês Cruz, the designer, and TNQ Books and Journals, the typesetter, for being so helpful and friendly while writing, preparing, and producing this project.

Dr. Mohammad Miransari
AbtinBerkeh Scientific Ltd. Company, Isfahan, Iran

Use of Biotechnology in Soybean Production Under Environmental Stresses

1

M. Miransari
AbtinBerkeh Scientific Ltd. Company, Isfahan, Iran

Introduction

The production of food for the increasing world population is among the most important goals of research work, globally. However, the production of crop plants is facing some restriction worldwide, including the limitation of agricultural fields and the presence of environmental stresses. The soybean is an important source of food, protein, and oil and is able to develop a symbiotic association with the nitrogen (N)-fixing bacterium, *Bradyrhizobium japonicum*, to acquire most of its essential nitrogen for growth and yield production (Davet, 2004; Schulze, 2004).

Soybeans are the number one economic oil seed crop, and the processed soybeans are the major source of vegetable oil in the world. Soybeans also contain metabolites including saponins, isoflavone, phytic acid, goitrogens, oligosaccharides, and estrogens (Sakai and Kogiso, 2008; Ososki and Kennelly, 2003). Soybean products are used worldwide because of their benefits, such as decreasing cholesterol, controlling diabetes and obesity, cancer prevention, and improving kidney and bowel activities (Friedman and Brandon, 2001).

It has been indicated that only 10% of agricultural fields are not under stress, and the remaining parts of the world are subjected to stress. Accordingly, it is important to find methods, techniques, and strategies including biotechnology, which may result in the alleviation of stresses and increases in soybean growth and yield production under stress. Biotechnology is a tool contributing to sustainable agriculture; different biotechnological techniques can be used to increase plant resistance under stress. The biological techniques, which are used for the improvement of plant tolerance under stress, include the use of molecular breeding, tissue culture, mutagenesis, and transformation of genes. Some of the most important details related to the use of biotechnology on soybean growth and yield production are presented in the following sections.

Soybean, *Bradyrhizobium japonicum*, Stress, and Biotechnology

Soybean response under stress is indicated by the following equation: $Y = HI \times WUE \times T$ in which HI is harvest index, WUE is water use efficiency, and T is the rate of transpiration (Turner et al., 2001). If the reduction of water is controlled by plan, it results

Environmental Stresses in Soybean Production. http://dx.doi.org/10.1016/B978-0-12-801535-3.00001-2

in the increase of WUE. The following are traits in plant control T: leaf area, root depth and density, phenology, developmental plasticity, water potential, regulation of osmotic potential, heat tolerance, and sensitivity of photoperiod. Plant stress physiology is a useful tool for improving plant tolerance under stress. However, mimicking the field environment must be among the important views of future research for the development of tolerant crop plants under multistressful conditions (Chen and Zhu, 2004; Luo et al., 2005).

Under abiotic stresses, different cellular and genetic mechanisms are activated to make the plant tolerate the stress. However, because more details have yet to be indicated on plant response under stress, a more detailed understanding of plant physiology and molecular biology under stress for the successful transformation of plants is essential (Umezawa et al., 2002). For example, the use of genetic, mutagenic, and transgenic approaches has been really useful for a better understanding related to plant response under salinity stress and hence for the production of more tolerant plants under stress (Foolad, 2004). It was accordingly indicated that if a single gene, which controls the antiport protein of Na^+/H^+ vacuolar or plasma membrane is overexpressed in Arabidopsis and tomatoes, their tolerance under greenhouse salinity is enhanced (Zhang and Blumwald, 2001; Shi et al., 2003).

The important point about using biotechnological techniques, used for improving legume resistance under stress, is the large genome size of some legumes. However, to make the use of such techniques easier and investigate the process of nodulation and legume response under stress, the two legume models including *Lotus japonicus* and *Medicago trancatula* have been used. The properties, including a smaller genome size and diploid genomes, autogenously nature, generation at reasonable time, and production of seed in a prolific manner, make such legumes suitable choices as model legumes (Cook, 1999).

Ever since, effective genetic and genomic tools have been developed and used, including their genome sequencing, the isolation of sequence tags, and the establishment of a genetic map for each legume (Dita et al., 2006). The increasing data related to the genomic and genetics and the high genetic similarity between legumes make the two legume species suitable for genetic research under different conditions, including stress. However, most of the research related to stress has been conducted using *Arabidopsis* as a model plant. The similarities and differences between *Arabidopsis* and legumes are significant.

Research work has indicated that it is possible to develop tolerant legumes under salinity stress using genetic tools (Foolad, 2004; Bruning and Rozema, 2013). The other important parameters affecting the expression of genes under different conditions, including stress, are the transcription factors. If their activity is modified, it is possible to produce legumes, which are tolerant under stress (Shinozaki and Yamaguchi-Shinozaki, 2000). Ethylene-responsive element-binding factors are among the most interesting transcription factors, and over 60 of them have been indicated in *M. trancatula* and their closely related drought-responsive element-binding (DREB) and cyclic adenosine monophosphate responsive element binding proteins (Yamaguchi-Shinozaki and Shinozaki, 2005).

Such transcription factors are responsive to stresses such as cold, drought, wounding, and pathogen infection. The other important class of transcription factors is the

WRKY, which are able to modify the response of plant stress genes, including receptor protein kinases, the genes of cold and drought, and the basic leucine zipper domain regulating the activity of genes such as Glutathione STranferase and PR-1 (Chen and Singh, 1999; Yamaguchi-Shinozaki, and Shinozaki, 2005). A transcription-like factor, *Mtzpt2-1*, affecting plant tolerance under salinity stress, has also been found in *M. Trancatula* (Zhu, 2001; Dita et al., 2006).

In molecular breeding, the DNA regions, which are the cause of agronomical traits in crop plants (molecular marker), are determined and used for improving crop response under stress. Although tissue culture is a method to produce tissues by organogenesis and embryogenesis, it is not yet a suitable method for legumes as it is not an efficient method for the production of transgenic legume plants. In the method of mutagenesis, mutants with the favorite traits are produced and diversity is created, which is the main goal of breeding. Improving crop response using gene transfer by *Agrobacterium tumefaciens* is a reality, and it is now possible to produce transgenic legumes, although in some cases legume response may not be high. Using such a method, DNA is inserted into the embryogenic or organogenic cultures (Vasil, 1987; Buhr et al., 2002).

Proteomics is a useful tool for the evaluation of plant response under stress, because the levels of mRNA are not sometimes correlated with the accumulation of proteins. A large number of research work has investigated the behavior of plant proteins under stress (Lee et al., 2013; Hirsch, 2010). The other important reason for the use of proteomic in parallel with metabolome and transcriptome is for understating the complete details related to gene activity and molecular responses controlling complex plant behavior. Such an approach has been investigated in *M. trancatula* response to environmental stimuli and in the metabolic alteration, during the process of biological N fixation by *L. japonicus*.

The soybean is not a tolerant plant under environmental stresses. The WRKY type is among the transcription factors, which is able to regulate different plant activities such as plant growth and development. However, its effect on plant response under stress is not known. Accordingly, Zhou et al. (2008) investigated the effects of *GmWRKY* by identification of 64 related genes, which were activated under abiotic stress. The effects of three induced genes under stress, including *GmWRKY13*, *GmWRKY21,* and *GmWRKY54*, on the response of plant under stress were investigated using transgenic *Arabidopsis*.

Calnexin is a chaperone protein, localized in the endoplasmic reticulum, controlling the quality and folding of other proteins. The expression of calnexin in soybean seedlings was evaluated under osmotic stress using a protein fraction of total membrane by immunoblot analysis (Nouri et al., 2012). The concentration of protein increased in soybean seedlings during the early growth stage, under nonstressed conditions. However, when the 14-d old seedlings were treated with 10% polyethylene glycol, the expression of protein decreased. There was a similar response by soybeans under other types of stress including treating with ABA, salinity, drought, and cold, which was also correlated with a decrease in soybean root length. However, the concentration of calnexin did not change in rice under stress. In conclusion the authors indicated that under stress the concentration of calnexin significantly decreases in the developing roots of soybeans.

The autophagy-related genes (ATGs) may have important roles in plant development, starvation, and senescence. Plant hormones controlling the expression of starvation-related genes can also affect the activity of ATGs. Okuda et al. (2011) investigated the effects of starvation on the expression of ATGs and the ethylene-related genes in young soybean seedlings. The expression of *GmATG4* and *GmATG8i* genes increased in a starvation medium, but remained unchanged in a sucrose and nitrate medium. The authors also found that ethylene insensitive 3, which is a transcription of ethylene, increases in soybean seedlings when subjected to sever starvation stress. The authors accordingly made the conclusion that under starvation stress the expression of *GmATG8i* and ethylene-related genes are stimulated, indicating the role of ethylene under such a stress.

Although the major production sites of reactive oxygen species are the intercellular sites, it has also been illustrated that a plasma membrane NADPH oxidase and a diamine oxidase in the cell wall are also activated under drought and osmotic stresses, respectively (Luo et al., 2005; Jung et al., 2006). Interestingly, H_2O_2 induced the activity of catalase gene CAT1, under osmotic stress; H_2O_2 is also able to activate the pathway mitogen-activated protein kinase in *Arabidopsis* under osmotic stress (Shao et al., 2005a,b). Future research work is essential to realize the role of calcium signaling and the cross talk of reactive oxygen species, H_2O_2, and mitogen-activated protein kinase under osmotic stress.

Salinity

Salinity adversely affects plant growth and yield production and causes several physiological responses in plant. However, just a few plant species can tolerate the stress of salinity in the root environment for a long time. Just 1% of plants are able to tolerate the stress of salinity in their root medium, under the concentration of at least 200 mM NaCl, and complete their life cycle (Rozema and Flowers, 2008). The salinity issue has existed in agriculture for 1000 years (Jacobsen and Adams, 1958). Salinity is the source of different issues in different parts of the world, including arid and semiarid areas, and it is becoming more severe with the issue of climate change (Rozema and Flowers, 2008). The percent of the affected land by salinity is equal to 20 (FAO, 2008).

Soil salinity is the result of salt presence in the soil by natural and anthropogenic causes. If the salinity of a soil is more than 4 dS/m (equal to 40 mM NaCl), the soil is saline. Although most plants are sensitive to the salinity stress, some plant species are able to tolerate the stress of salinity. Using such tolerant plant species is useful for both reclaiming the salty fields and for the use of saline water for the irrigation of crop plants. This is especially significant in the areas with little water (Munns, 2002; Munns and Tester, 2008).

The rate of fresh water in the world is equal to 1% (>0.05% dissolved salt), the seawater is equal to 97% (<3% dissolved salt), and the other part is of intermediate salinity. A large part of fresh water is used by human beings (with a higher rate of demand related to the population growth) for especially agriculture (70%), industry (20%), and domestic (10%) use. Accordingly, the use of tolerant plants may be a useful solution to the issue of salinity and the scarceness of fresh water. Although a few species of crop plants can survive under saline conditions, more attempts are essential for the production of tolerant plant species under saline conditions. It is because salinity tolerance is a complex trait controlled by multiple genes, resulting in different mechanisms of salinity tolerance in plant. A promising method is domesticating the wild species of salt-tolerant plants (Bruning and Rozema, 2013).

The interesting point about using the seawater is the presence of micronutrients, which are essential for plant growth in the water. Accordingly, the need for the use of fertilization in the fields irrigated with salt water decreases. However, the rate of the most essential nutrient for plant growth, N, is little in the seawater. Legumes are among the sensitive plant species under salinity stress. The process of biological N fixation is more sensitive to the salinity stress than plant biomass production (Rozema and Flowers, 2008).

According to plant tolerance under salinity stress, plants are classified into four categories: (1) sensitive plants: 80% of biomass production at the salinity of 3 dS/m (30 mM NaCl) related to the control treatment, (2) moderately sensitive: 80% biomass production at the salinity of 6 dS/m (60 mM NaCl), (3) moderately tolerant: 80% biomass production at the salinity of 11 dS/m (110 mM NaCl), and (4) tolerant: 80% biomass production at the salinity of 16 dS/m (160 mM NaCl, 30% of sweater salinity). According to the graph, which relates plant salinity tolerance to salt concentration, less biomass reduction is resulted in tolerant plants with increasing the salt concentration, compared with the less tolerant plants (Munns and James, 2003).

Most legumes are sensitive or moderately sensitive under salinity stress (at the salinity of 3–6 dS/m, produce 80% of biomass related to nonstress conditions). Similar to the other crop plants, the morphology and physiology of legumes is adversely affected by salinity stress; however, the other important point is that the process of symbiosis is also affected by the stress, influencing plant growth and yield production. The process of symbiosis is more sensitive to the stress of salinity than the two symbionts by themselves (Bordeleau and Prévost, 1994).

Different stages of symbiotic association are affected by salinity, including the initiation of process and the production and functionality of nodules according to the following (Fig. 1.1): (1) formation of plant root hairs, regardless of the process of symbiosis; (2) the process of the signaling exchange between the two symbionts and the growth and activity of rhizobium; (3) curling of the root hairs, including the physical attachment of rhizobium to the root hairs, resulting in the subsequent alteration of root hair morphology and the production of infection thread; and (4) formation of nodules, including nodule number and weight, and nodule functionality (Swaraj and Bishnoi, 1999; Miransari and Smith, 2007).

Under salinity stress the growth of legumes and the number of root hairs decreases, and the morphology and physiology of root hairs is also adversely affected (Fig. 1.1). The adverse effects of salinity on the production of root nodules is also affecting the development of root hairs, as such a process is essential for the production of root nodules. Under stress a high number of rhizobium is essential for the successful inoculation of host roots, because usually under salinity stress the number of rhizobium decreases. However, it has been indicated that rhizobium is more tolerant under salinity stress than the host plant, and there is a significant interaction between rhizobium

Non-saline	Saline	Nodulation phase	Adverse effects of salinity (references)	Species and references
		Formation of root hairs	Decreased plant growth and the number of root hairs, root hairs deformed (Tu, 1981; Miransari and Smith, 2009)	Soybean (Tu, 1981; Miransari and Smith, 2009)
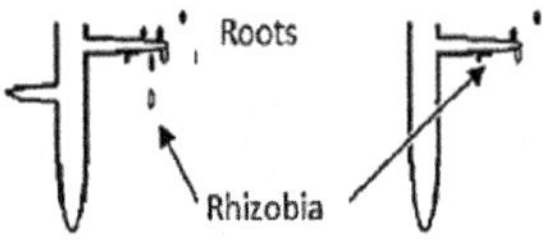 Roots / Rhizobia		Signa communication, Rhizobia growth and attachment to the roots	Decreased possibilities of signaling exchange between the symbionts, decreased rhizobial growth (Tu, 1981; Miransari and Smith, 2009)	Soybean (Tu, 1981; Singleton and Bohlool, 1984; Miransari and Smith, 2009)
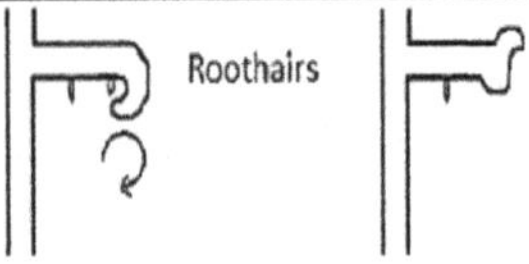 Roothairs		Curling of root hairs and formation of infection thread	Decreased curling, decreased expansion of root hairs, reduced formation of infection thread, unregulated release of bacteria from infection thread into the cells (Tu, 1981; Miransari et al., 2006; Miransari and Smith, 2009)	Soybean and faba bean (Tu, 1981; Schubert et al., 1990; Abd-Alla, 1992; Miransari and Smith, 2009)
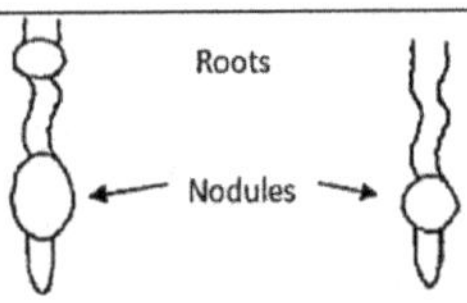 Roots / Nodules		Formation and functioning nodules (N fixation)	Decreased nodule number and nodule weight, increased nodule weight, alteration of nodule morphology, nodule necrosis, increased concentrations of proteins and sugars in nodules, high concentration of salt, decreased content of leghemoglobin, decreased activity of acetylene reduction (Yousef and Sprent, 1983; Singleton and Bohlool, 1984; Soussi et al., 1999; Miransari and Smith, 2009)	Chickpea, faba bean, soybean, pea and bean (Tu, 1981; Schubert et al., 1990; Abd-Alla, 1992; Soussi et al., 1998; 1999; Begum et al., 2001; Miransari and Smith, 2009)

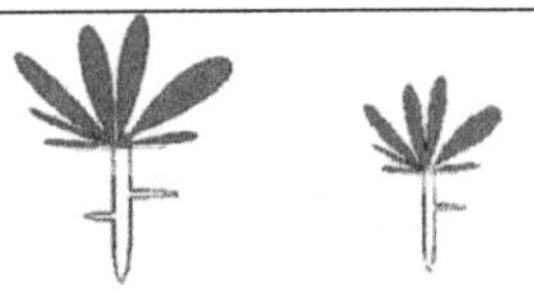

Figure 1.1 Different phases of nodulation and functioning and the effects of salinity (NaCl) (Bruning and Rozema, 2013).
With kind permission from Elsevier. License number 3771420931052.

and salinity stress. This indicates that the rhizobium species determines the bacterial tolerance under salinity stress (Zhang and Smith, 1995; Hashem et al., 1998).

Legumes fed with N fertilization are more tolerant under the stress of salinity compared with the legumes, which acquire their N by the process of biological N fixation. For example, the chickpea is among the legumes, which are the most sensitive under salinity stress, and some of the genotypes are not even able to tolerate the salinity stress at 25 mM under hydroponic conditions (Flowers et al., 2010a,b).

The understanding of mechanisms related to the salt tolerance of legumes have become easier with the progress in molecular techniques and the presence of two model legumes, *L. japonicus* and *M. truncatula*. Accordingly, the following may be used as the real approach for the improvement of legume tolerance under salinity stress: (1) finding the tolerant species among the wild species, (2) researching the salinity tolerance of different plant species, and (3) enhancing the salinity tolerance of plants using the genetic modification methods (Abshukor et al., 1988; Soussi et al., 1999).

According to Bruning and Rozema (2013), the first genetic improvement under salinity stress was done by Verdoy et al. (2006). They inserted a gene from another legume, *Vigna aconitifolia*, which was able to produce proline in *M. truncatula* and hence increase its level in plant nodule. Accordingly, plant tolerance under stress increased as a smaller decrease in N fixation resulted in improved plants under stress related to the control plants. Some other approaches, which may result in enhanced tolerance of legumes under salinity stress, have been indicated in the following.

(1) Using the tolerant species of rhizobium, particularly if isolated from stress conditions, is among the useful methods that can be used for increasing the efficiency of N fixation by legumes and rhizobium under salinity stress. It has been indicated that the salt tolerance of rhizobium is higher than the legume host plant. (2) Using the signal molecule genistein has also alleviated the stress of salinity on the growth and yield of soybeans. Miransari and Smith (2007, 2008, 2009) found that pretreatment of rhizobium with the signal molecule genistein, which is essential for the chemotactic approach of bacteria toward the host plant roots, under field and greenhouse conditions can significantly alleviate the stresses such as salinity, acidity, and suboptimal root zone heat. (3) Treating the soils with minerals such as nitrogen, calcium, and boron can improve the salt tolerance of fertilized legumes. (4) Using plant microbes, including arbuscular mycorrhizal fungi and plant growth-promoting rhizobacteria, can enhance the salt tolerance of different crop plants including legumes under salinity stress. Accordingly, (1) the modification of a cotransporter, which affects plant activity under salinity stress (Verdoy et al., 2006), and (2) the use of the tripartite symbiosis including rhizobium, mycorrhizal fungi, and soybean host plants are among the useful strategies and techniques for improving soybean growth under salinity stress.

Yoon et al. (2009) investigated the effects of methyl jasmonate (MeJa) on the alleviation of salinity stress in soybeans using hydroponics medium. The soybean seedlings were subjected to salinity stress by the concentrations of 60 mM NaCl 24 h after using MeJa at 20 and 30 μM. Salinity stress significantly decreased soybean growth, gibberellins concentration, the rate of photosynthesis, and water loss from plant, but significantly increased the rate of abscisic acid (ABA) production and proline content. MeJa alleviated the adverse effects of salinity by increasing the rate of ABA, plant growth, photosynthetic rate, chlorophyll content, plant water efficiency, and proline content. The important effects of jasmonates, including methyl jasmonate and jasmonic acid in plants, are by affecting (1) root growth, (2) seed germination, (3) ripening of fruits, (4) and fertility (Parthier, 1990; Pozo et al., 2004).

Jasmonate is an important signaling molecule affecting plant response under biotic and abiotic stresses, and its biosynthetic pathway is catabolized by Allene oxide cyclase (AOC: EC5.3.99.6). Accordingly, Wu et al. (2011) isolated six AOC genes from soybeans, which were randomly located on chromosomes 1, 2, 8, 13, 18, and

19. Real-time PCR indicated that such genes are specifically activated with complex patterns in different plant tissues under stress. The overexpression of *GmAOC1* and *GmAOC5* in plants resulted in plant tolerance under salinity and oxidative stresses, respectively. Such a large diversity of the AOC family, developed with time, may result in the adaptive responses of soybeans under stress.

The ionic and osmotic effects of salinity stress in plants were discriminated by Umezawa et al. (2002) using a modified technique of cDNA-amplified fragment length polymorphism (AFLP). Soybean seedlings were subjected to the stress using 100 mM NaCl and polyethylene glycol 6000 (12%, w/v) for 24 h. The activation of inositol-1-phosphate synthase gene indicated plant salt tolerance under salinity stress. The number of transcripts was dependent on ionic effects, which was a function of salinity stress, and osmotic effect, which was a function of both salinity and drought stress. The number of activated genes in response to ionic effects was higher in the roots than the aerial part. A set of redox enzymes and transcription factors may have important roles in soybean tolerance under salinity stress by affecting the ionic effects.

Soybean genotypes are different in their tolerance under salinity stress (Wang and Shannon, 1999), and it is being indicated which regions of DNA make soybeans tolerant under salinity stress. The combination of plant natural genetic with the physiological mechanisms and the genetic modification make it likely to produce soybean plants for use in saline agriculture (Flowers, 2004).

The family of BURP domain proteins compromises a set of plant-specific proteins, with a unique BURP domain at the C-terminus. However, there is not much research on the functions and subcellular localization of such proteins. The expression of RD22 gene under stress indicates its ability of enhancing plant tolerance under stress. Using different methods of genetic modification (cells and *in planta*), Wang et al. (2012) investigated how the expression of an RD22 protein in soybean (GmRD22) may alleviate salinity and osmotic stresses in soybeans.

Following the subcellular localization of *GmRD22*, the authors indicated that the gene is able to interact with a peroxidase in the cell, and its expression in *Arabidopsis thaliana* and genetically-modified (GM) rice resulted in the enhanced rate of lignin production under salinity stress. The authors accordingly indicated that by regulating the peroxidase activity, and hence strengthening the integrity of the cell wall, the gene is able to increase plant tolerance under salinity and osmotic stresses (Wang et al., 2012).

It has been indicated that under salinity, osmotic, and oxidative stresses the *GmPAP3* gene in soybeans is induced, indicating that the gene has a likely role in soybean response under abiotic stresses (Li et al., 2008). The main location of *GmPAP3* is in the mitochondria as the major site of reactive oxygen species. When the transgenic of *A. thaliana* with *GmPAP3* was subjected to the stresses such as salinity (NaCl) and drought (polyethylene glycol), root elongation significantly increased related to the wild type. The authors accordingly made the conclusion that *GmPAP3* is able to alleviate the adverse effects of salinity and osmotic stresses on the growth of soybeans by increased scavenging of reactive oxygen species.

Under osmotic stress the production of antioxidant enzymes and osmolytes, which are able to scavenge reactive oxygen species, is enhanced (Chen and Zhu, 2004;

Nakagami et al., 2005; Shao et al., 2005a; Suzuki et al., 2005; Jung et al., 2006). However, it has also been indicated that reactive oxygen species are able to act as a signaling pathway under biotic stresses, and their production results in a plasma membrane NADPH oxidase (Jung et al., 2006).

Xu et al. (2011) investigated the effects of salinity on the seed germination of barley at both the physiological and proteomic level. A salt-tolerant and a salt-sensitive genotype of soybean were subjected to the stress of salinity at 100 mmol/L until the radical was grown from the seed. Although the rate of seed germination was not affected by salinity stress, germination was done at a 0.3 and 1 day later in the salt-tolerant and salt-sensitive genotype, respectively, related to the control treatment. Under salinity the rate of ABA increased; however, the rate of gibberellic acid and isopentenyladenosine decreased. Although the rate of auxin increased in the tolerant genotype, it remained unchanged in the sensitive genotype. The proteins, which can have a role in the salt tolerance of soybean seeds, were indicated, including 20S proteasome and glutathione *S*-transferase.

Hamayun et al. (2010a) investigated the effects of salt on the growth and hormonal activity of soybeans, including ABA, gibberellins, salicylic acid, and jasmonic acid. Under the salt levels of 70 and 140 mM, plant parameters including plant biomass and height, rate of chlorophyll, number of pods, weight of 100 seeds, and yield decreased. The endogenous concentration of gibberellins and salicylic acid decreased under stress, while the concentrations of ABA and jasmonic acid increased under the stress. In conclusion the authors indicated that such hormonal effects are among the most important parameters regulating soybean growth under salinity stress. With respect to the above mentioned details, it is likely to enhance the tolerance of legumes, including soybeans under salinity stress, if the related morphological and physiological responses are evaluated and accordingly the related modifications in the genetic combination of plants are done.

Drought

Drought stress is among the most prevalent stresses, as one-third of the world population are subjected to such a stress. With the increasing level of CO_2 in the atmosphere and the issue of global changing, the stress can become even more severe and frequent. Under drought stress, 40% of soybean yield decreases (Specht et al., 1999). Plants use different mechanisms to alleviate the drought stress, including drought avoidance and drought tolerance (Turner et al., 2007). As a result, a plant completes its life cycle during the period of sufficient water, although under such conditions the plant growth stage is not long and the plant produces some seeds.

For example, in the Southern United States, the early mature genotypes are planted during March–April, producing flowers in late April and early May and pods at late May. Accordingly, a plant completes its life cycle before the period of drought in July–August. Under drought tolerance, the plant is able to avoid the stress by the production of osmolytes, solutes and osmoprotectants, or protoplasmic tolerance; as a result the plant will be able to maintain its cellular water potential and continue its metabolism (Boyko and Kovalchuk, 2008; Manavalan et al., 2009).

Plant drought resistance can be determined if the plant is subjected to stress each year for a long period of time. For conducting such an experiment, a field with the little rate of water potential, with specified soil properties, and an acceptable drought level is essential each year. However, most of the time, achieving such conditions is not easy. If the precise measurement of soybean response under drought stress is favored, new facilities for the determination of precipitation is essential. For example, such facilities have been used in the National Center for Plant Gene Research (NCPGR) in China for the determination of complex traits controlling plant drought tolerance (Granier et al., 2006; Yue et al., 2006).

Li et al. (2013) found that the expression of the drought gene in *Arabidopsis* activates the production of aldehyde oxidase, resulting in the production of ABA. Such responses make the plant tolerate the drought stress. The insertion of the gene in soybeans decreased water loss by reducing the size of stomata and the rate of water loss from the plant leaf by alleviating leaf wilting and increasing the rate of relative water content.

Under drought stress the damage of cellular membrane decreased in the genetically modified soybean due to the following: (1) production of malondialdehyde, (2) decreased rate of electrolyte leakage, (3) increased activities of antioxidants, and (4) higher production of proline. The seeds of GM soybeans also increased by 21% under drought stress related to the wild type. The authors accordingly indicated that the drought gene is able to enhance drought tolerance in GM soybeans by increasing the production of ABA, which can activate the expression of the stress-related genes and results in the activation of a complex of biochemical and physiological tolerance responses.

The water deficit is among the most important stress symptoms resulting from drought, salinity, and suboptimal root zone temperature. Under water stress a set of hormonal and signaling pathways are activated including the production of ABA, which results in the activation of the related genes under the stresses of salt, drought, and cold (Lawrence et al., 2001; Cooke et al., 2003; Chen and Zhu, 2004; Luo et al., 2005; Jung et al., 2006; Shao et al., 2006; Yang and Zhang, 2006; Yao et al., 2006).

The other important signaling pathway, which is activated under water stress, is the calcium pathway (Chen and Zhu, 2004). The source of calcium in the related pathway is from the cellular storage. Under hyperosmotic stresses the rate of inositol 1.4,5-trisphosphate (IP3) phospholipase increases, which is suppressed by C inhibitors (Carafoli, 2002; Shao et al., 2007a,b). IP3 is able to activate the channels of vacuolar calcium, resulting in the release of calcium under stress (Nakagami et al., 2005).

Using polyethylene glycol solutions (8% and 16%), Hamayun et al. (2010b) investigated the effects of pre- and postflowering drought stress (during a 14-day period) on the growth, yield, and hormonal responses of soybeans. The drought stress significantly decreased soybean growth and yield especially at the preflowering stage by the 16% treatment. The level of bioactive plant hormones, gibberellins, decreased with increasing the level of drought stress; however, the level of other plant hormones, including jasmonic, salicylic, and abscisic, increased under drought stress. Accordingly, in conclusion the authors indicated that the adverse effects of drought on plant growth at the preflowering stage is more significant than the postflowering stage, and the increased production of plant hormones under drought stress is among plant physiological responses to tolerate the stress.

In a review by Manavalan et al. (2009), they evaluated the methods that can be used for improving soybean tolerance under drought stress using the combination of physiology, genomics, and molecular breeding. Drought stress is prevalent in different parts of the world and is considered as a constraint to soybean production. The most used method for the plantation of soybeans under drought stress is the selection of the most tolerant genotype and testing its tolerance in different growing sites. However, such a method may not be recommended because it is (1) labor intensive, (2) time-consuming, (3) the yield is a highly quantitative trait with little rate of heritability, and (4) yield is under the influence of environmental factors and soil heterogeneity.

The method of indirect selection has not been successful due to its repeatability and lack of suitable phenotyping, especially for the traits, which are related to the root. The use of genetic modification has been tried for a large number of crop plants, and the production of tolerant soybean lines under drought stress is in progress. A high rate of research has been done related to finding the traits affecting soybean drought tolerance. Using the following techniques it has become more likely to develop, tolerant-soybean genotypes under drought stress: (1) the complete soybean genome, (2) molecular breeding, (3) genetic map, (4) different tools including genetic and genomics ones, and (5) genetic modification (Manavalan et al., 2009).

Wang et al. (2004) evaluated the effects of drought stress on the physiological–biochemical responses of different soybean genotypes. Drought stress resulted in the damage of cellular membrane and the increased production of free proline and malonaldehyde (MDA); however, the production of superoxide dismutase (SOD) increased at the earlier growth stage and decreased at the later growth stage. The above mentioned parameters were correlated with plant response under drought stress, and in the tolerant genotypes the production of free proline increased and the membrane damage decreased. However, there was not a correlation between MDA and SOD in soybean genotypes under drought stress.

Brassinolide (BR) is among the newest plant hormones regulating different plant activities. Zhang et al. (2008) evaluated the effects of BR on soybean growth under water deficit by determining the rate of plant photosynthesis, distribution of assimilates, and production of antioxidant enzymes and seed yield. BR was applied to foliage at 0.1 mg/L at the start of blooming. The plants were treated with control (80% field capacity) and drought stress (35% drought stress) at the initiation of pod production. Biomass production and seed yield were increased by using BR under both treatments. Under drought stress the translocation of ^{14}C from the plant leaf was suppressed by drought stress; however, BR was able to increase the process of translocation under both treatments.

The stress decreased chlorophyll content and the assimilation rate, but BR was able to increase both parameters under the stress. The other parameters, including the leaf water potential, ribulose-1,5-bisphosphate carboxylase activity, and the maximum quantum yield of PS II, was enhanced by the BR treatment. Under drought stress, BR treatment also had positive effects on the concentration of sugars and proline and the activity of superoxide dismutase and peroxidase in the soybean leaf. However,

BR decreased the leaf electrical conductivity and the malondialdehyde concentration under the stress. The authors accordingly indicated that BR is able to alleviate the adverse effects of drought stress on soybean growth and decreases the yield loss under the stress.

Acidity

Under acidic conditions, aluminum (Al) toxicity is among the parameters limiting plant growth and yield production. Using the proteomic method, Zhen et al. (2007) investigated the soybean proteins, which are responsive under Al stress in a soybean genotype, which is Al resistant. Using 50 μM AlCl$_3$, one-week-old soybean seedlings were treated for 24, 48, and 72 h. Using the two-dimensional electrophoresis the proteins were isolated from the roots. The analyses indicated that under the stress, 21 proteins were upregulated, 13 were induced, and 5 were downregulated. The authors also indicated that different cellular activities related to gene regulation, stress, protein folding, and cellular signaling and metabolisms, as plant responses under stress, can be used for the characterization of plant roots under Al stress. The proteomic method can be used for the determination of proteins, which are activated under Al toxicity.

Using nontolerant and tolerant soybean genotypes, Ermolayev et al. (2003) investigated the eight genes, which make soybean genotypes tolerant under Al stress. Among such genes, only one gene, phosphoenolpyruvate carboxylase (PEPC), was expressed in the roots of tolerant soybean genotypes, related to nontolerant genotype under the stress. The authors also isolated two genes with the increased expression in the roots of the tolerant genotype, which were inserted in *Arabidopsis* to indicate their functional analysis.

Under the stress only 6% of the wild type genotype survived, while in the GM soybean, 86% were able to survive the stress. The GM soybean genotypes showed a decreased rate of Al concentration in the roots, especially in the zones of division and elongation, when the plants were subjected to stress. A high number of root hairs were produced in the GM plants under stress, indicating the ability of such plants to tolerate the Al stress.

The other important factor affecting plant growth and yield production under acidity stress is the high concentration of H$^+$ affecting plant activities and microbial growth. Under high rate of soil H$^+$, the availability of different macronutrients, such as nitrogen and phosphorous, and hence their uptake by the plant decreases; however, the availability of macronutrients such as iron and zinc increases, which sometimes may be toxic to plants (Miransari, 2013).

For the evaluation of acidity stress on plant growth and microbial activities, Miransari and Smith (2007) conducted field and greenhouse experiments. In the field experiments, favorite levels of elemental sulfur, as a safe method of decreasing soil pH, were used to decrease soil pH to the favorite levels. The elemental sulfur was mixed with the soil and after a certain time (24 d), soybean seeds, which had been treated with the inoculums of *B. japonicum*, pretreated with genistein (as the plant to bacterium signal during the process of biological N fixation), were planted in a 2-year period.

Soil acidity increased by using elemental sulfur, the process of nodulation and N fixation, and soybean growth and yield were investigated during the 2 years of research work. Due to the decreased soil pH, soybean growth and the process of biological N fixation were affected; however, genistein was able to alleviate the stress partially or completely by increasing the rate of biological N fixation and soybean growth and yield. The authors accordingly made the conclusion that it is likely to increase soybean tolerance under acidity stress by using the signal molecule genistein.

The experiments were also conducted under greenhouse conditions (Miransari and Smith, unpublished data) to evaluate the effects of acidity on the growth and nitrogen fixation of soybeans and the symbiotic *B. japonicum*. The soybean seeds were planted in a sterilized surface and were inoculated with *B. japonicum* inoculums pretreated with genistein at different concentrations including control, 5, and 20 μM. Using HCl 1N and Hoagland nutrient solutions the soybean seedlings were treated with the favorite levels of acidity and were fertilized with their essential nutrients for growth. Genistein was able to alleviate the acidity stress in some cases by increasing plant growth and enhancing the process of nodulation and biological N fixation.

Cold

The soybean is not a tolerant plant species under cold stress, and its growth and yield as well as the growth and activities of its symbiotic rahizobium decreases. However, research work has indicated that it is possible to increase soybean tolerance under suboptimal root zone heat using the signal molecule genistein (Zhang and Smith, 1995, 1996). For example, Miransari and Smith (2008) indicated that it is possible to alleviate the stress of suboptimal root zone heat on the growth and nodulation of processes using genistein. Accordingly, they examined the effects of suboptimal root zone heat on the growth and activities of the soybean and its symbiotic rhizobium, *B. japonicum*, conditions by stimulating the field conditions in the greenhouse.

Intact soil samples with different soil textures were collected from the field using an aluminum cylinder. The samples were placed in the greenhouse and were subjected to different soil heat at 14, 19, and 25°C using a compressor and thermostat. Soybean seeds were sterilized and planted in vermiculite; seven-day-old seedlings were placed in the aluminum cylinder, and each seedling was inoculated with *B. japonicum* (pretreated with genistein at 0, 5, 10, and 20 μM) at 1 mL containing 10^8 cells. Genistein at 5 and 20°C significantly increased soybean growth and nodulation at 14°C. The effect of soil texture was also significant on the growth and nodulation of soybeans under suboptimal root zone heat (Miransari and Smith, 2008).

A promoter fragment with bases −1058 to −664 was found to be the cause of response under cold stress, which was detected one hour following the cold treatment. However, 24h following the cold treatment, such a response was suppressed by a transcriptional repressor, which was bound to a *cis*-element in the region −1403 to −1058. Physiological responses in *Arabidopsis* illustrated that the gene resulted in a higher fresh weight and osmolality related to the wild type. The authors accordingly indicated that it is likely to enhance the tolerance of crop plants under stress using such a gene, especially if combined with the appropriate promoter (Chen et al., 2009).

GmWRKY21-transgenic *Arabidopsis* indicated tolerance to cold stress, but *GmWRKY54* induced salt and drought tolerance in the plant, and the overexpression of *GmWRKY13* resulted in the increased sensitivity of plant under salt and mannitol stress. However, compared with the wild-type plant, sensitivity to abscisic acid decreased. Accordingly, the authors made the conclusion that the three *GmWRKY* genes make variable responses in plants under abiotic stresses, and *GmWRKY13* can affect both the development of plant lateral roots and plant response under stress.

DREB is among the transcription factors with important roles in plants under stress. Chen et al. (2009) investigated the effects of cold affecting soybean growth by isolating a DREB orthologue, *GmDREB3*. The authors found that subjecting the seedlings to the cold stress for 0.5 h resulted in the induction of *GmDREB3*. However, when the plant was subjected to high salt concentration, drought, salinity, and ABA, the gene was not expressed. Analyzing the promoter of *GmDREB3* indicated its cold-induced activity.

The main function of L-asparaginase (EC 3.5.1.1) is to catalyze the amide of L-asparagine, resulting in the release of NH_4^+ and aspartate. Cho et al. (2007) isolated a cDNA sequence, which was induced under suboptimal root zone heat and was able to activate L-asparaginase in a soybean leaf. The L-asparaginase cDNA, with the full length and *GmASP1*, contained a frame of 1258 bp activating a protein with 326 amino acids. It has been indicated that the genome of soybeans contained two *GmASP1*, ABA, suboptimal root zone heat, and NaCl and no drought and heat stress results in the induction of *GmASP1* mRNA. As a result the authors indicated the likely role of *GmASP1* during the early stage of suboptimal root zone heat.

The transcription of MYB type has a binding domain of conserved MYB DNA with 50 amino acids regulating different plant activities, growth, and development under different conditions including stress. Liao et al. (2008) obtained 156 *GmMYB* genes and examined their expression under treatments including ABA, salt, cold, and drought, and 43 of such genes were expressed. The genes, including *GmMYB76*, *GmMYB92*, and *GmMYB177*, were selected for analyses. It was accordingly indicated that the threes genes are able to bind to *cis* elements **CCG GAA AAA AGG AT** and **TAT AAC GGT TTT TT**, however, with a different affinity. Under cold and salt stress the genetically modified *Arabidopsis* indicated a higher tolerance. However, their response under ABA treatment decreased at germination stage related to the wild type. A subset of responsive genes was affected by the three genes although they were also able to regulate a subset of some other genes. The authors accordingly indicated that such genes are able to affect soybean responses under stress by the possible regulation of stress-responsive genes. With respect to the above mentioned details, it is possible to increase soybean growth and nodulation under suboptimal root zone heat using the biotechnological techniques and strategies.

Rhizobium in the Absence of Host Plant

The other important question is how rhizobium may survive under stress when its host plant is not present. The bacteria, which are more likely to persist under stress in the absence of their host plants, are the dormant ones, and the bacteria, which are in the stationary growth stage, and the actively growing bacteria are not able to survive. The bacteria in the forms of nonspores are able to survive under stress by the production of biofilms (Fujishige et al., 2006, 2008), which are a community of surface-attached rhizobium (single or multiple species) in a self-produced extracellular matrix. In a biofilm, bacteria are protected from a predator and also can survive abiotic stresses more efficiently, which is due to their extracellular matrix and their decreased rate of metabolic activities.

Under drought stress, the viability of rhizobia attached to the soil particles decreased in 9 weeks time; however, 10^4 cells were detected in the soil months later. In contrast, when the vegetative cells of *Bacillus* sp., as a gram positive and with the ability to form spores, were subjected to the drought stress (exactly similar conditions), the spores were produced. Accordingly, although rhizobia are not able to form spores, they can maintain their viability under severe stress conditions. It has been estimated that rhizobia are able to survive in the soil in a 4–5 year period; however, their rate of survival is much higher in the rhizosphere than the bulk soil (Hirsch, 1996, 2010).

Rhizosphere is a favorite environment to the rhizobia compared with the bulk soil, which is a desert by comparison. Root products can support the survival or high rhizobial number; for example, each gram of root exudate is equal to the rate of 50–100 mg, supporting the rhizobium population of 2×10^{10}. It is also the case for nonlegumes with the ability of supporting a high number of rhizobium. For example, the internal tissues of rice roots contained 10^6 bacterial cells, *Rhizobium leguminosarum* bv. *trifolii* with the ability of nodulating *Trifolium alexandrinum* L., indicating their symbiotic potential (Yanni et al., 1997). In contrast, most of the rhizobia isolated from the soil are not symbiotically active, indicating that the loss of plasmid results in the enhanced survival of rhizobia in the soil. Some of the rhizobia are also not able to nodulate the host plant, as they do not have plasmid-borne but just have symbiotic genes on their chromosome. However, if such bacteria acquire a plasmid, they will be able to fix N biologically.

Several strains of rhizobium are tolerant under salt or osmotic stress and hence are able to tolerate the severe stress of water deficiency. Such kind of responses can overlap with the responses to the other types of stress. However, there is not much detail related to the genetic response of rhizobium under stress, and the details on the signals regulating rhizobium response under stress are scarce. When the free-living bacteria face the drought stress the following changes result: (1) the osmotic potential increases, due to the enhanced concentration of salt; (2) water activity decreases; (3) because the DNA may be damaged, the transcriptional and translation of DNA decreases; and (4) there is a leakage of cellular membrane. To avoid such adverse effects the bacteria increase the production rate of carbohydrates or osmoprotectants (Arkhipova et al., 2007; Hamwieh et al., 2011; Vassilev et al., 2012; Evans and Wallenstein, 2014).

Among the useful methods for the evaluation of plant and microbial response to salinity stress is to investigate the alteration of genetic combination when the plant or microbe is subjected to the stress. Under such conditions the stress genes are activated and enable the bacteria and the plant to tolerate the stress. Such genes include the ones essential for (1) the regulation of transcription, (2) the repair of DNA, (3) regulation of cell cycling, (4) assembly of pili proteins and flagellin, (5) uptake of cations, and (6) transport of different molecules including sucrose (Parkinson and Kofoid, 1992; Craig et al., 2004; Danhorn, and Fuqua, 2007; Hamwieh et al., 2011).

For example, analyzing the genetic combination of *B. japonicum* under drought stress interestingly indicated that the genes, which are essential for the assembly of pili proteins including *ctpA* and *pilA* and *pilA2*, are expressed. Pili, especially the type IV, is essential for the production of biofilm. It is likely that the genes, which are essential for the bacterial response under stress, are also activated for the production of biofilm. Under stresses such as drought the production of biofilm can make the bacteria tolerate the stress. For example, the presence of exopolysaccharide in the biofilm matrix enhances bacterial tolerance under the drought stress. The loss of functionality in the exopolysaccharide inhibits the production of bacterial biofilm (Yildiz and Schoolnik, 1999; Danese et al., 2000; Fujishige et al., 2006; Wang et al., 2008).

In gram positive bacteria such as *Bacillus subtilis*, which are able to produce spores under stress, the sigma factors are activated, indicating the process of sporulation and nutrient deficiency. A high number of responses are also resulted in the biofilms of gram negative bacteria under nutrient deficiency. Under nutrient-deficient conditions, the center of large biofilms is nutrient starved, and the biofilms may be highly dispersed looking for nutrients (Fig. 1.2). Different research work has indicated that rhizobia are able to produce biofilms more quickly under nutrient-deficient conditions, including N and P, than nutrient-sufficient conditions. Most genes, mentioned above, can indirectly affect the process of symbiosis by membrane leakage and nutrient deficiency (Fujishige et al., 2006).

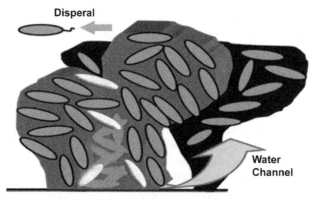

Figure 1.2 A biofilm with water channels and cells with different functions: pink (light gray in print versions), cells with oxygen deficiency; speckled, nutrient-deficient cells; yellow (gray in print versions), cells containing intact membrane (Hirsch, 2010).
With kind permission from Springer. License number 3771430609292.

Conclusion and Future Perspectives

Soybean yield and growth and its symbiosis with *B. japonicum* are subjected to environmental stresses worldwide. The use of different methods, techniques, and strategies have indicated that it is possible to enhance soybean growth and yield as well as its symbiosis with the N fixing bacteria, *B. japonicum*, under stress conditions. For example, the use of biological techniques, resulting in the modification of plant and bacterial genetic combination, is among the most used and effective methods for the alleviation of stress. It has also been indicated that because during the stress the initial stages of the symbiosis process, including the signaling communication between the two symbionts, are adversely affected, the use of signal molecules such as genistein can alleviate the stress. Research work has been so far partially successful in finding the methods, strategies, and techniques, which may alleviate the stress. However, more research is essential in this respect for the production of tolerant soybeans and *B. japonicum* and their practical use under field conditions.

References

Abd-Alla, M.H., 1992. Nodulation and nitrogen fixation in faba bean (*Vicia faba* L.) plants under salt stress. Symbiosis 12, 311–319.

Abshukor, N., Kay, Q., Stevens, D., Skibinski, D., 1988. Salt tolerance in natural-populations of *Trifolium repens* L. New Phytologist 109, 483–490.

Arkhipova, T., Prinsen, E., Veselov, S., Martinenko, E., Melentiev, A., Kudoyarova, G., 2007. Cytokinin producing bacteria enhance plant growth in drying soil. Plant and Soil 292, 305–315.

Begum, A.A., Leibovitch, S., Migner, P., Zhang, F., 2001. Specific flavonoids induced nod gene expression and pre-activated nod genes of *Rhizobium leguminosarum* increased pea (*Pisum sativum* L.) and lentil (*Lens culinaris* L.) nodulation in controlled growth chamber environments. Journal of Experimental Botany 52, 1537–1543.

Bordeleau, L.M., Prévost, D., 1994. Nodulation and nitrogen fixation in extreme environments. In: Symbiotic Nitrogen Fixation. Springer Netherlands, pp. 115–125.

Boyko, A., Kovalchuk, I., 2008. Epigenetic control of plant stress response. Environmental and Molecular Mutagenesis 49, 61–72.

Bruning, B., Rozema, J., 2013. Symbiotic nitrogen fixation in legumes: perspectives for saline agriculture. Environmental and Experimental Botany 92, 134–143.

Buhr, T., Sato, S., Ebrahim, F., Xing, A., Zhou, Y., Mathiesen, M., et al., 2002. Ribozyme termination of RNA transcripts down-regulate seed fatty acid genes in transgenic soybean. The Plant Journal 30, 155–163.

Carafoli, E., 2002. Calcium signaling: a tale for all seasons. Proceedings of the National Academy of Sciences of the United States of America 99, 1115–1122.

Chen, W.Q., Singh, K.B., 1999. The auxin, hydrogen peroxide and salicylic acid induced expression of the Arabidopsis GST6 promoter is mediated in part by an ocs element. Plant Journal 19, 667–677.

Chen, W.Q.J., Zhu, T., 2004. Networks of transcription factors with roles in environmental stress response. Trends in Plant Science 9, 591–596.

Chen, M., Xu, Z., Xia, L., Li, L., Cheng, X., Dong, J., Wang, Q., Ma, Y., 2009. Cold-induced modulation and functional analyses of the DRE-binding transcription factor gene, GmDREB3, in soybean (*Glycine max* L.). Journal of Experimental Botany 60, 121–135.

Cho, C., Lee, H., Chung, E., Kim, K., Heo, J., Kim, J., Chung, J., Ma, J., Fukui, K., Lee, D., Kim, D., Chung, Y., Lee, J., 2007. Molecular characterization of the soybean L-asparaginase gene induced by low temperature stress. Molecules and Cells 23, 280–286.

Cook, D.R., 1999. *Medicago truncatula*—A model in the making!. Commentary Current Opinion in Plant Biology 2, 301–304.

Cooke, J.E.K., Brown, K.A., Wu, R., Davis, J.M., 2003. Gene expression associated with N-induced shifts in resource allocation in poplar. Plant, Cell and Environment 26, 757–770.

Craig, L., Pique, M.E., Tainer, J.A., 2004. Type IV pilus structure and bacterial pathogenicity. Nature Reviews Microbiology 2, 363–378.

Danese, P.N., Pratt, L.A., Kolter, R., 2000. Exopolysaccharide production is required for development of *Escherichia coli* K-12 biofilm architecture. Journal of Bacteriology 182, 3593–3596.

Danhorn, T., Fuqua, C., 2007. Biofilm formation by plant-associated bacteria. Annual Review of Microbiology 61, 401–422.

Davet, P., 2004. Microbial Ecology of the Soil and Plant Growth. Science Publishers, Enfield, NH.

Dita, M.A., Rispail, N., Prats, E., Rubiales, D., Singh, K.B., 2006. Biotechnology approaches to overcome biotic and abiotic stress constraints in legumes. Euphytica 147, 1–24.

Ermolayev, V., Weschke, W., Manteuffel, R., 2003. Comparison of Al-induced gene expression in sensitive and tolerant soybean cultivars. Journal of Experimental Botany 54, 2745–2756.

Evans, S.E., Wallenstein, M.D., 2014. Climate change alters ecological strategies of soil bacteria. Ecology Letters 17, 155–164.

FAO, 2008. The State of Food and Agriculture. Rome, Italy. Available at: http://www.fao.org/docrep/011/i0100e/i0100e00.htm.

Flowers, T.J., 2004. Improving crop salt tolerance. Journal of Experimental Botany 55, 307–319.

Flowers, T.J., Galal, H.K., Bromham, L., 2010a. Evolution of halophytes: multiple origins of salt tolerance in land plants. Functional Plant Biology 37, 604–612.

Flowers, T.J., Gaur, P.M., Gowda, C.L., Krishnamurthy, L., Samineni, S., Siddique, K.H., et al., 2010b. Salt sensitivity in chickpea. Plant, Cell & Environment 33, 490–509.

Foolad, M.R., 2004. Recent advances in genetics of salt tolerance in tomato. Plant Cell, Tissue and Organ Culture 76, 101–119.

Friedman, M., Brandon, D.L., 2001. Nutritional and health benefits of soy proteins. Journal of Agriculture and Food Chemistry 49, 1069–1086.

Fujishige, N.A., Kapadia, N.N., De Hoff, P.L., Hirsch, A.M., 2006. Investigations of *Rhizobium* biofilm formation. FEMS Microbiology Ecology 56, 195–205.

Fujishige, N.A., Lum, M.R., De Hoff, P.L., Whitelegge, J.P., Faull, K.F., Hirsch, A.M., 2008. *Rhizobium* common *nod* genes are required for biofilm formation. Molecular Microbiology 67, 504–595.

Granier, C., Aguirrezabal, L., Chenu, K., Cookson, S.J., Dauzat, M., Hamard, P., et al., 2006. PHENOPSIS, an automated platform for reproducible phenotyping of plant responses to soil water deficit in *Arabidopsis thaliana* permitted the identification of an accession with low sensitivity to soil water deficit. New Phytologist 169, 623–635.

Hamayun, M., Khan, S., Khan, A., Khan, Z., Iqbal, I., Sohn, E., Khan, M., Lee, I., 2010a. Effect of salt stress on growth attributes and endogenous growth hormones of soybean cultivar Hwangkeumkong. Pakistan Journal of Botany 42, 3103–3112.

Hamayun, M., Khan, S.A., Shinwari, Z.K., Khan, A.L., Ahmad, N., Lee, I.J., 2010b. Effect of polyethylene glycol induced drought stress on physio-hormonal attributes of soybean. Pakistan Journal of Botany 42, 977–986.

Hamwieh, A., Tuyen, D.D., Cong, H., Benitez, E.R., Takahashi, R., Xu, D.H., 2011. Identification and validation of a major QTL for salt tolerance in soybean. Euphytica 179, 451–459.

Hashem, F.M., Swelim, D.M., Kuykendall, L.D., Mohamed, A.I., Abdel-Wahab, S.M., Hegazi, N.I., 1998. Identification and characterization of salt- and thermotolerant Leucaena-nodulating Rhizobium strains. Biology and Fertility of Soils 27, 335–341.

Hirsch, P.R., 1996. Population dynamics of indigenous and genetically modified rhizobia in the field. New Phytologist 133, 159–171.

Hirsch, A.M., 2010. How rhizobia survive in the absence of a legume host, a stressful world indeed. In: Symbioses and Stress. Springer Netherlands, pp. 375–391.

Jacobsen, T., Adams, R.M., 1958. Salt and silt in ancient Mesopotamian agriculture. Science 128, 1251–1258.

Jung, H.W., Lim, C.W., Hwang, K.B., 2006. Isolation and functional analysis of a pepper lipid transfer protein III (CALTPIII) gene promoter during signaling to pathogen, abiotic and environmental stresses. Plant Science 170, 258–266.

Lawrence, S.D., Cooke, J.E., Greenwood, J.S., Korhnak, T.E., Davis, J.M., 2001. Vegetative storage protein expression during terminal bud formation in poplar. Canadian Journal of Forest Research 31, 1098–1103.

Lee, J., Lei, Z., Watson, B.S., Sumner, L.W., 2013. Sub-cellular proteomics of Medicago truncatula. Frontiers in Plant Science 4.

Li, W., Shao, G., Lam, H., 2008. Ectopic expression of GmPAP3 alleviates oxidative damage caused by salinity and osmotic stresses. New Phytologist 178, 80–91.

Li, Y., Zhang, J., Zhang, J., Hao, L., Hua, J., Duan, L., Zhang, M., Li, Z., 2013. Expression of an Arabidopsis molybdenum cofactor sulphurase gene in soybean enhances drought tolerance and increases yield under field conditions. Plant Biotechnology Journal 11, 747–758.

Liao, Y., Zou, H., Wang, H., Zhang, W., Ma, B., Zhang, J., Chen, S., 2008. Soybean GmMYB76, GmMYB92, and GmMYB177 genes confer stress tolerance in transgenic Arabidopsis plants. Cell Research 18, 1047–1060.

Luo, G., Wang, H., Huang, J., Tian, A., Wang, Y., Zhang, J., Chen, S., 2005. A putative plasma membrane cation/proton antiporter from soybean confers salt tolerance in Arabidopsis. Plant Molecular Biology 59, 809–820.

Manavalan, L., Guttikonda, K., Tran, L., Nguyen, H., 2009. Physiological and molecular approaches to improve drought resistance in soybean. Plant and Cell Physiology 50, 1260–1276.

Miransari, M., Balakrishnan, P., Smith, D., Mackenzie, A.F., Bahrami, H.A., Malakouti, M.J., Rejali, F., 2006. Overcoming the stressful effect of low pH on soybean root hair curling using lipochitooligosaccharides. Communications in Soil Science and Plant Analysis 37, 1103–1110.

Miransari, M., Smith, D.L., 2007. Overcoming the stressful effects of salinity and acidity on soybean nodulation and yields using signal molecule genistein under field conditions. Journal of Plant Nutrition 30, 1967–1992.

Miransari, M., Smith, D., 2008. Using signal molecule genistein to alleviate the stress of suboptimal root zone temperature on soybean-Bradyrhizobium symbiosis under different soil textures. Journal of Plant Interactions 3, 287–295.

Miransari, M., Smith, D.L., 2009. Alleviating salt stress on soybean (Glycine max (L.) Merr.) – Bradyrhizobium japonicum symbiosis, using signal molecule genistein. European Journal of Soil Biology 45, 146–152.

Miransari, M., 2013. Soil microbes and the availability of soil nutrients. Acta Physiologiae Plantarum 35, 3075–3084.

Munns, R., 2002. Comparative physiology of salt and water stress. Plant, Cell & Environment 25, 239–250.

Munns, R., James, R.A., 2003. Screening methods for salinity tolerance: a case study with tetraploid wheat. Plant and Soil 253, 201–218.

Munns, R., Tester, M., 2008. Mechanisms of salinity tolerance. Annual Review of Plant Biology 59, 651–681.

Nakagami, H., Pitzschke, A., Hirt, H., 2005. Emerging MAP kinase pathways in plant stress signaling. Trends in Plant Science 10, 339–346.

Nouri, M., Hiraga, S., Yanagawa, Y., Sunohara, Y., Matsumoto, H., Komatsu, S., 2012. Characterization of calnexin in soybean roots and hypocotyls under osmotic stress. Phytochemistry 74, 20–29.

Okuda, M., Nang, M., Oshima, K., Ishibashi, Y., Zheng, S., Yuasa, T., Iwaya-Inoue, M., 2011. The ethylene signal mediates induction of *GmATG8i* in soybean plants under starvation stress. Bioscience, Biotechnology, and Biochemistry 75, 1408–1412.

Ososki, A.L., Kennelly, E.J., 2003. Phytoestrogens: a review of the present state of research. Phytotherapy Research 17, 84–869.

Parthier, B., 1990. Jasmonates: hormonal regulators or stress factors in leaf senescence? Journal of Plant Growth Regulation 9, 57–63.

Pozo, M.J., Van Loon, L.C., Pieterse, C.M., 2004. Jasmonates-signals in plant-microbe interactions. Journal of Plant Growth Regulation 23, 211–222.

Parkinson, J.S., Kofoid, E.C., 1992. Communication modules in bacterial signaling proteins. Annual Review of Genetics 26, 71–112.

Rozema, J., Flowers, T., 2008. Crops for a salinized world. Science 322, 1478–1480.

Sakai, T., Kogiso, M., 2008. Soyisoflavones and immunity. Journal of Medical Investigation 55, 167–173.

Schubert, S., Schubert, E., Mengel, K., 1990. Effect of low pH of the root medium on proton release, growth, and nutrient uptake of field beans (*Vicia faba*). Plant and Soil 124, 239–244.

Schulze, J., 2004. How are nitrogen fixation rates regulated in legumes? Journal of Plant Nutrition and Soil Science 167, 125–137.

Shao, H., Liang, Z., Shao, M., Wang, B.C., 2005a. Changes of some physiological and biochemical indices for soil water deficits among 10 wheat genotypes at seedling stage. Colloids and Surfaces B: Biointerfaces 42, 107–113.

Shao, H., Liang, Z., Shao, M., 2005b. Adaptation of higher plants to environmental stresses and stress signal transduction. Acta Ecologica Sinica 25, 1871–1882.

Shao, H., Liang, S., Shao, M., 2006. Osmotic regulation of 10 wheat (*Triticum aestivum* L.) genotypes at soil water deficits. Colloids and Surfaces B: Biointerfaces 47, 132–139.

Shao, H., Jiang, S., Li, F., Chu, L., Zhao, C., Shao, M., Zhao, X., Li, F., 2007a. Some advances in plant stress physiology and their implications in the systems biology era. Colloids and Surfaces B: Biointerfaces 54, 33–36.

Shao, H.B., Chu, L.Y., Wu, G., Zhang, J.H., Lu, Z.H., Hu, Y.C., 2007b. Changes of some anti-oxidative physiological indices under soil water deficits among 10 wheat (*Triticum aestivum* L.) genotypes at tillering stage. Colloids and Surfaces B: Biointerfaces 54, 143–149.

Shi, H.Z., Lee, B.H., Wu, S.J., Zhu, J.K., 2003. Overexpression of a plasma membrane Na^+/H^+ antiporter gene improves salt tolerance in *Arabidopsis thaliana*. Nature Biotechnology 21, 81–85.

Shinozaki, K., Yamaguchi-Shinozaki, K., 2000. Molecular responses to dehydration and low temperature: differences and cross-talk between two stress signaling pathways. Current Opinion in Plant Biology 3, 217–223.

Singleton, P.W., Bohlool, B.B., 1984. Effect of salinity on nodule formation by soybean. Plant Physiology 74, 72–76.

Soussi, M., Ocana, A., Lluch, C., 1998. Effects of salt stress on growth, photosynthesis and nitrogen fixation in chick-pea (*Cicer arietinum* L.). Journal of Experimental Botany 49, 1329–1337.

Soussi, M., Lluch, C., Ocana, A., 1999. Comparative study of nitrogen fixation and carbon metabolism in two chick-pea (*Cicer arietinum* L.) cultivars under salt stress. Journal of Experimental Botany 50, 1701–1708.

Specht, J.E., Hume, D.J., Kumudini, S.V., 1999. Soybean yield potential—a genetic and physiological perspective. Crop Science 39, 1560–1570.

Suzuki, N., Rizhsky, L., Liang, H., Shuman, J., Mittler, R., 2005. Enhanced tolerance to environmental stress in transgenic plants expressing the transcriptional coactivator multiprotein bridging factor 1c (MBF1C). Plant Physiology 139, 1313–1322.

Swaraj, K., Bishnoi, N.R., 1999. Effect of salt stress on nodulation and nitrogen fixation in legumes. Indian Journal of Experimental Biology 37, 843–848.

Turner, N.C., Wright, G.C., Siddique, K.H.M., 2001. Adaptation of grain legumes (pulses) to water limited environments. Advances in Agronomy 71, 193–231.

Turner, N., Abbo, S., Berger, J.D., Chaturvedi, S., French, R.J., Ludwig, C., et al., 2007. Osmotic adjustment in chickpea (*Cicer arietinum* L.) results in no yield benefit under terminal drought. Journal of Experimental Botany 58, 187–194.

Tu, J.C., 1981. Effect of salinity on rhizobium-root-hair interaction nodulation and growth of soybean. Canadian Journal of Plant Science 61, 231–239.

Umezawa, T., Mizuno, K., Fujimura, T., 2002. Discrimination of genes expressed in response to the ionic or osmotic effect of salt stress in soybean with cDNA-AFLP. Plant, Cell & Environment 25, 1617–1625.

Vasil, I.K., 1987. Developing cell and tissue culture systems for the improvement of cereal and grass crops. Journal of Plant Physiology 128, 193–218.

Vassilev, N., Eichler-Löbermann, B., Vassileva, M., 2012. Stress-tolerant P-solubilizing microorganisms. Applied Microbiology and Biotechnology 95, 851–859.

Verdoy, D., De la Pena, T.C., Redondo, F.J., Lucas, M.M., Pueyo, J.J., 2006. Transgenic *Medicago truncatula* plants that accumulate proline display nitrogen-fixing activity with enhanced tolerance to osmotic stress. Plant, Cell & Environment 29, 1913–1923.

Wang, D., Shannon, M.C., 1999. Emergence and seedling growth of soybean cultivars and maturity groups under salinity. Plant and Soil 214, 117–124.

Wang, Q., Xu, X., Ma, Y., Wu, S., 2004. Influences of drought stress on physiological and biochemical characters of different soybean varieties in flowering period. Agricultural Research in the Arid Areas 23, 98–102.

Wang, P., Zhong, Z., Zhou, J., Cai, T., Zhu, J., 2008. Exopolysaccharide biosynthesis is important for *Mesorhizobium tianshanense*: plant host interaction. Archives of Microbiology 189, 525–530.

Wang, H., Zhou, L., Fu, Y., Cheung, M.Y., Wong, F.L., Phang, T.H., Sun, Z., Lam, H.M., 2012. Expression of an apoplast-localized BURP-domain protein from soybean (GmRD22) enhances tolerance towards abiotic stress. Plant, Cell & Environment 35, 1932–1947.

Wu, Q., Wu, J., Sun, H., Zhang, D., Yu, D., 2011. Sequence and expression divergence of the AOC gene family in soybean: insights into functional diversity for stress responses. Biotechnology Letters 33, 1351–1359.

Xu, X.Y., Fan, R., Zheng, R., Li, C.M., Yu, D.Y., 2011. Proteomic analysis of seed germination under salt stress in soybeans. Journal of Zhejiang University Science B 12, 507–517.

Yang, J.C., Zhang, J.H., 2006. Grain filling of cereals under soil drying. New Phytologist 169, 223–236.

Yao, Y., Ni, Z., Du, J., Wang, X., Wu, H., Sun, Q., 2006. Isolation and characterization of 15 genes encoding ribosomal proteins in wheat (*Triticum aestivum* L.). Plant Science 170, 579–586.

Yanni, Y.G., Rizk, R.Y., Corich, V., Squartini, A., Ninke, K., Philip-Hollingsworth, S., Orgambide, G., de Bruijn, F., Stoltzfus, J., Buckley, D., Schmidt, T.M., Mateos, P.F., Ladha, J.K., Dazzo, F.B., 1997. Natural endophytic association between *Rhizobium leguminosarum* bv. *trifolii* and rice roots and assessment of its potential to promote rice growth. Plant and Soil 194, 99–114.

Yildiz, F.H., Schoolnik, G.K., 1999. *Vibrio cholerae* O1 El Tor, identification of a gene cluster required for the rugose colony type, exopolysaccharide production, chlorine resistance, and biofilm formation. Proceedings of the National Academy of Sciences of the United States of America 96, 4028–4033.

Yousef, A.N., Sprent, J.I., 1983. Effects of NaCl on growth nitrogen incorporation and chemical-composition of inoculated and NH_4NO_3 fertilized *Vicia faba* (L.) plants. Journal of Experimental Botany 34, 941–950.

Yamaguchi-Shinozaki, K., Shinozaki, K., 2005. Organization of cis-acting regulatory elements in osmotic- and cold-stress-responsive promoters. Trends in Plant Science 10, 88–94.

Yoon, J., Hamayun, M., Lee, S., Lee, I., 2009. Methyl jasmonate alleviated salinity stress in soybean. Journal of Crop Science and Biotechnology 12, 63–68.

Yue, B., Xue, W., Xiong, L., Yu, X., Luo, L., Cui, K., et al., 2006. Genetic basis of drought resistance at reproductive stage in rice: separation of drought tolerance from drought avoidance. Genetics 172, 1213–1228.

Zhang, F., Smith, D.L., 1995. Preincubation of *Bradyrhizobium japonicum* with genistein accelerates nodule development of soybean at suboptimal root zone temperatures. Plant Physiology 108, 961–968.

Zhang, F., Smith, D.L., 1996. Genistein accumulation in soybean (*Glycine max* [L.] Merr.) root systems under suboptimal root zone temperatures. Journal of Experimental Botany 47, 785–792.

Zhang, H.X., Blumwald, E., 2001. Transgenic salt-tolerant tomato plants accumulate salt in foliage but not in fruit. Nature Biotechnology 19, 765–768.

Zhang, M., Zhai, Z., Tian, X., Duan, L., Li, Z., 2008. Brassinolide alleviated the adverse effect of water deficits on photosynthesis and the antioxidant of soybean (*Glycine max* L.). Plant Growth Regulation 56, 257–264.

Zhen, Y., Qi, J., Wang, S., Su, J., Xu, G., Zhang, M., Miao, L., Peng, X., Tian, D., Yang, Y., 2007. Comparative proteome analysis of differentially expressed proteins induced by Al toxicity in soybean. Physiologia Plantarum 131, 542–554.

Zhou, Q., Tian, A., Zou, H., Xie, Z., Lei, G., Huang, J., Wang, C., Wang, H., Zhang, J., Chen, S., 2008. Soybean WRKY-type transcription factor genes, GmWRKY13, GmWRKY21, and GmWRKY54, confer differential tolerance to abiotic stresses in transgenic *Arabidopsis* plants. Plant Biotechnology Journal 6, 486–503.

Zhu, J.K., 2001. Plant salt tolerance. Trends in Plant Science 6, 66–71.

Soybean Production Under Flooding Stress and Its Mitigation Using Plant Growth-Promoting Microbes

2

S. Tewari, N.K. Arora
Babasaheb Bhimrao Ambedkar University, Lucknow, Uttar Pradesh, India

Introduction

Abrupt changes in climatic conditions are posing a potential threat to biodiversity (Eigenbrod et al., 2014). A rapid rise in the levels of carbon dioxide (CO_2), methane (CH_4), and other potent greenhouse gases due to industrial revolution are also one of the major factors for inducing global warming and changing precipitation schemes (Hao et al., 2010). These changing conditions have not only impacted biotic factors, but have also severely affected several abiotic stressors like drought, salinity, flooding, cold, and high temperature (Beck et al., 2007; Parvaiz and Satyawati, 2008; Manavalan et al., 2009; Bita and Greats, 2013; Tewari et al., 2016). Among all of the abiotic factors, flooding is one of the major environmental stress factors that has a devastating effect on crop growth and ultimately causes reduction in yield and production (Normile, 2008). Flooding is usually triggered by heavy and unpredictable rainfall, which has increased globally since the 1950s as a result of climate change (Bailey-Serres et al., 2012; Oh et al., 2014). Flooding harshly affects the productivity of farmland, because most agriculturally vital crops are incapable of tolerating such stress (Setter and Waters, 2003).

Flooding or waterlogging is the saturation of soil with water. In waterlogged soil, the water table is very high, affecting the normal biological activities (Jackson and Colmer, 2005). Flooded conditions decrease root development, hence reducing the crop's ability to absorb water and nutrients and tolerate drought stress during the season. Flooding is one of the major stresses in certain parts of the world, especially in rain-fed ecosystems, where poor drainage exists. About 10% of the global agricultural area is suffering from the constraints of flooding stress. Yield loss in various crops suffering from waterlogging varies between 15% and 80%, depending on the plant species, soil type, and time duration of the stress (Patel et al., 2014). The impact of flooding stress is observed on a variety of crop plants, but major grain loss has been observed in the case of the miracle crop "soybean" (*Glycine max* L.).

The soybean is one of the most important legume crops, as it contains a high amount of protein content. The soybean is vulnerable to various abiotic stresses, including flooding stress, limiting growth and yield (Hou and Thseng, 1991). Soybean crops are

Environmental Stresses in Soybean Production. http://dx.doi.org/10.1016/B978-0-12-801535-3.00002-4

commonly intolerant of waterlogged stress (Tougou et al., 2012). The production and grain yield of soybean crops is particularly affected by flooding stress, especially during vegetative and seed germination stages (Githiri et al., 2006). Exposure of soybean crops to flooding stress immediately damages the plant due to the rapid imbibition of water by the cotyledons, negatively affecting root growth (Nakayama et al., 2004). It has been estimated that up to a 25% reduction in soy crop yield is due to flooding injuries in Asia, North America, and other regions of the world (Mustafa and Komatsu, 2014). A reduction in soybean yield by 17–43% at the vegetative stage and 50–56% at the reproductive stage, due to flooding stress, has been observed by Oosterhuis et al. (1990). This stress leads to a shift toward alternative pathways of energy generation. Flooding results in the scarcity of O_2, resulting in a change from aerobic to anaerobic respiration (Voesenek et al., 2006).

An increase in the commercial value of soybeans in the international market has resulted in an increase in the cultivated area in a range of climatic conditions and soils, including flooded soils that occur in many areas of the world (Beutler et al., 2014). Therefore it is important to recognize and understand the mechanism of the flooding responses in soybeans in order to improve production and yield of this very important crop. But flooding responses in soybeans are neither well categorized nor characterized. Generally the studies on soybean seedlings in flooded soil showed that it obstructs hypocotyl pigmentation, root elongation, scavenges reactive oxygen species (ROS), scavenges glycolysis, protein storage, and reduces defense against diseases (Russell et al., 1990). The main aim of this chapter is to highlight the impact of flooding stress on soybean crops and to give possible remedies for ameliorating such constraints by utilizing plant growth-promoting rhizobacteria (PGPR).

Impact of Flooding Injuries on Soybean Growth

The cultivation of soybeans in flooded or waterlogged soils of the world is affected with low production and yields (Sosbai, 2012). For cultivation of soybean crops with higher yields, it is essential to have optimum or sufficient water content. But the water level, particularly in rain-fed systems, can show fluctuations resulting in excess or flooding stress. The effects of water-deficient conditions (also called as drought stress) has been clearly described by several workers in their studies worldwide (Dogan et al., 2007; Lobato et al., 2008; Mastrodomenico et al., 2013); however, the impact of flooding or excess water stress on soybean crops is reviewed in this chapter.

Flooding stress is demonstrated in two ways in soybeans: the first is physical injury and the second is anaerobic stress (Rizal and Karki, 2011). The physical injury happens due to the quick absorption of water, collapse of seed structure, and outflow of internal seed contents, which ultimately leads to the hampering of seed germination. The excessive amount of water that results in flooding stress generates a hypoxic situation, ie, insufficient oxygen (O_2) concentration for maintaining normal or healthy respiration in roots. Flooding or waterlogged conditions decreases the amount of O_2 in the soil, to reach roots, and it hampers soybean growth. Once the hypoxia sets in, a change from aerobic to anaerobic path takes place, bringing changes in the respiratory metabolism of roots. Abrupt changes in respiratory metabolism in turn produce toxic metabolites like lactic acid ($C_3H_6O_3$) and ethanol (C_2H_5OH), which increase the activity of fermentative enzymes (Kolb and Joly, 2009; Borella et al., 2014). Hypoxia further leads to decreased biological nitrogen fixation, as nodules need sufficient O_2 for carrying aerobic respiration and supplying adenosine triphosphate, which is essential for maintaining nitrogenase activity.

Furthermore, the concentration of CO_2 in flooded soil is usually high, and as a result, biomass and elongation of soybean roots get inhibited (Grable, 1966; Boru et al., 2003). Apart from reduced N_2 fixation, flooding also affects biochemical and physiological processes in soybeans, such as the decrease in photo assimilation and reduced photosynthetic activity, inhibiting the absorption of carbon, nitrogen, and other macronutrients and increasing the absorption of iron, leading to iron toxicity (Davanso et al., 2002; Pires et al., 2002). Flooding stress results in highly complex effects, depending on the duration of stress and developmental stage of the plant (Schöffel et al., 2001). It also affects vegetative and reproductive stages in soybean crops. Other effects like abscission and yellowing of leaves at the lower nodes, reduced dry weight of the plant, decreased seed yield, and short stunted growth of crops are also observed (Scott et al., 1989). Soybeans flooded at the vegetative stage are also reported to have reduced leaf area, dry weight, and plant height (Griffin and Saxton, 1988; Linkemer et al., 1998). Griffin and Saxton (1988) reported that soybean crops flooded with excessive water for 4 days at V6 (early flowering) stage showed severe chlorosis and stunting symptoms (Table 2.1). These researchers also stated that crop growth rate has been regularly affected when the flooding stress was applied for more than 2 days.

The sowing time of soybeans overlaps with the rainy season in many parts of the world. Of copious stresses during its growth series, stress due to flooding during germination is damaging to seeds, seedling growth, crop formation, and crop yield (Hou and Thseng, 1992). Flooding significantly decreases soybean growth and yield parameters including germination, plant length, dry weight, yield, and photosynthetic activity. Naeve (2002) reported that soybean crops get slightly affected when flooding was applied for more than 48 h, but flooding above this duration can even lead to delayed germination and reduced plant growth, which in some cases may even destroy the complete stand. Due to flooding injuries, several phenotypic and genotypic changes take place in soy crops, which are discussed in the next section.

Germination

There are several reports that account for the negative correlation between germination percentage and flooding stress (Yaklich and Abdul-Baki, 1975; Maryam and Nasreen, 2012). When soy seeds are planted under nonstress conditions (optimum conditions), they start to absorb water, imbibe, swell, and germinate within 1 or 2 days. But the germination of seeds is delayed when fields are sometimes saturated with water from heavy rains and poor drainage. Flooding causes mechanical damage on the soybean seeds and prohibits germination. Wuebker et al. (2001) reported, when seeds were flooded 3 days after the start of imbibition, a significant drop in germination

Table 2.1 Impact of Flooding Stress on Different Growth Stages of a Soybean Crop

Growth/ Reproductive Stages	Description of Growth/ Reproductive Stages	Symptoms Caused Due to Flooding	References
Vegetative Growth Stages			
V_1	One unrolled trifoliate leaf	Reduced weight	Oosterhuis et al. (1990)
V_2	Second trifoliate stage	Reduced plant growth	Linkemer et al. (1998)
V_3	Two unrolled trifoliate leaves	Reduced leaf area and photosynthetic activity	Yordanova and Popova (2007)
V_4	Trifoliate stage	Reduced dry weight	Maryam and Nasreen (2012)
V_5	Fifth trifoliate stage	Reduced plant growth	Cho and Yamakawa (2006a,b)
V_6	Early flowering stage	Chlorosis and stunted symptoms; diminished plant growth	Griffin and Saxton (1988)
Reproductive Stages			
R_1	Beginning flowering stage	Decreased shoot and dry matter	Youn et al. (2008)
R_2	Full flowering stage	Reduced biomass	Oosterhuis et al. (1990)
R_3	Beginning pod stage	Reduced seed yield	Van Toai et al. (1994)
R_4	Full pod stage	Reduced grain mass	Schöffel et al. (2001)
R_5	Grain filling stage	Reduced seed yield	Singh and Singh (1995)

percentage occurred and seed injury was observed. These workers also stated that flooding stress at 15°C is more severe on seed growth than at 25°C. The flooded soil reduces seed development and growth. Shanmugasundaram (1980) accounted that continuous rainfall may adversely affect seed germination by increasing anaerobic conditions, and a 28% reduction was reported in plant establishment due to flooding immediately after sowing.

Root Length and Shoot Length

Plant growth is generally affected by changes in soil water content. Oxygen content can be altered by flooding stress. A hypoxic situation caused by water stress results in root damage due to insufficient allocation of water, minerals, nutrients, and hormones. This insufficiency in nutrient and water uptake further leads to shoot damage (Vartapetian and Jackson, 1997; Jackson and Ricard, 2003). Hence flooding stress causes damage of both root length and shoot length in soybean crops. The symptoms of flooding stress, which appear on soybean shoots, are wilting of the leaves (Kramer, 1951). There is solid evidence that proves that extended periods of root anoxia results in a decline in leaf area. Flood tolerance in plants is strongly correlated to root length and to root surface area. Sallam and Scott (1987) reported significant correlation between root length, surface area, dry weight, and flood tolerance in soybean crops. Oosterhuis et al. (1990) displayed a reduction in dry weight/biomass production in soybeans at V4 (trifoliate) and R2 (full flowering) stages. Bacanamwo and Purcell (1999) observed that a decrease in soybean dry matter was 34% in flooded soil. Youn et al. (2008) reported that water stress at the R1 (beginning flowering) growth stage decreased shoot and dry matter accumulation in soybean crops in comparison to a control. They also observed a reduction of root dry mass by 41–45% in wild types and 62–67% in supernodulating mutants just after drainage, and 51–64% in wild types and 64–75% in supernodulating mutants 30 days after drainage. In vivo field trials were conducted in flooded soil by Cho and Yamakawa (2006b); they observed that shoots were more starved than roots due to nutrient deficiency. Maintaining proper root and shoot function is very essential for the sustainable growth and development of the plants.

Grain Yield

An essential factor for the evaluation of soybean tolerance under flooding stress is the yield and production of high-quality, marketable seeds (Van Toai et al., 1994). Linkemer et al. (1998) reported a significant decline in pod number, branch number, pods per node, and seed size after 7 days of flooding at various vegetative and reproductive stages. It was observed that waterlogging reduced the seed yield by 93%, 67%, and 30% at the R3 (beginning pod), R1, R5 (grain filling stage), and V2 (second trifoliate) stages, respectively. A decrease in yield can be attributed to reduced pod growth and seed size, brought about by nitrogen stress and reduced photosynthesis (Oosterhuis et al., 1990).

A reduction in pod number in soybeans was also established by Sullivan et al. (2001). They accounted for a reduction in plant height for 3, 5, and 7-day floods at early vegetative growth stages. They also stated that a reduction in soybeans ranged from 20–93% after 6 days of flooding. Several workers reported a reduction in soybean yields on flooding, ranging in duration from 24 hours to 14 days (Scott et al., 1989; Singh and Singh, 1995).

Field studies were conducted by Cho and Yamakawa (2006a) on flooded soil, taking three soybean cultivars: Pungsan–namulkong, Sobaeg–namulkong, and Saebyeolkong. It was observed that seed yields of soybeans were significantly affected by the flood duration in all the three cultivars. As the duration of flooding was increased, seed yield decreased. Except Saebyeolkong, all the three cultivars showed significant seed yield reduction on exposing the plant to 3 days of flood treatment. Results also highlighted that the seed yield reduced by 38%, 44%, and 66% in Saebyeolkong, Sobaeg–namulkong, and Pungsan–namulkong (in comparison to the control, ie, soybean sown under non-flooded conditions), respectively on the ninth day of flooding. The number of pods per m^2 also decreased as the flooding duration was increased. Several workers reported that the soybean is more sensitive against the excessive water in soil during the early reproductive stage rather than on the vegetative stages (Griffin and Saxton, 1988; Kwon and Lee, 1988; Heatherly and Pringle, 1991; Choi et al., 1995). Beutler et al. (2014)

accounted that soybean grain yield reduced progressively as the increase in flooding during the R2 and R5 stages resulted in a 17% and 29% reduction after 16 days, and 41% and 36% reduction after 32 days of water stress, respectively.

Schöffel et al. (2001) showed a decreased number of pods per plant at the R4 stage on the 10th day of flooding in pot trails. These workers found no difference in grain mass in R2 and R4 stages when flooding occurred for 10 days. Whereas Rhine et al. (2010) conducted a field experiment in flooded soil and obtained yield reductions from 20% to 39% in the different soybean cultivars when subjected during the R5 stage on the eighth day of flooding. These workers also reported that flooding at the R5 stage resulted in a more significant reduction of soybean grain yield compared with flooding at the R2 stage.

Nodulation

Long-lasting nitrogen fixation and good nodulation are the two key factors, which can bring high seed yield and increased protein content in the seeds of soy crops. Soybean seeds hold high protein content of about 35–40%, depending on the weight of the seed. Ohyama et al. (2013) showed that seed yield of soybeans is directly proportional to the nitrogen assimilation in plants. The author also showed that soybean seed yield exhibits linear correlation with the amount of N_2 accumulation. Nagumo et al. (2010) reported that soybean seeds have four times higher protein content than any of the cereals. Due to a high amount of protein, soybean seeds require nearly 70–90 kg of nitrogen per hectare. Approximately 20% of total nitrogen is assimilated by soy crop until the initial flowering stage (R1 stage), and 80% of nitrogen is assimilated during the reproductive stage. Hence the uninterrupted assimilation of nitrogen after the R1 stage is necessary for respectable growth and high seed yield in soybean farming (Ohyama et al., 2013). The development and formation of nodules is severely affected by flooding stress. Exposure of plants to flooding stress for about a week is sufficient to reduce leaf nitrogen content at early vegetative stages (Sullivan et al., 2001). The reduction in nitrogen concentration in flooded soybean plants has also been identified as the major limiting factor for plant growth (Pankhurst and Sprent, 1975; Sugimoto and Satou, 1990; Bacanamwo and Purcell, 1999). Reduction in nitrogen has been attributed to decreased nodulation, reduced nitrognase activity, and increased ethylene production (Sprent and Gallacher, 1976; Bennett and Albrecht, 1984; Sung, 1993). Miao et al. (2012) conducted a pot experiment with soybean cultivars (Hefeng 50 and Kenfeng 16) under flooding stress and observed that nodule number of Hefeng 50 and Kenfeng 16 decreased by 84% and 64% in comparison to the control, respectively. They also reported that nodule numbers are significantly reduced at the flowering stage, pod bearing stage, and then at grain filling stages. Robredo et al. (2011) accounted that prolonged stress can restrict the ability of plants for nitrogen assimilation, thereby inhibiting nitrogen metabolism. Hence it was elucidated that flooding can affect nitrogen fixation of soybeans, since the total nitrogen content in soybean tissues is considerably lower under flooding stress conditions as compared to normal conditions (Jin et al., 2005).

Photosynthetic Activity

Flooded soybeans show a decrease in photosynthetic activity (Mutava et al., 2015). Cho and Yamakawa (2006a) reported that the leaves and the branches of soybeans are the first parts to respond to flooding stress. It was observed that flooding significantly decreased N uptake in soybean leaves and branches. The longer exposure of plants to water stress leads to an observable drop in photosynthesis and chlorophyll (Yordanova and Popova, 2007). Flooding brought reduction in leaf number, leaf area, canopy height, and dry weight at maturity in soybean crops (Sallam and Scott, 1987; Scott et al., 1989). The decrease in photosynthetic activity with longer exposure to flooding may be caused by the reduction in chlorophyll, transpiration, and Ribulose-1, 5-biphosphate (RuBP) carboxylase activity. These collective effects against flooding had brought a decline in the rate of crop growth, net assimilation, and leaf expansion of plants. An increase in water stress results in decrease of net CO_2 assimilation per unit area of leaf, mainly due to stomatal closure dipping RuBP production (Jackson and Attwood, 1996). Decreases in CO_2 assimilation eventually reduce dry weight of soybean plants (Trought and Drew, 1980). Kawase (1981) stated that prolonged flooding may result in the swelling of cells in the cortex, swelling of the stem base, or hypocotyl and hypertrophy that are often accompanied by the collapse of cells. Extended flooding may also lead to leaf chlorosis, epinasty, and plant death (Kramer and Boyer, 1995). Reduction in nutrient content in stressed soil can decrease photosynthesis and dry weight content (Wilson, 1988). The photosynthetic rate declined in the waterlogged soybean, and it decreased greatly by the longer flooding treatments.

Impact of Flooding Stress on Soybean Proteome

Several studies on the flooding responsive mechanisms in soybeans using proteomics have shown that several proteins regulating glucose degradation, sucrose accumulation, signal transduction, cell wall relaxing, and alcohol fermentation were altered under flooding stress (Komatsu et al., 2012, 2015). Whereas proteins related to production of energy increased, proteins involved in maintaining cell structure and protein folding were reduced in response to flooding stress (Nanjo et al., 2013). Apart from this, proteomic studies identified a number of calcium (Ca)-binding proteins that might have played important roles in flooding stress response.

As already discussed, flooding forces soybean crops to shift from aerobic to anaerobic respiration. This metabolic shift helps plants to regenerate NAD$^+$ through alcoholic (C_2H_5OH) fermentation by producing flood-inducible proteins that are involved in fermentation that brings drastic increase in alcohol dehydrogenase (ADH) activity in soybeans (Komatsu et al., 2011a). The repressed energy metabolisms accelerate the depletion of energy, resulting in retarded growth, and render flooded plants vulnerable to flooding stress.

Flooding stress primarily damages the mitochondrial electron transport chain (ETC), resulting in additional ROS generation. Proteomic and metabolomics in combination have been effectively used to study the effect of flooding on soybean mitochondria

(Komatsu et al., 2011b). Flooding causes problem in the ETC in the hypocotyls and roots of soybean seedlings. Inner membrane carrier proteins and proteins related to complexes III, IV, and V of the ETC were found to be reduced, while metabolites and proteins linked to tricarboxylic acid cycle (TCA) and γ-amino butyrate (GABA) shunt were amplified under flooding stress, resulting in high NADH production. In addition, succinate semialdehyde dehydrogenase and GABA were significantly increased by flooding stress, as was 2-oxoglutarate dehydrogenase, suggesting that the GABA shunt is involved in the replenishment of intermediates required for energy production that have been depleted by flooding stress (Hossain and Komatsu, 2014).

Mass spectroscopy (MS)-based quantitative proteomics have been exploited for profiling soybean root tips under flooding stress. It was observed that proteins involved in fermentation, glycolysis, nucleotide, and cell metabolism was amplified, whereas proteins involved in cell organization and amino acid metabolism were decreased. Few other proteins, including phosphatidylinositol-4-phosphate 5-kinases, sucrose-binding protein, actins, and alpha-tubulins, were found specifically in the root tip region. The addition of sucrose-binding proteins in flooded soybean root tips explains an enhanced sucrose accumulation (Nanjo et al., 2010).

Nanjo et al. (2010) conducted a gel-free MS-based proteomic study and observed regulation of 20S proteasome subunits in flooded soybeans. A different expression of 20S proteasome subunit may thus affect the quantity as well as the activity of the 26S proteasome, thereby altering flooding tolerance. Differentially expressed ROS scavenger proteins, superoxide dismutase, and cytosolic ascorbic peroxidase (cAPX) decreased in response to flooding. A significant decrease in cAPX 2 proteins on exposure to flooding was observed during proteonomic screening of six different soybean varieties (Shi et al., 2008). An abundance of cAPX 2 transcripts was also found to be decreased significantly after flooding, as did the APX activity. Hence it was suggested that cAPX 2 plays a significant role in flood-induced stress response of young soybean seedlings.

Komatsu et al. (2013) conducted proteomic studies on flooded cotyledons of soybeans and observed a decrease in the number of calcium oxalate crystals in the cotyledon. Calcium oxalate crystals control physiological calcium stages in plant cells (Franceschi and Nakata, 2005). The proteomic analysis of cotyledon displayed the HSP70 protein. This protein is involved in many biological works like protein folding and translocation, as well as protein degradation. These all proteonomic and metabolomics findings related to the change in soybean proteome in response to flooding give a clear insight into the complex metabolic process of injuries in soybeans.

Mitigation of Flooding Stress in Soybeans Utilizing Plant Growth-Promoting Rhizobacteria

When flooding occurs, plant roots become hypoxic. Oxygen limited response results in the synthesis of enzyme 1-Aminocyclopropane-1-Carboxylate (ACC) synthase along with several stress proteins (Li et al., 2012). The stressed plant consequently manufactures more ACC in their roots. Since this freshly prepared ACC cannot be converted back to ethylene (in the roots) because ethylene synthesis requires oxygen, this ACC is transferred to the shoots, where there is an oxygenic environment, and this ACC can be converted to ethylene (Bradford and Yang, 1980; Else and Jackson, 1998; Tewari and Arora, 2013). In soybean plants, phytohormone indole acetic acid (IAA) induces ethylene production, and inhibitory effects of high IAA on root growth are also facilitated by ethylene (Grichko and Glick, 2001a). The ethylene production can be reduced by inhibiting ethylene synthesis enzymes. PGPR-producing ACC deaminase can be used to convert ACC (precursor of ethylene) into α-ketobutyrate and ammonia, thus reducing the levels of ethylene under water stress conditions (Grichko and Glick, 2001b).

The inoculated plants produce low ethylene, thereby reducing the damage of elevated ethylene (Grichko and Glick, 2001b). The synthesis of ethylene by waterlogged plants results in wilting, necrosis, chlorosis, and reduced biomass yield. However, the application of ACC deaminase-producing PGPRs can protect plants from these damages (Grichko and Glick, 2001a; Barnawal et al., 2012; Li et al., 2013; Glick, 2014).

Husen et al. (2011) also reported the role of ACC deaminase and the IAA-producing strain of *Pseudomonas* for enhancing the growth and yield of soybean crops under stress conditions. Certain transgenic plants expressing ACC deaminase genes have demonstrated tolerance toward flooding stress. The adverse effects of root hypoxia on plant development can also be reduced in transgenic crops. Plants that have the ACC-deaminase gene under the regulation of the rolD promoter are guarded to the highest extent (Grichko and Glick, 2001b; Dimkpa et al., 2009). These workers also observed the effects of inoculating ACC deaminase PGPR on tomato seedlings exposed to waterlogged stress. Seeds of tomato plants were treated either with *Enterobacter cloacae* CAL2 or different strains of *Pseudomonas putida* like UW4, ATCC17399, pRKACC, and ATCC17399/pRK415. It was found that tomato plants inoculated with ACC deaminase-producing PGPR strains showed significant tolerance to flooding stress, suggesting the role of bacterial ACC deaminase in reducing the effects of ethylene stress.

It has been observed that flooding affects nodule development and nitrogenase activity more intensely. A physical barrier located in nodular parenchymal cells regulate the diffusion of O_2 within the nodules. The absence of O_2 in the nodules increases their diffusion resistance under stress conditions, resulting in the decreased activity of nodules (Day and Copeland, 1991). Under such conditions the use of aerobic bacteria, which are able to utilize nitrogenous oxides, as terminal electron acceptors causes their survival and growth under the periods of anoxia. Accordingly, rhizobia will be able to survive in the soil and biologically fix the atmospheric N (Zablotowicz et al., 1978). Nodule formation and nodule development can bring an onset of nitrogen fixation where leghemoglobin is present. Leghemoglobin can provide oxygen for the growth of bacteria under rigorous conditions (Shleev et al., 2001). A positive relationship between leghemoglobin content and rhizobia efficiency was also observed by Dong (1999). Soybean crops inoculated with the bacterium *Bradyrhizobium elkanii* and *Bradyrhizobium japonicum* can play a significant role in enhancing N_2 fixation in plants and improving plant growth under soil flooding conditions (Scholles and Vargas, 2004; Beutler et al., 2014). Kadempir et al. (2014) also observed the positive effects of inoculating *B. japonicum* for enhancing the nutritional status of the plant along with increased N_2 uptake and flooding amelioration in soybean plants. Another free-living, motile, aerobic bacterium that can thrive in flooded conditions is *Azospirillum*, which can promote plant growth and development at various stages (Bhattacharyya and Jha, 2012; Sahoo et al., 2014). The application of *Azospirillum* on seeds has demonstrated beneficial effects on diverse plants both in the greenhouse and in field trials (Saikia et al., 2013). Additional research work is required to find out the possible role of beneficial rhizobacterial species in alleviating flooding stress, as very little work has been done on this aspect until now.

The presence of mycorrhizae in flooded habitats has been documented, but the available data is very limited and needs thorough understanding. Sondergaard and Laegaard (1977) stated that 0–96% of the aquatic plant roots, when examined thoroughly in four oligotrophic lakes, showed mycorrhizal colonization. Taimer and Clayton (1985) also stated that submerged aquatic plants extensively colonized with arbuscular mycorrhizal (AM) fungi contained 20% more phosphorus on a dry weight basis than nonmycorrhizai plants. Anderson et al. (1984) observed no functional mycorrhizae (presence of arbuscules) in plants growing in the wettest habitats (shallow emergent zone) along a soil moisture gradient, but some plants growing in moist to wet habitats were mycorrhizal. AM colonization has been documented in rice, soybeans, Populus, Salix, and Nyssa spp. growing in flooded soils (Keeley, 1980; Dhillion and Ampompan, 1992). AM fungi also have been reported in both fresh (Sengupta and Chaudhuri, 1990) and saline (Ragupathy et al., 1990) tropical wetlands. Certain species of AM fungi have acclimatized themselves to maintain survivability under water stress conditions (Turner et al., 2000; Landwehr et al., 2002). The AM species *Glomus geosporum* is generally prevalent in European salt marshes (Landwehr et al., 2002; Carvalho et al., 2004). It has been recognized that AM fungi obtained from rice fields were showing excessive colonization in soybean fields (Isobe et al., 2011). These AM fungi were quite suitable for the mitigation of flooding damage in soybeans. It is also essential to emphasize soybean cultivars. Some cultivars have wide aerenchymatous tissue and/or adventitious roots, thus leading to a higher percentage of air inside the roots (Shimamura et al., 2003; Thomas et al., 2005). Soy plants that grow into high secondary aerenchymatous tissue and adventitious roots could willingly form symbiotic relationships with AM fungi, even under flooding or water stress conditions. Hattori et al. (2013) reported that soy crops can form symbiosis with AM fungi and rhizobia (*B. japonicum*), and the symbiotic nutrient acquisition controls the seed yield and growth of this crop grown under water stress conditions. Colonization of AM fungi by *Glomus intraradices* has significantly impacted flood tolerance of *Pterocarpus officinalis* seedlings by improving P acquisition and plant growth development in leaves. Waterlogging induces nodule formation both on adventitious roots and drowned parts of the stem (Fougnies et al., 2007). AM fungi also play a significant role in increased phosphorus uptake and nitrate reduction (Harley and Smith, 1983). AM fungi upsurge plant water uptake and lessen wilting during water or flooding stress. Water and nutrient assimilation is believed to be improved during flooding because the fungal mycelium spreads the root system, creating a better pool of water and nutrients accessible to the plant (Harley and Smith, 1983). Snellgrove et al. (1982) observed that AM fungi increased the transport of the total photosynthate by 7%.

These authors also suggested that plants with AM fungi can pay off for this loss of photosynthate by lowering the percentage of dry matter and increasing the assimilation rate per unit dry weight of the leaf.

It has been observed that mycorrhizal (*Casuarina equisetifolia*) seedlings could better adapt to flooding than noninoculated or control seedlings. The reason behind this could be due to increased O_2 diffusion and the removal of ethanol through greater development of adventitious roots, aerenchymatous tissue, and hypertrophies lenticels on the root area and sunken part of the stem (Rutto et al., 2002). The AM inoculated *Aster tripolium* plants showed increased tolerance to flooding by enhanced osmotic adjustment and accumulation of proline in plant tissues (Neto et al., 2006). However, further studies are required on the role of AM fungi on root nodule formation under flooding stress conditions in soybean plants.

Conclusion

Rainfall is one of the most significant factors required for the proper growth and development of crop plants, particularly in developing countries. However, an excessive amount of rainfall has its impact in the form of flooding, which has severe impact on the growth of crop plants, including soybean crops cultivated in rain-fed regions. Flooding stress causes significant damage to the morphological and physiological features of soybean crops and hence reduces grain yield severely. A proteomic study also stated the downregulation of certain proteins involved in flooding stress. Numerous investigators try to develop stress-tolerant soybean crops to alleviate these losses. Upgrading such genetic breeds is not a simple task, and success is also not confirmed. Hence the use of PGPR to augment the effect of flooding stresses can be a more sustainable approach to enhance the yield of soybean crops in flood-affected regions. There is an emergent need to hunt more flood-tolerant PGPR and AM strains displaying plant growth promotion and stress (water) ameliorating abilities for better prospects of soybean crops. Though very large amounts of work have been done on plant growth promotion and stress amelioration on a wide variety of crops under salinity, drought, temperature, and pH stress but very little data is available on the subject of flooding stress. Certain PGPRs and AM fungi, including ACC deaminase-producing bacteria, *Pseudomonas*, *Azospirillum*, *Rhizobium*, and *Bradyrhizobium*, have been known for their role in enhancing soybean growth under flooding stress. A combination of these PGPRs, along with AM fungi in form of bioformulations, could be a novel step in the alleviation of flooding-impacted plants. Also, there are reports available where authors have used the foliar spray of inorganic nutrients and chemical plant growth regulators to minimize the effect of flooding. Hence efforts should be taken to use the mixed inoculum of liquid inorganic nutrients in combination with PGPRs and AM fungi to lessen the injuries of flooding in crop plants. It has been already mentioned in some sections of the review that during flooding, stressed plants cope or tolerate these effects by enhancing proline concentration and by osmoting adjustment in cells. Researchers can also explore proline, glutamine, and trehalose producing beneficial strains for osmotic adjustment that can help stressed plants to cope with the effect of flooding stress. Enhanced production of such a commercially important oil seed crop will serve as a boon not only in raising the economy of developing countries, but also the potentialities of soybean can be utilized by the agroindustries, due to its miscellaneous uses worldwide. Taking the present clues available, rigorous future exploration is required mainly on field assessment and the application of potential flood-tolerant PGPR microorganisms.

Acknowledgments

The authors are thankful to Professor RC Sobti Vice Chancellor BBA University for providing relentless support.

References

Anderson, R.C., Liberta, A.E., Dickman, L.A., 1984. Interaction of vascular plants and vesicular-arbuscular mycorrhizal fungi across a soil moisture-nutrient gradient. Oecologia 64, 111–117.

Bacanamwo, M., Purcell, L.C., 1999. Soybean dry matter and N accumulation responses to flooding stress, N sources and hypoxia. Journal of Experimental Botany 50, 689–696.

Bailey-Serres, J., Fukao, T., Gibbs, D.J., Holdsworth, M.J., Lee, S.C., Licausi, F., et al., 2012. Making sense of low oxygen sensing. Trends in Plant Science 17, 129–138.

Barnawal, N., Bharti, D., Maji, C.S., Chanotiya, A., Kalra, A., 2012. 1-Aminocyclopropane-1-carboxylic acid (ACC) deaminase-containing rhizobacteria protect *Ocimum* sanctum plants during waterlogging stress via reduced ethylene generation. Plant Physiology & Biochemistry 58, 227–235.

Beck, E.H., Fettig, S., Knake, C., Hartig, K., Bhattarai, T., 2007. Specific and unspecific responses of plants to cold and drought stress. Journal of Biosciences 32, 501–510.

Bennett, J.M., Albrecht, S.L., 1984. Drought and flooding effects on N_2 fixation, water relations, and diffusive water resistance of soybean. Agronomy Journal 76, 735–740.

Beutler, A.N., Giacomeli, R., Albertom, C.M., Silva, V.N., da Silva, Neto, G.F., Machado, G.A., Santos, A.T.L., 2014. Soil hydric excess and soybean yield and development in Brazil. AJCS 8, 1461–1466.

Bhattacharyya, P.N., Jha, D.K., 2012. Plant growth-promoting rhizobacteria (PGPR): emergence in agriculture. World Journal of Microbiology & Biotechnology 28, 1327–1350.

Bita, C.E., Greats, T., 2013. Plant tolerance to high temperature in a changing environment: scientific fundamentals and production of heat stress-tolerant crops. Frontiers of Plant Science 4, 273.

Borella, J., do Amarante, L., Santos Colares de Oliveira, D., Barneche de Oliveira, A.C., Braga, E.J.B., 2014. Waterlogging-induced changes in fermentative metabolism in roots and nodules of soybean genotypes. Scientia Agricola 71, 499–508.

Boru, G., Vantoat, T., Alves, J., Hua, D., Knee, M., 2003. Responses of soybean to oxygen deficiency and elevated root-zone carbon dioxide concentration. Annals of Botany 91, 447–453.

Bradford, K.J., Yang, S.F., 1980. Xylem transport of l-aminocyclopropane-1-carboxylic acid, an ethylene precursor, in waterlogged tomato plants. Plant Physiology 65, 322–326.

Carvalho, L.M., Correia, P.M., Martins-Loucao, M.A., 2004. Arbuscular mycorrhizal fungal propagules in a salt marsh. Mycorrhiza 14, 165–170.

Cho, J.W., Yamakawa, T., 2006a. Effects on growth and seed yield of small seed soybean cultivars of flooding conditions in paddy field. Journal-Faculty of Agriculture Kyushu University 51, 189–193.

Cho, J.W., Yamakawa, T., 2006b. Tolerance differences among small seed soybean cultivars against excessive water stress conditions. Journal-Faculty of Agriculture Kyushu University 51, 195–199.

Choi, K.J., Lee, L.S., Kwon, Y.W., 1995. Physiological response of soybean under excessive soil water stress during vegetative growth period. Korean Journal of Crop Science 40, 595–599.

Davanso, V.M., Souza, L.A., Medri, M.E., Pimenta, J.A., Bianchini, E., 2002. Photosynthesis, growth and development of *Tabebuia avellanedae* Lor. ex Griseb. (Bignoniaceae) in flooded soil. Brazilian Archives of Biology Technology 45, 375–384.

Day, D.A., Copeland, L., 1991. Carbon metabolisms and compartmentation in nitrogen fixing legume nodules. Plant Physiology & Biochemistry 29, 185–201.

Dhillion, S.S., Ampompan, L., 1992. The influence of inorganic nutrient fertilization on the growth, nutrient composition and vesicular-arbuscular mycorrhizal colonization of pre-transplant rice (*Oryza sativa*) plants. Biology and Fertility of Soils 13, 85–91.

Dimkpa, C., Weinand, T., Asch, F., 2009. Plant–rhizobacteria interactions alleviate abiotic stress conditions. Plant Cell and Environment 32, 1682–1694.

Dogan, E., Kirnak, H., Copur, O., 2007. Effect of seasonal water stress on soybean and site specific evaluation of CROPGRO Soybean model under semi-arid climatic conditions. Agricultural Water Management 90, 56–62.

Dong, Z., 1999. Soybean Yield Physiology. China Agriculture Press, pp. 103–107.

Eigenbrod, F., Gonzalez, P., Dash, J., Steyl, I., 2014. Vulnerability of ecosystems to climate change moderated by habitat intactness. Global Change Biology 21, 275–286.

Else, M.A., Jackson, M.B., 1998. Transport of 1-aminocyclopropane- 1-carboxylic acid (ACC) in the transpiration stream of tomato (*Lycopersicon esculentum*) in relation to foliar ethylene production and petiole epinasty. Australian Journal of Plant Physiology 25, 453–458.

Fougnies, L., Renciot, S., Muller, F., Plenchette, C., Prin, Y., de Faria, S.M., Bouvet, J.M., Sylla, S.N., Dreyfus, B., Ba, A.M., 2007. Arbuscular mycorrhizal colonization and nodulation improve tolerance in *Pterocarpus officinalis* Jacq. seedlings. Mycorrhiza 17,159–166.

Franceschi, V.R., Nakata, P.A., 2005. Calcium oxalate in plants: formation and function. Annual Review of Plant Biology 56, 41–71.

Githiri, S.M., Watanabe, S., Harada, K., Takahashi, R., 2006. QTL analysis of flooding tolerance in soybean at an early vegetative growth stage. Plant Breeding 125, 613–618.

Glick, B.R., 2014. Bacteria with ACC deaminase can promote plant growth and help to feed the world. Microbiological Research 169, 30–39.

Grable, A.R., 1966. Soil aeration and plant growth. Advances in Agronomy 18, 57–106.

Grichko, V.P., Glick, B.R., 2001a. Ethylene and flooding stress in plants. Plant Physiology & Biochemistry 39, 1–9.

Grichko, V.P., Glick, B.R., 2001b. Amelioration of flooding stress by ACC deaminase-containing plant growth-promoting bacteria. Plant Physiology & Biochemistry 39, 11–17.

Griffin, J.L., Saxton, A.M., 1988. Response of solid–seeded soybean to flood irrigation. II. Flood duration. Agronomy Journal 80, 885–888.

Hao, X.Y., Han, X., Ju, H., Lin, E.D., 2010. Impact of climatic change on soybean production: a review. Ying Yong Sheng Tai Xue Bao 21, 2697–2706.

Harley, J.L., Smith, S.E., 1983. Mycorrhizal Symbiosis. Academic Press, London.

Hattori, R., Matsumura, A., Yamawaki, K., Tarui, A., Daimon, H., 2013. Effects of flooding on arbuscular mycorrhizal colonization and root-nodule formation in different roots of soybeans. Agricultural Sciences 4, 673–677.

Heatherly, L.G., Pringle, H.C., 1991. Soybean cultivars response to flood irrigation of clay soil. Agronomy Journal 83, 231–236.

Hossain, Z., Komatsu, S., 2014. Potentiality of soybean proteomics in untying the mechanism of flood and drought stress tolerance. Proteomes 2, 107–127.

Hou, F.F., Thseng, F.S., 1991. Studies on the flooding tolerance of soybean seed: varietal differences. Euphytica 57, 169–173.

Hou, F.F., Thseng, F.S., 1992. Studies on the screening technique for pregermination flooding tolerance in soybean. Japanese Journal of Crop Science 61, 447–453.

Husen, E., Wahyudi, A.T., Suwanto, A., Giyan, 2011. Growth enhancement and disease reduction of soybean by 1-aminocyclopropane-1-carboxylate deaminase-producing Pseudomonas. American Journal of Applied Sciences 8, 1073–1080.

Isobe, K., Maruyama, K., Nagai, S., Higo, M., Maekawa, T., Mizonobe, G., Drijber, R.A., Ishii, R., 2011. Arbuscular mycorrhizal fungal community structure in soybean roots: comparison between Kanagawa and Hokkaido, Japan. Advances in Microbiology 1, 13–22.

Jackson, M.B., Attwood, P.A., 1996. Roots of willow (*Salix viminalis* L.) show marked tolerance to oxygen shortage in flooded soils and in solution culture. Plant and Soil 187, 37–45.

Jackson, M.B., Colmer, T.D., 2005. Response and adaptation by plants to flooding stress. Annals of Botany 96, 501–505.

Jackson, M.B., Ricard, B., 2003. Physiology, Biochemistry and Molecular Biology of Plant Root Systems Subjected to Flooding of the Soil. In: De Koon, H., Visser, E.J.W. (Eds.), Root Ecology. Springer, Berlin, pp. 193–213.

Jin, J., Wang, G.H., Liu, X.B., Pan, X.W., Herbert, S.J., 2005. Phosphorus regulates root traits and phosphorus uptake to improve soybean adaptability to water deficit at initial flowering and full pod stage in a pot experiment. Soil Science and Plant Nutrition 51, 953–960.

Kadempir, M., Galeshi, S., Soltani, A., Ghaderifar, F., 2014. The effect of flooding and nutrition levels on reproductive growth stages of aerenchyma formation and ethylene production in soybean (*Glycine max* L). International Journal of Advanced Biological and Biomedical Research 2, 487–495.

Kawase, M., 1981. Effects of ethylene on aerenchyma development. American Journal of Botany 68, 651–658.

Keeley, J.E., 1980. Endomycorrhizae influence growth of blackgum seedlings in flooded soils. American Journal of Botany 67, 6–9.

Kolb, R.M., Joly, C.A., 2009. Flooding tolerance of *Tabebuia cassinoides*: metabolic, morphological and growth responses. Flora 204, 528–535.

Komatsu, S., Deschamps, T., Hiraga, S., Kato, M., Chiba, M., Hashiguchi, A., Tougou, M., Shimamura, S., Yasue, H., 2011a. Characterization of a novel flooding stress-responsive alcohol dehydrogenase expressed in soybean roots. Plant Molecular Biology 77, 309–322.

Komatsu, S., Nanjo, Y., Nishimura, M., 2013. Proteomic analysis of the flooding tolerance mechanism in mutant soybean. Journal of Proteomics 79, 231–250.

Komatsu, S., Yamamoto, A., Nakamura, T., Nouri, M.Z., Nanjo, Y., Nishizawa, K., Furukawa, K., 2011b. Comprehensive analysis of mitochondria in roots and hypocotyls of soybean under flooding stress using proteomics and metabolomics techniques. Journal of Proteome Research 10, 3993–4004.

Komatsu, S., Hiraga, S., Yanagawa, Y., 2012. Proteomics techniques for the development of flood tolerant crops. Journal of Proteome Research 11, 68–78.

Komatsu, S., Sakata, K., Nanjo, Y., 2015. Omics techniques and their use to identify how soybean responds to flooding. Journal of Analytical Science Technology 6, 9.

Kramer, P.J., 1951. Causes and injury to plants resulting from flooded soil. Plant Physiology 26, 722–736.

Kramer, P.J., Boyer, J.S., 1995. Water Relations of Plants and Soils. Academic Press, San Diego, CA.

Kwon, Y.W., Lee, M.K., 1988. Physiological responses of soybean plants to flooding at the vegetative growth and the flowering stages. Research Report RDA 31, 289–300.

Landwehr, M., Hildebrandt, U., Wilde, P., Nawrath, K., Toth, T., Biro, B., Bothe, H., 2002. The arbuscular mycorrhizal fungus *Glomus geosporum* in European saline, sodic and gypsum soils. Mycorrhiza 12, 199–211.

Li, G., Meng, X., Wang, R., Mao, G., Han, L., Liu, Y., Zhang, S., 2012. Dual-level regulation of ACC synthase activity by MPK3/MPK6 cascade and its downstream WRKY transcription factor during ethylene induction in Arabidopsis. PLoS Genetics 8, e1002767.

Li, J., McConkey, B.J., Cheng, Z., Guo, S., Glick, B.R., 2013. Identification of plant growth-promoting rhizobacteria-responsive proteins in cucumber roots under hypoxic stress using a proteomic approach. Journal of Proteomics 84, 119–131.

Linkemer, G., Board, J.E., Musgrave, M.E., 1998. Waterlogging effects on growth and yield components in late–planted soybean. Crop Science 38, 1579–1584.

Lobato, A.K.S., Costa, R.C.L., Oliveira Neto, C.F., Santos Filho, B.G., Cruz, F.J.R., Freitas, J.M.N., Cordeiro, F.C., 2008. Morphological changes in soybean under progressive water stress. International Journal of Botany 4, 231–235.

Manavalan, L.P., Guttikonda, S.K., Tran, L.S., Nguyen, H.T., 2009. Physiological and molecular approaches to improve drought resistance in soybean. Plant and Cell Physiology 50, 1260–1276.

Maryam, A., Nasreen, S., 2012. A review: water logging effects on morphological, anatomical, physiological and biochemical attributes of food and cash crops. International Journal of Water Resources and Environment Sciences 1, 113–120.

Mastrodomenico, A.T., Purcell, L.C., King, C.A., 2013. The response and recovery of nitrogen fixation activity in soybean to water deficit at different reproductive developmental stages. Environmental and Experimental Botany 85, 16–21.

Miao, S., Shi, H., Jian, J., Judong, L., Xiaobing, L., Guanghua, W., 2012. Effects of short-term drought and flooding on soybean nodulation and yield at key nodulation stage under pot culture. Journal of Food, Agriculture, and Environment 10, 819–824.

Mustafa, G., Komatsu, S., 2014. Quantitative proteomics reveals the effect of protein glycosylation in soybean root under flooding stress. Frontiers of Plant Science 18, 627.

Mutava, R.N., Prince, S.J.K., Syed, N.H., Song, L., Valliyodan, B., Chen, W., Nguyen, H.T., 2015. Understanding abiotic stress tolerance mechanisms in soybean: a comparitive evaluation of soybean response to drought and flooding stress. Plant Physiology and Biochemistry 86, 109–120.

Naeve, S., 2002. Flooded Fields and Soybean Survival. University of Minnesota. Minnesota Crop eNews, July 14. http://www.extension.umn.edu/cropEnews/2002/02MNCN26.htm.

Nagumo, Y., Sato, T., Hattori, M., Tsuchida, T., Hosokawa, H., Takahashi, Y., Ohyama, T., 2010. Effect of sigmoidal releasing-type coated urea fertilizer and ridge tillage on nitrogen accumulation and rate of side-wrinkled seeds in soybean cultivated in rotated paddy fields under poor drainage conditions. Japanese Journal of Soil Science and Plant Nutrition 81, 360–366.

Nakayama, N., Hashimoto, S., Shimada, S., Takahashi, M., Kim, Y.H., Oya, T., Arihara, J., 2004. The effect of flooding stress at the germination stage on the growth of soybeans in relation to initial seed moisture content. Japanese Journal of Crop Science 73, 323–329.

Nanjo, Y., Nakamura, T., Komatsu, S., 2013. Identification of indicator proteins associated with flooding injury in soybean seedlings using label-free quantitative proteomics. Journal of Proteome Research 12, 4785–4798.

Neto, D., Carvalho, L.M., Cruz, C., Martins-Loução, M.A., 2006. How do mycorrhizas affect C and N relationships in flooded *Aster tripolium* plants? Plant and Soil 279, 51–63.

Nanjo, Y., Skultety, L., Ashraf, Y., Komatsu, S., 2010. Comparative proteomic analysis of early-stage soybean seedlings responses to flooding by using gel and gel-free techniques. Journal of Proteome Research 9, 3989–4002.

Normile, D., 2008. Reinventing rice to feed the world. Science 321, 330–333.

Oh, M.W., Nanjo, Y., Komatsu, S., 2014. Gel-free proteomic analysis of soybean root proteins affected by calcium under flooding stress. Frontiers of Plant Science 5, 559.

Ohyama, T., Minagawa, R., Ishikawa, S., Yamamoto, M., Hung, N.V.P., Ohtake, N., Sueyoshi, K., Sato, T., Nagumo, T., Takahashi, Y., 2013. Soybean Seed Production and Nitrogen Nutrition. Intech, pp. 115–157.

Oosterhuis, D.M., Scott, H.D., Hampton, R.E., Wullschleger, S.D., 1990. Physiological response of two soybean (*Glycine max* L. Merr) cultivars to short-term flooding. Environmental and Experimental Botany 30, 85–92.

Pankhurst, C.E., Sprent, J.I., 1975. Surface features of soybean root nodules. Protoplasma 85, 85–98.

Parvaiz, A., Satyawati, S., 2008. Salt stress and phyto biochemical responses of plants. Plant Soil and Environment 54, 89–99.

Patel, P.K., Singh, A.K., Tripathi, N., Yadav, D., Hemantaranjan, A., 2014. Flooding: abiotic constraint limiting vegetable productivity. Advances in Plants and Agricultural Research 1, 00016.

Pires, J.L.F., Soprano, E., Cassol, B., 2002. Adaptações morfofisiológicas da soja em solo inundado. Pesq Agropec Bras 37, 41–50.

Ragupathy, S., Mohankumar, V., Mahadevan, A., 1990. Occurrence of vesicular arbuscular mycorrhizae in tropical hydrophytes, 2010 mycorrhizae in tropical hydrophytes. Aquatic Botany 36, 287–291.

Rhine, M., Stevens, G., Shannon, G., Wrather, A., Sleper, D., 2010. Yield and nutritional responses to waterlogging of soybean cultivars. Irrigation Science 28, 135–142.

Rizal, G., Karki, S., 2011. Alcohol dehydrogenase (ADH) activity in soybean (*Glycine max* [L.] Merr.) under flooding stress. Electronic Journal of Plant Breeding 2, 50–57.

Robredo, A., Perez-Lopez, U., Mirznda-Apodaca, J., Lacuesta, M., Mena-Petite, A., Munoz-Rueda, A., 2011. Elevated CO_2 reduces the drought effect on nitrogen metabolism in barley plants during drought and subsequent recovery. Environmental and Experimental Botany 71, 399–408.

Russell, D.A., Wong, D.M.L., Sachs, M.M., 1990. The anaerobic response of soybean. Plant Physiology 92, 401–407.

Rutto, K.L., Mizutani, F., Kadoya, K., 2002. Effect of root-zone flooding on mycorrhizal and non-mycorrhiozal peach (*Prunus persica* Batsch) seedlings. Scientia Horticulturae 94, 285–295.

Sahoo, R.K., Ansari, M.W., Pradhan, M., Dangar, T.K., Mohanty, S., Tuteja, N., 2014. Phenotypic and molecular characterization of efficient native *Azospirillum* strains from rice fields for crop improvement. Protoplasma. http://dx.doi.org/10.1007/s00709-013-0607-7.

Saikia, S.P., Bora, D., Goswami, A., Mudoi, K.D., Gogoi, A., 2013. A review on the role of *Azospirillum* in the yield improvement of non-leguminous crops. African Journal of Microbiology Research 6, 1085–1102.

Sallam, A., Scott, H.D., 1987. Effects of prolonged flooding on soybeans during vegetative growth. Soil Science 144, 61–66.

Scholles, D., Vargas, L.K., 2004. Viability of soybean inoculation with *Bradyrhizobium* strains in flooded soil. Revista Brasileira de Ciência do Solo 28, 973–979.

Schöffel, E.R., Saccol, A.V., Manfron, P.A., Medeiros, S.L.P., 2001. Excesso hídrico sobre os componentes do rendimento da cultura da soja. Ciência Rural 31, 7–12.

Scott, H.D., DeAngulo, J., Daniels, M.B., Wood, L.S., 1989. Flood duration effects on soybean growth and yield. Agronomy Journal 81, 631–636.

Sengupta, A., Chaudhuri, S., 1990. Vesicular arbuscular mycorrhiza (VAM) in pioneer salt marsh plants of the Ganges river delta in West Bengal (India). Plant and Soil 122, 111–113.

Setter, T.L., Waters, I., 2003. Review of prospects for germplasm improvement for waterlogging tolerance in wheat, barley and oats. Plant and Soil 253, 1–34.

Shanmugasundaram, S., 1980. The role of AVRDC I the improvement of soybean and mungbean for the developing tropical countries. In: Grain Legume Production in Asia, pp. 137–166.

Shi, F., Yamamoto, R., Shimamura, S., Hiraga, S., Nakayama, N., Nakamura, T., Yukawa, K., Hachinohe, M., Matsumoto, H., Komatsu, S., 2008. Cytosolic ascorbate peroxidase 2 (cAPX 2) is involved in the soybean response to flooding. Phytochemistry 69, 1295–1303.

Shimamura, S., Mochizuki, T., Nada, Y., Fukuyama, M., 2003. Formation and function of secondary aerenchyma in hypocotyl, roots, and nodules of soybean (*Glycine max*) under flooded conditions. Plant and Soil 251, 351–359.

Shleev, S.V., Rozov, F.N., Topunov, A.F., 2001. A method for producing multiple forms of metleghemoglobin reductase and leghemoglobin components from lupine nodules. Applied Biochemistry and Microbiology 37, 195–200.

Singh, K.D., Singh, N.P., 1995. Effect of excess soil water and nitrogen on yield, quality, and N-uptake of soybean (*Glycine max*. (L) Merrill). Annals of Agricultural Research 16, 151–155.

Snellgrove, R.C., Splittstoesser, W.J.E., Stribley, D.P., Tinker, P.B., 1982. The distribution of carbon and the demand of the fungal symbiont in leek plants with vesicular arbuscular mycorrhizas. New Phytologist 92, 75–87.

Sondergaard, M., Laegaard, S., 1977. Vesicular-arbuscular mycorrhiza in some aquatic vascular plants. Nature 268, 232–233.

Sosbai, 2012. Sociedade Sul-Brasileira de Arroz Irrigado Arroz irrigado: recomendações técnicas da pesquisa para o Sul do Brasil. Sosbai, Itajaí, Brazil. p. 17.

Sprent, J.I., Gallacher, A., 1976. Anaerobiosis in soybean root nodules under water stress. Soil Biology & Biochemistry 8, 317–320.

Sugimoto, H., Satou, T., 1990. Excess moisture injury of soybean cultivated in upland field converted from paddy. Japanese Journal of Crop Science 59, 727–732.

Sullivan, M., Van Toai, T.T., Fausey, N., Beuerlein, J., Parkinson, R., Soboyejo, A., 2001. Evaluating on-farm flooding impacts on soybean. Crop Science 41, 93–100.

Sung, F.J.M., 1993. Waterlogging effects on nodule nitrogenase and leaf nitrate reductase activities in soybean. Field Crops Research 35, 183–189.

Taimer, C.C., Clayton, J.S., 1985. Effects of vesicular-arbuscular mycorrhizas on growth and nutrition of a submerged aquatic plant. Aquatic Botany 22, 377–386.

Tewari, S., Arora, N.K., 2013. Plant growth promoting rhizobacteria for ameliorating abiotic stresses triggered due to climatic variability. Climate Change and Environmental Sustainability 1 (2), 95–103.

Tewari, S., Arora, N.K., Miransari, M., 2016. Plant growth promoting rhizobacteria to alleviate soybean growth under abiotic and biotic stresses. In: Miransari, M. (Ed.), Abiotic and Biotic Stresses in Soybean Production. Academic Press, Elsevier, USA, pp. 131–156.

Thomas, A.L., Guerreiro, S.M.C., Sodek, L., 2005. Aerenchyma formation and recovery from hypoxia of the flooded root system of nodulated soybean. Annals of Botany 96, 1191–1198.

Tougou, M., Hashiguchi, A., Yukawa, K., Nanjo, Y., Hiraga, S., Nakamura, T., Nishizawa, K., Komatsu, S., 2012. Responses to flooding stress in soybean seedlings with the alcohol dehydrogenase transgene. Plant Biotechnology 29, 301–305.

Trought, M.C.T., Drew, M.C., 1980. The developmental of waterlogging damage in young wheat plants in anaerobic solution cultures. Journal of Experimental Botany 31, 1573–1585.

Turner, S.T., Amon, J.P., Schneble, R.M., Friese, C.F., 2000. Mycorrhizal fungi associated with plants in ground-water fed wetlands. Wetland 20, 200–204.

Van Toai, T.T., Beuerlein, J.E., Schmitthenner, A.F., St Martin, S.K., 1994. Genetic variability for flooding tolerance in soybean. Crop Science 34, 1112–1115.

Vartapetian, B.B., Jackson, M.B., 1997. Plant adaptations to anaerobic stress. Annals of Botany 79, 3–20.

Voesenek, L.A.C.J., Colmer, T.D., Pierik, R., Millenaar, F.F., Peeters, A.J.M., 2006. How plants cope with complete submergence. New Phytologist 170, 213–226.

Wilson, J.B., 1988. A review of evidence on the control of shoot: root ratio, in relation to models. Annals of Botany 61, 433–449.

Wuebker, E.F., Russell, E.M., Kenneth, K., 2001. Flooding and temperature effects on soybean germination. Crop Science 41, 1857–1861.

Yaklich, R.W., Abdul-Baki, A.A., 1975. Variability in metabolism of individual axes of soybean seeds and its relationship to vigor. Crop Science 15, 424–426.

Youn, J.T., Van, K., Lee, J.E., Kim, W.H., Yun, H.T., Kwon, Y.U., Ryu, Y.H., Lee, S.H., 2008. Waterlogging effects on nitrogen accumulation and N_2 fixation of supernodulating soybean mutants. Journal of Crop Science and Biotechnology 11, 111–118.

Yordanova, R.Y., Popova, L.P., 2007. Flooding-induced changes in photosynthesis and oxidative status in maize plants. Acta Physiologiae Plantarum 29, 535–541.

Zablotowicz, R.M., Eskew, D.L., Focht, D.D., 1978. Denitrification in *Rhizobium*. Canadian Journal of Microbiology 24, 757–776.

Soybean Tillage Stress

3

M. Miransari
AbtinBerkeh Scientific Ltd. Company, Isfahan, Iran

Introduction

The soybean (*Glycine max* (Merr.) L), as a leguminous plant, is an important source of protein and oil. The high rate of oil content (18–22%) makes it a suitable crop for the oil industry. The oil contains a high-quality oil with a suitable rate of unsaturated fatty acids, especially the essential linoleic acid. The soybean is also a great source of protein, with the highest rate of seed protein (35–45%) among the legumes and with favorable amino acids. It is accordingly used as a source of a cholesterol-free diet in human nutrition (Fecák et al., 2010).

The soybean is able to acquire most of its essential nitrogen (N) by developing a symbiotic association with its specific rhizobium, *Bradyrhizobium japonicum*. The bacteria are able to fix the atmospheric N and reduce it to ammonia for the use of their host plant. Such an association is of economic and environmental significance, because it significantly decreases the use of N chemical fertilization. Although N chemical fertilization is a quick method of providing crop plants with their essential N, due to the high solubility of N compounds, it is subjected to leaching and hence results in the contamination of the environment. However, compared with N chemical fertilization, biological N fixation is of high efficiency and can greatly contribute to plant growth and yield production with economic and environmental significance (Miransari, 2011; Miransari et al., 2013).

Soil tillage is an important agronomical practice affecting soil and plant properties. Accordingly, different types of soil tillage are used in the field to make it suitable for planting crops. The most usual methods of soil tillage are conventional tillage (CT), reduced tillage (RT), and no tillage (NT), affecting the physical, chemical, and biological properties of soil. The major reason for cultivating the soil is to make it suitable for the germination of plant seeds and the growth of seedlings and crop plants (Vetsch and Randall, 2002).

Among the most used tillage practices for soybean production is conservation tillage, especially in the United States, Canada, and Western Europe. The related tillage practices are (1) RT, including disking, mulching, and strip tillage; and (2) NT, including direct seeding. It has been indicated that some tillage practices right before planting soybeans is beneficial. Among the most important benefits of NT is the increased water use efficiency and soil health under different climatic conditions, soil types, and agronomical practices. The benefits of NT become evident when it is used continuously and with the crop residue kept on the soil surface. It has been shown that it is possible to produce wheat using NT practice in India, which reduces the rate of production expenses (Malik et al., 2004).

Environmental Stresses in Soybean Production. http://dx.doi.org/10.1016/B978-0-12-801535-3.00003-6

Mulching can increase soybean yield by decreasing soil temperature and water evaporation from the soil. Accordingly, soil water efficiency increases and the plant faces less water deficit (Hatfield et al., 2001). It is important to use tillage practices, which have more benefits and less expenses. It is also pertinent to find the most suitable practices and strategies related to soybean production singly or rotated with the other crop plants, resulting in the sustainability of soybean production as affected by soil tillage, straw mulch, water requirement, and crop establishment.

However, soil tillage also affects soil aeration, soil water, soil reduction and oxidation, soil organic matter, and soil microbial activities. Such alteration of soil properties affects soil microbial growth and activities and hence soil productivity. It is also possible to alleviate the adverse effects of stresses such as compaction, suboptimal root zone temperature, drought, salinity, and heavy metal on the growth of crop plants and soil microbes using the appropriate method of soil tillage (West and Post, 2002; Six et al., 2002; Alvarez and Steinbach, 2009).

Soil tillage and rotation practices are essential to enhance the quality of the soil and hence for the long-term crop production. Alteration of soil tillage and rotation practices result in different soil physical, chemical, and biological properties and hence alter the soil quality index (Ding et al., 2011). If improper tillage practices are used, the quality of soil, including the rate of organic matter, decreases and hence soil efficiency and productivity decreases (Thomas et al., 2007).

The benefits of using the appropriate agronomical practices are (1) maintaining the structure of soil, (2) increased water use efficiency, and (3) enhanced crop production. The positive effects of NT on soil quality enhance the diversity of plant biomass, resulting in higher soil moisture and cooler temperature, decreased soil bulk density, more efficient microbial activity, improved structure of soil, increased soil N, C, and organic content, enhanced cation exchange capacity, and decreased C/N rate (Benitio, 2010; Jina et al., 2011; Aziz et al., 2013).

However, among the most suitable indicators of soil quality are soil biological properties, as such properties are responsive and sensitive to different soil tillage and rotation practices (Kennedy and Papendick, 1995). Soil organic matter is also an important soil parameter, significantly affecting soil physical, chemical, and biological properties. Soil physical properties are also among the important factors affecting the efficiency and productivity of soil (de Moraes et al., 2016).

In this chapter the most important findings, including the most recent ones related to the effects of soil stresses on soybean growth and its symbiotic association with *B. japonicum*, and how such adverse effects may be alleviated using different tillage practices, are presented and analyzed. Such details can be used for the more efficient production of soybeans under different conditions including stress.

Tillage and Soybean Yield Production

Soil tillage affects crop production by affecting soil properties. Mazzoncini et al. (2008) investigated the effects of NT and CT on the production of durum wheat (*Triticum durum* Desf.) and soybeans in a long-term experiment in the field, using a 2-year rotation under Mediterranean conditions. During the 16-year (1990–2005) experiment, NT resulted in a wheat yield of 3.97 Mg/ha, which was 8.9% less than CT at 4.36 Mg/ha. There were significant differences among different tillage practices in 6 out of the 16 seasons.

However, the differences were negligible when the seeds were planted early, with suitable control of weeds and with just rainfall deficiency during the grain-filling period. Similarly for soybeans, NT resulted in less grain yield (2.60 Mg/ha) compared with CT (3.08 Mg/ha), and there were significant differences among different tillage practices in just 4 out of 16 seasons. There was a higher rate of weeds under NT soybeans related to wheat. Tillage did not have any effects on the N concentration of wheat and soybeans. However, although soybean P was not affected by tillage, wheat P (grain and straw) was higher under NT than CT.

Although in some research it has been indicated that there were not significant differences among different tillage practices and under different climatic conditions (West et al., 1996; Yin and Al-Kaisi, 2004), some research has shown a higher rate of soybean grain yield under CT compared with NT (Elmore, 1987; Mazzoncini et al., 2008). The effects of tillage practices are higher on soybean and wheat yield production than their N and P uptake.

According to statistics from Canada (2000), just about 4.8% of agricultural fields were under NT in Quebec, Canada, related to the 18.5% of agricultural fields under CT. It was during the 1980s and early 1990s that Italy started to use minimum tillage and NT instead of CT. It is because such tillage practices are economically more beneficial, can improve the structure of soil, and hence decrease the erodibility of soil, increasing the rate of organic matter and hence the rate and the availability of soil nutrients. However, the Italian farmers were not willing to use such new methods of tillage practices due to the related issues. There was also a need for the increased soil filtration under the Mediterranean climate with a high rate of rainfall, which is achieved by using CT. Among the most important reasons for the reduction of crop yield under NT compared with CT is the decreased rate of temperature and the excess of soil moisture at planting, especially when the crop residues are kept on the surface. This is more evident on the soils with a lower rate of infiltration (Mazzoncini et al., 2008).

Kihara et al. (2011) investigated the rate of biological N fixation in soybeans as affected by RT and CT under the conditions of subhumid Western Kenya. The different use of crop residues was also among the treatments in soybean corn rotation and intercropping practices. The rate of soybean N biological fixation ranged from 41–65%, and it was significantly higher in RT (55.6%) than CT (46.6%). The rate of biological fixation in the upper part of the soybean was equal to 26–48 kg/ha with intercropping and 53–82 kg/ha using rotation. The yearly N uptake by soybean grain was higher in RT (−9 to −32 kg N/ha) than CT (−40 to −60 kg N/ha). The use of P fertilization significantly increased nodule weight by 3–16 times compared with the control treatment. To decrease the rate of soil N mining, soybean residues must be kept on the soil surface at harvest. The use of RT+NT increases the rate of biologically fixed N by soybeans related to CT. The use of chemical P fertilization can improve the weight of nodules (Kihara et al., 2011).

Under the cold and wet soil conditions the process of seed germination is delayed, resulting in a lower growth rate of seedlings and hence decreased rate of grain yield production at the harvest. Gesch et al. (2012) suggested and tested a method for the alleviation of cold and wet conditions early in the season by using the polymer-coated seeds. Because under NT soil gets warm and dry later than CT, the use of early growing genotypes may be a good idea.

Such polymer coatings are activated at the time of increased temperature, and hence can protect the seeds from cold and wet stresses under NT. Accordingly, in a 2-year experiment, the effects of such polymer coatings were tested on two different soybean genotypes, planted at an early time (early- to mid-April) and at the usual time (mid-May). The rate of seed germination was only significantly increased in 2005 by seed coatings at 51% and 35% in genotype 1 and 2, respectively, compared with the control treatment.

However, in 2006 the seed-coating treatment under an unusually dry climate decreased the rate of seed germination, although the yield was not affected. A decreased rate of initial osmotic moisture potential increased the delay of seed germination by the seed coating treatment. Although genotype 1 resulted in a significantly higher rate of soybean yield, the effect of planting date was not significant on the yield of soybeans. The authors accordingly indicated that it is possible to enhance the rate of seed germination early in the season under NT using the seed-coating treatment (Gesch et al., 2012).

Under NT the weed seeds are on the soil surface; however, when using CT, they are more homogeneously distributed in the soil depth. Accordingly, the weed species, which require soil depth for their germination and growth, germinate more efficiently under CT than NT. However, under NT, because of less disturbance of agricultural practices, more annual weed grasses and perennial weeds are found on the soil surface. Under NT, more chemicals must be used to avoid the growth of herbicide resistant weeds. Using the resistant corn and soybean genotypes is among the most useful methods for controlling the crop plants from pathogens; however, tillage can also control such pathogens. Due to the crop residue presence of pathogens, especially during the winter, NT can be a more favorable tillage practice for the growth of pathogens (Whalen et al., 2007).

Whalen et al. (2007) evaluated the growth of corn and soybeans, under different soil tillage practices including CT and NT, by investigating the rate of weeds and pathogens. During the first 2 years the rate of annual grass was higher than the rate of annual broadleaf weeds under NT compared with CT; however, the broadleaf weeds occupied a higher surface area under CT related to NT. Although the rate of crop damage by pathogens was less than the economic thresholds, during the 5-year of the experiment (2000–2004), CT resulted in a higher rate of silage corn and soybean yields, compared with NT. The authors accordingly indicated that the decreased rate of yield under NT, related to CT, was due to less root growth, resulting from the less favorable properties of the seed growth medium rather than the effects of biotic stresses such as pathogens.

West and Post (2002) investigated the effects of different agronomical practices, including tillage and rotation on the sequestration of atmospheric CO_2. A meta-analysis of 67 data sets from long-term agricultural experiments was conducted. According to the results, on average, the rate of C sequestration increased by $57 \pm 14\,g\,C/m^2/year$ when NT was used instead of CT, although this was not the case for wheat-follow. A more complex use of rotation enhanced C sequestration by $20 \pm 12\,g\,C/m^2/year$, although this was not the case when corn–soybean rotation was used instead of continuous corn. About 5–10 years is essential for the highest rates of C sequestration and 15–20 years for equilibrium. However, when using a new complex of rotation, about 40–60 years will be essential for the equilibrium of soil organic carbon. The estimation of carbon sequestration rates for different crop plants and agronomical practices in this research can be used for modeling usages to determine the potentials of C sequestration (West and Post, 2002).

Different methods, practices, and strategies can be used to improve the efficiency of agricultural fields including the use of conservation tillage such as NT, use of soil amendments, organic farming, increasing soil organic matter, and keeping crop residue on the soil surface (Seufert et al., 2012; Abdollahi et al., 2015; de Moraes et al., 2016).

Among such methods, NT is among the most important strategies, essential for the sustainable production and for the future demand of the increasing world population. The beneficial effects of NT include the enhancement of soil quality by (1) decreasing the erodibility of soil, (2) increasing the rate of organic matter, (3) improving the structure of soil, (4) enhancing biodiversity, and (5) increasing biological activity (Anken et al., 2004; Lal, 2007; Alvarez and Steinbach, 2009; Babujia et al., 2010).

Different parameters including the rate, source, number of use, crop strategies, and the growing season affect plant response to organic amendments. The favorable effects of organic amendments on soil properties increase the rate of soil organic matter and are also a suitable source of nutrients for plant and soil microbes. For example, treating the agricultural fields, which had received manure, increased soybean grain yield under NT. The rate of soil C also increased by 22% in such fields (Singer et al., 2008).

Singer et al. (2008) evaluated the effects of soil tillage and compost on soybean growth, nutrient uptake, and water use efficiency of soil. The rotation was a corn-soybean-wheat-clover, using compost during the fall, and started in 1998 in the field treated with NT, chisel plow (CT), and moldboard plow (MP) from 1988. There was a significant interaction between NT and compost; however there was not any significant differences between MP and CT. The rate of crop yield under different tillage practices was similar in 2003, but the rate of crop yield under NT was 15% higher than MP and CT in 2004 and 8% less in 2005. The average of yield was similar across the 3 years.

The effect of compost was just significant in 2004 by increasing the crop yield at 9% related to the control treatment and an average of 2% across the 3 years. Compost and tillage practices did not have significant effects in soil water from 0 to 90 cm. NT resulted in a higher P and K uptake at 18% and 16% related to MP and CT, and compost resulted in a 16% and 13% higher rate of P and K, respectively, in 2004. Compost also resulted in a lower rate of soybean grain protein compared with the control treatment. However, there was a lower rate of oil at 2 % and 1% in the control rather than the compost treatment in 2003 and 2004, respectively (Singer et al., 2008).

Soybean, Stress, and Tillage

Soybeans and Soil Compaction

The use of agricultural machinery, especially under a high rate of moisture, results in the compaction of soil. Under soil compaction, the growth of crop plants and hence the rate of yield production decreases, which is due to the following reasons: (1) increased rate of soil bulk density, (2) decreased structure of soil, (3) decreased rate of macropores, (4) increased rate of micropores, (5) emission of greenhouse gases from the soil, (6) decreased rate of oxygen in the soil, (7) decreased root growth, (8) decreased biological activity, (9) decreased water and nutrient uptake by plants, (10) pollution of the environment, and (11) increased rate of nutrient loss (Franchini et al., 2012; Miransari, 2013a; Nawaz et al., 2013; Soane and van Ouwerkerk, 2013; Chen et al., 2014).

The effects of soil tillage on soil properties are a function of agricultural practices, time, and space (Wang and Shao, 2013), resulting in a high variability of results. Such details indicate the importance of long-term experimental data so that the results can be evaluated more precisely. Different methods, as indicated in the following, have been suggested to alleviate the adverse effects of soil compaction on the properties of soil and hence plant growth and yield production. Such methods and strategies can improve the properties of soil and hence plant growth and yield production by (1) improving the structure of soil, (2) increasing the rate of soil organic matter, (3) decreasing soil bulk density, (4) enhancing the rate of biopores by increasing root growth, (5) decreasing the emission rate of greenhouse gases, and (6) enhancing biological activity (Calonego and Rosolem, 2010; Miransari et al., 2008, 2009a,b; Miransari, 2013a; de Moraes et al., 2016).

(1) The use of different tillage practices such as the use of chiseling and subsoiling: although such methods may alleviate the stress of compaction, such stress alleviation is not for a long time and is time- and labor-consuming. (2) The use of proper crop rotation is among the methods that can alleviate the stress of A by increasing the rate of soil biopores and biological activities. Compared with the pores resulting from soil tillage, the biopores produced by plant roots are continuous and long and can favorably increase plant roots by facilitating the circulation of water and air, including oxygen in the soil. They can also be used by roots of the subsequent crop (Calonego and Rosolem, 2010).

(3) Using NT, which can decrease the use of agricultural machinery in the field, increases the rate of crop residue on the soil surface and hence improves the properties of the soil and increases plant growth and yield production (DeLaune and Sij, 2012; Hansen et al., 2012). (4) The increased use of soil organic matter in the field by, for example, using manure, which can improve the structure of soil and enhance soil biological activities (Hazarika et al., 2009). (5) Using biological methods such as the use of soil microbes including arbuscular mycorrhizal (AM) fungi (Miransari et al., 2008, 2009a,b; Miransari, 2013a).

Among the suggested methods, the combined use of NT, organic matter, and biological methods may be the most effective for alleviating the stress of compaction, because they are economically and environmentally of significance. Miransari et al. (2007, 2008, 2009a,b) investigated the biological method of using AM fungi under field and greenhouse conditions. Using a tractor, the field soil was compacted to the favorable levels, and corn seeds were inoculated with the inoculums of mycorrhizal fungi at seeding in a 2-year experiment. Corn plants were irrigated during the growing season, and different soil and plant parameters were determined to find the effects of AM fungi on corn growth and yield as affected by soil compaction.

Different soil and plant parameters, including soil bulk density, soil resistance at a certain moisture, plant leaf weight, and corn grain yield, were determined. The results indicated that the tractor compacted the soil and increased soil bulk and soil resistance. Stress decreased plant growth and yield production, as the symptoms of stress including plant pale color, decreased plant height, and root cluster growth were evident. However, AM fungi were able to alleviate the stress, although with increasing the level of stress the effectiveness of mycorrhizal fungi on the alleviation of soil compaction decreased.

The experiments were also conducted under growth chamber and greenhouse conditions, and wheat and corn seeds were planted in 10-kg pots and were inoculated with mycorrhizal fungi in a 2-year experiment. Different soil and plant parameters were determined during the season. According to the results, the use of weights, which were released from a 20-cm height, compacted the soil, as higher soil bulk densities and soil resistance under soil compaction treatments resulted. However, the fungi were able to alleviate the stress by increasing plant growth parameters (leaf and root growth) and grain yield.

Soybean and Drought

Drought stress adversely affects plant growth and yield production by decreasing the uptake of water and nutrients by plants. Plants use different morphological and physiological mechanisms to alleviate drought stress, including the alteration of leaves and root growth and the production of different metabolites such as proline, which is able to adjust the cellular water potential. Different methods have been used to alleviate the stress of drought on plant growth and yield production including the use of tolerant genotypes, the increased use of organic matter in the field, the use of appropriate tillage and rotation strategies, and the use of soil microbes (Marino et al., 2007; Silvente et al., 2012).

The use of organic matter is among the most useful methods of alleviating drought stress on plant growth and yield production. It is because soil organic matter (1) absorbs a high rate of water, which becomes readily available to plants; (2) is a source of different nutrients for plant and microbial use; (3) enhances the structure of soil; (4) decreases the rate of water loss from the soil surface; (5) adjusts the temperature of soil; and (6) increases the population and activity of soil microbes (Peralta and Wander, 2008; He et al., 2009; Zhu and Cheng, 2011; Miransari, 2011, 2013b).

Among the most useful methods of soil tillage on drought stress is NT, because under no tillage the presence of crop residue on the soil surface can act as a favorable source of organic matter, with respect to the above mentioned details. Soil organic matter can significantly enhance the properties of soil and hence increase plant tolerance under stress. Under NT the microbial decomposition of soil organic matter decreases because soil organic matter is not incorporated in the soil and hence it is decomposed in a longer time (Fernández et al., 2008; Hammerbeck et al., 2012).

Using a pot experiment, Xue et al. (2013) investigated the effects of mulching (control, 3750, 7500, 11,000, and 14,750 kg/ha) on the morphological, biochemical, and physiological properties of soybean under water stress. Soybean plants were subjected to drought stress at the beginning of blooming as the plants were not watered for 15 days and were sampled every 5 days for the content of proline and malondialdehyde. Mulch treatments significantly improved different morphological parameters of soybean growth including plant height, the growth of the aerial part, the leaf area, and the number of leaves per plant.

The physiological parameters enhanced by the mulch treatments, compared with the control treatment, were the rate of photosynthesis, the rate of transpiration, the conductance of stomata, and the intercellular concentration of CO_2. The parameters related to the gas exchange, including the rate of photosynthesis as well as the rate of transpiration, were the most affected by the highest rate of mulch treatment (11,000 kg/ha). The rate of malondialdehyde in plant cells decreased by the mulch treatments, alleviating the adverse effect of drought on the membrane damage.

There was a linear and significant correlation between the rate of mulch treatments and the rate of free proline contents, as the higher rates of mulch treatment resulted in the higher rates of cellular proline content. The authors accordingly indicated that it is possible to alleviate the adverse effects of drought stress on soybean growth and physiology using the mulch treatments, which is a good idea for planting soybeans under drought conditions (Xue et al., 2013).

In a factorial experiment, on the basis of completely randomized block design with four replicates, Fecák et al. (2010) investigated the effects of different tillage practices including CT, RT (to a depth of 10 cm), and NT and different rates of N fertilization including 25 and 50 kg/ha on the growth of soybeans in 2006–2008. The most significant parameter on the rate of soybean yield was the climatic conditions. The highest rate of soybean yield was related to 2008 at 2.77 t/ha, followed by 2.34 t/ha in 2006, and 1.98 t/ha in 2007. The seed-filling stage was the most sensitive under drought conditions, decreasing the rate of soybean grain yield. Compared with tillage and N fertilization treatments, the effects of climatic conditions were also significant on soybean protein and oil.

Seed protein was highly and negatively correlated with the amount of rainfall; however, the correlation of seed oil and rainfall was positive and significant. The effects of tillage was significant on soybean grain yield, as the highest was resulted by CT at 2.60 t/ha, followed by RT (2.39 t/ha), and NT (2.11 t/ha). Accordingly, the authors indicated that using NT may not be a favorable method of tillage for soybean production on heavy soils. However, the use of starter N fertilization had significant effects on the grain yield of soybeans, and 25 kg/ha resulted in the highest soybean grain yield, which was 0.05 t/ha higher than the rate of soybean grain yield using N fertilization at 50 kg/ha (Fecák et al., 2010).

Soybean and Acidity

Soil acidity is an important stress factor negatively affecting crop yield and root growth. Under lower concentrations of base cations, particularly calcium, the high concentration of aluminum root growth is adversely affected and hence water and nutrient uptake and the subsequent crop yield decreases (Marsh and Grove, 1992; Miransari and Smith, 2007), especially under tropical and subtropical conditions.

Under such climatic conditions the use of NT with crop rotation is among the most useful practices enhancing the sustainability of agriculture and soil nutrient efficiency by decreasing the erodibility of soil. The rate of agricultural fields under NT was at 20 million hectares in 2008. This type of tillage strategy is especially practiced in areas of Brazil with high concentrations of Al. Surface liming is one of the most efficient strategies for the amelioration of soil acidity; although it is a quick method for the alleviation of surface acidity, a longer time is essential for the amelioration of subsoil acidity, especially with a high rate of acidity and Al^{3+} (Caires et al., 2008).

The most important parameters affecting the movement of lime into the deeper depths of soil are the properties of the soil, the rate of surface pH and liming, climate, cropping strategies, and the use of acidic fertilization (Blevins et al., 1978; Oliveira and Pavan, 1996; Conyers et al., 2003; Caires et al., 2006). Subsoil acidity is an important factor negatively affecting crop yield production, especially in areas with water deficiency (Marsh and Grove, 1992).

Different soil chemical parameters are affected by soil tillage, affecting plant growth and yield production; however, the presence of crop residues on the soil surface (soil organic matter) under NT may alleviate the adverse effects of aluminum on plant growth and yield production (Blevins et al., 1978; Bayer et al., 2000; Rhoton, 2000). It is mainly because the increased rate of cation exchange capacity by soil organic matter can increase the soil capacity for the absorption and hence exchange of Ca^{2+}, which also alleviates the adverse effects of Al^{3+} on root growth (Alva et al., 1986). Soil organic matter on the soil surface can also decrease the rate of evaporation and enhance the rate of soil moisture, increasing the availability of nutrients under acidic conditions. It has been indicated that the adverse effects of acidity on plant growth and yield production is alleviated under NT; however, more detail for the related causes is essential (Caires et al., 2008).

The use of NT is rapidly increasing in Brazil, and due to the presence of soil acidity, lime is surface broadcast and is not incorporated. Caires et al. (2008) investigated the effects of previously and newly applied liming under NT on the yield and root growth of corn, wheat, and soybeans in a 3-year field experiment. Treatments included (1) a control (no lime), (2) 3 t/ha liming in 2000, (3) 6 t/ha liming in 1993, and (4) liming in 1993 and reliming in 2000. Corn was planted in 2000–2001 and soybeans in 2001–2003 and 2002–2003 without a water deficit, and wheat was planted in 2003 with water deficiency during the vegetative growth and the initiation of flowering.

The liming treatment in 2000 increased the soil pH and the rate of exchangeable Ca^{2+}; however, it decreased the rate of exchangeable Al^{3+} in the 0–5 cm of soil surface. The 1993 liming treatment alleviated the soil pH stress compared with the control treatment and decreased Al concentration to the soil depth of 60 cm. The previously treated plots with liming, which had been relimed in 2000, had a higher pH to the soil depth of 10 cm just 1 year after liming; however, it was to the 60 cm depth 3 years after liming. These results indicate that the slight acidity of surface soil resulted in a deeper treatment of the liming in 2000 (Caires et al., 2008).

The density of root length and corn and wheat grain yields was not affected by surface liming. However, there was a 100% increase in the wheat root length density at the depths of 0–10 and 10–20 cm by the 2000 liming treatment on the previous lime plots, resulting in a 210% increase in wheat grain yield. Soil exchangeable Al^{3+} at concentrations higher than 3 mmol (+)/dm^3 decreased root growth rate. There was a high and significant correlation between wheat grain yield and root length per soil surface area. Accordingly, the authors indicated that under the conditions of high rainfall and NT the Al toxicity decreases; however, under water deficit conditions, Al adversely affected root growth and crop yield (Caires et al., 2008).

Soybean and Suboptimal Root Zone Temperature

NT and crop rotation are among the most usual types of tillage used for corn (*Zea mays* L.) and soybean production in the Midwest of the United States. However, latitude and cropping strategies determine the benefits of NT. Accordingly, Wilhelm and Wortmann (2004), in a 16-year period, investigated the effects of precipitation and seasonal temperature affecting different tillage strategies including disking, plowing, chiseling, subsoiling, ridge tillage, and NT, as well as the growth and yield of corn and soybean rotated and continuously grown under the rain-fed conditions of southern Nebraska.

Under the higher temperatures of summer, a lower rate of grain yield was produced by corn and soybeans. Less rainfall in the spring and higher rainfall in the summer resulted in higher corn grain yield. Although corn grain yield was affected by tillage and rotation practices, soybean yield was affected by just rotation. The rate of corn grain yield was less under no till than under plowing. There were significant interactions between tillage and year on the yield of both crop plants, and there were less advantages of plowing under the warmer conditions of summer.

Under the favorable conditions the response of soybean to chisel was less than the other tillage practices. The yield of soybean (2.6 vs. 2.4 Mg/ha) and corn (7.1 vs. 5.8 Mg/ha) was higher under rotation, related to the continuous cropping. Cooler springs resulted in higher corn yield under rotation. However, for soybeans the benefits of rotation were not a function of weather. Although corn yield under rotation and tillage was affected by seasonal temperature and rainfall, soybean yield was just as affected by seasonal temperature under different tillage practices.

The benefits of corn rotated with soybeans are higher than continuous corn, because of different advantages such as the need for less N fertilization and a lower rate of N deep leaching. Meese et al. (1991) incanted that when corn was rotated with soybeans, there were not significant differences with the other types of soil tillage; however under monocropping, the yield of corn and soybeans was less than the other tillage practices. Similar results were obtained by Pedersen and Lauer (2004) as they found that soybeans rotated with corn resulted in higher yield compared with monocropping under no till. da Veiga et al. (2007) compared different tillage strategies with respect to their rate of compressibility and found that crop yield was higher under NT related to the other tillage practices.

Using crop models, it is possible to evaluate the effects of different tillage strategies on crop growth and yield production. A tillage model (Andales et al., 2000) was used for the evaluation of soybean growth and yield in Iowa, USA. The model was able to properly indicate soil bulk density, surface organic matter residue, and hydraulic conductivity in agreement with previous research. The soil temperature was appropriately calculated by the model at the 6 cm depth under NT ($R^2 = 0.81$), moldboard ($R^2 = 0.81$), and chisel plow ($R^2 = 0.72$) and also indicated cooler temperatures and hence delayed seed germination during early spring under NT. However, there were no significant differences in the temperature of the three different tillage practices.

There were great predictions of soybean biomass production and phenology in 1997, as for example, for pod production under NT moldboard, and for chisel plow the related determination values were equal to 0.95, 0.98, and 0.97, respectively. The model also properly indicated the related differences of different soybean components including canopy height and weight, leaf, stem and pod weight, and nodule number resulted by different tillage practices during the vegetative and reproductive growth stages. The validation of the model also indicated the proper yield production by soybeans under different tillage practices. A sensitivity analysis indicated that the yield and canopy weight calculated by the model were not significantly affected by different tillage practices. Due to a delayed seed germination, the model indicated that the yield under NT was less compared with the other tillage practices.

Aziz et al. (2013) investigated the effects of different tillage (two treatments) and rotation (three treatments) practices on soil quality using a factorial experiment on the basis of completely randomized block design from 2002 to 2007. Tillage practices of CT and NT and the rotation practices of corn-soybean-wheat-cowpea, corn-soybean, and continuous corn were used as the experimental treatments. From each soil depth of 0–7.5, 7.5–15, 15–22.5, and 22.5–30, 10 soil samples were collected and their physical, chemical, and biological properties were determined.

The related quality indices of physical, chemical, and biological, as well as the composite were calculated. The effect of NT was significant on soil physical, chemical, and biological properties. Accordingly, NT resulted in a higher soil quality index compared with conventional tillage. The authors indicated that soil biological properties are suitable indicators of soil quality in response to different soil tillage and rotation practices (Aziz et al., 2013).

In other research, de Moraes et al. (2016) investigated the long-term effects of tillage practices and cropping strategies on soil physical quality and hydraulic properties in Southern Brazil. The experiment was a 5×2 factorial on the basis of a completely randomized block design with four replicates, including five different tillage practices and two cropping strategies with an emphasis on the use of NT to improve the quality of soil.

The following soil tillage practices were used: (1) continuous NT for 11 years, (2) continuous NT for 24 years, (3) CT, (4) minimum tillage (MT) with chiseling soil each year, and (5) MT with chiseling soil every 3 years. The two cropping strategies consisted of two different rotations of 2 and 4 years with crop plants such as corn, soybeans, wheat, etc. in winter and summer. Soil physical properties including soil bulk density, rate and distribution of soil pores, soil moisture behavior, hydraulic conductivity, and infiltration rate were determined using the undisturbed soil samples from the depths of 0–0.10, 0.10–0.20, and 0.20–0.30.

Although the interaction of soil tillage and cropping strategies were not significant on soil physical quality and hydraulic conductivity, regardless of cropping strategies, the persistence of chiseling on soil physical properties were for less than 22 months to the depth of below 0.20 m. CT increased the soil macropores and decreased soil bulk density compared with the other soil tillage practices. CT also decreased macropores to unfavorable rates for plant growth and increased soil bulk density at the depths of less than 0.10 m.

Soil physical quality was improved by continuous NT at the deeper soil depths and increased the availability of soil water for plant use compared with CT and MT with chiseling each year. The authors accordingly indicated that the use of NT results in the improvement of soil physical quality and enhances the availability of soil water for plant use with time. The use of periodic chiseling for the alleviation of soil compaction is not recommended, because such effects do not last a long time (de Moraes et al., 2016).

The rate of agricultural fields in the world has increased from 45 million ha in 1999 to 111 million ha in 2009. The most important reasons for such rapid increase of agricultural fields under NT are (1) deceased erodibility of soil, (2) decreased use of fuel and labor, and (3) saving time (Lal, 2007). However, the compaction of soil is the most important concern of agricultural fields under NT in different parts of the world (Anken et al., 2004; Alvarez and Steinbach, 2009; Chen et al., 2014; de Moraes et al., 2016).

Soybeans, Tillage, and Microbial Activities

Biological properties are among the most important effects of soil tillage practices on soil properties. Microbes are an essential part of the soil environment with beneficial effects on the soil properties. It has been indicated that under extensive crop production, the method of direct seeding increases the number of soil microbes (Frey et al., 1999; Feng et al., 2003). However, soil tillage can also adversely affect soil microbes by decreasing the rate of soil moisture and the structure of soil, compaction of soil, porosity of soil, and decreasing the rate of food availability for soil microbes.

With respect to the properties of different plant species and the agronomical practices the diversity of soil microbes is different. Accordingly, crop plants that are rotated are healthier than the crop plants that are planted under monocropping, because crop rotation decreases the rate of unfavorable microbes (Li et al., 2006). A higher rate of soil microbial diversity increases the buffer capacity of soil under unfavorable conditions (Ibekwe et al., 2007; Gil et al., 2011).

The method of phospholipid fatty acid (PLFA) is suitable for the determination of soil microbial biomass and diversity in the soil. Using such a method, Gil et al. (2011) indicated that under long-term research conditions, the number of soil microbes significantly decreased using RT and monoculture. The rhizosphere of different plant species is different, which is mainly due to the production of different compounds affecting the microbial composition (Shen et al., 2011; Miransari, 2014). Accordingly, the availability of soil nutrients, which are used by plants and microbes, is affected by soil tillage and hence plant growth, and microbial activities are also affected. The use of rotation increased soil organic matter at 20% when corn was planted before soybeans, compared with soybean monocropping (Hinsinger et al., 2009; Yu et al., 2010).

Accordingly, Gil et al. (2011) had indicated that the rate of corn residue in the field was twice the rate of soybean residue, explaining the higher rate of soil organic matter in the rotation of corn and soybeans. The high rate of carbon in the corn residue is a suitable source of food for soil microbes affecting their activity and growth. However, the effect of preceding corn residue on the following soybeans in the rotation and the related microbial activities is a function of climate, soil type, and the other agronomical practices affecting the rate of soil organic matter. It has also been indicated that soil tillage affects the branched type of PLFA, which is indicative of gram positive bacteria. Soil tillage also resulted in the alteration of other fatty acids, which are indicative of gram negative bacteria, under NT and MT for soybean monocropping (Gil et al., 2011).

Teasdale et al. (2007) compared the effects of NT and organic farming on the grain yield of corn, soybeans, and wheat under erodible drought conditions in an experiment from 1994 to 2002. Different tillage practices were compared including (1) NT with the essential rate of N fertilization and herbicide, (2) an NT with cover crop, (3) an NT with vetch as organic mulch, and (4) organic farming using a chisel plow with manure and cover crops for nutrients weed control using postplanting cultivation.

The yield of corn decreased by 12% and 28% under NT with N fertilization and herbicide and organic farming, respectively, compared with standard NT. However, there were no differences between corn yields under NT with cover crop and standard NT. Corn yield was increased by 18% and 19% using organic farming and NT with N fertilization and herbicide, related to standard NT. However, corn yield was not different under standard NT and NT with cover crop.

Using different methods of soil, N testing including presidedress soil nitrate test indicated that the rate of N available to corn was higher under organic farming and NT with N fertilization and herbicide compared with the standard NT. The authors accordingly indicated that organic farming results in more benefits for corn production in the long term, related to standard NT; however, the disadvantage of organic farming is the process of weed controlling.

Al-Kaisi and Kwaw-Mensah (2007) investigated the interactions of soil tillage, N, and P on the yield of corn rotated with soybeans. The experiment was a split plot on the basis of completely randomized design in three replicates. Different tillage systems including NT, strip tillage, and chisel plow were devoted to the main plots, and in each tillage method, four rates of N (0, 85, 170, and 250 kg N/ha) as the sources of manure (total N) and chemical fertilizer (anhydrous ammonia) were used as the subplots using a corn-soybean rotation. There were no effects of tillage and N fertilizer higher than 85 kg/ha on using the two sources of N. There was a significant interaction effect of tillage and N on the rate of plant N and P uptake, particularly during the early plant growth using the two sources of N. Just 40% and 12% of N was utilized by plants using manure and chemical fertilization, respectively, during the corn 12th-leaf growth stage (V12). The rate of plant N and P uptake at V12 was equal to 44% and 37%, averaged for different tillage practices, N sources, and N rates. The authors accordingly indicated that the rate of plant N and P uptake was more affected by N rate than the tillage practices.

Yang et al. (2013) investigated the activity of soil microbes and the related carbon cycle in the rhizosphere and in the bulk soil as affected by long-term NT and crop residue under a 10-year wheat-soybean maize-winter rotation. The experiment was a factorial in which the conventional and NT were used as the experimental factors with or without crop residue. The microbial activity in the rhizosphere was much higher than the bulk soil. The rate of soil organic matter and microbial carbon under the experimental treatments of NT and without residue were significantly higher than the bulk soil.

In the presence of crop residue, soil microbes preferentially used such sources of food for their growth and activity. The effect of NT was significant on the use of carbon sources by soil microbes compared with conventional tillage. Accordingly, the authors indicated that soil microbial activities and the utilization of carbon sources are significantly affected by soil tillage practices, and NT resulted in the highest rate of soil microbial activities and carbon cycle.

The beneficial soil microbes have a wide range of activities in the soil including (1) alleviation of soil stresses, (2) fixation of nutrients such as N and P for the use of their host plant, (3) recycling of organic matter affecting the structure of soil, (4) interaction with the other soil microbes, (5) production of plant hormones, and (6) controlling pathogens (Compant et al., 2010; Miransari, 2011, 2013b, 2014; Faust and Raes, 2012).

The most active part of the soil is the rhizosphere, with a wide range of soil microbes interacting with each other and plant roots. The properties of the rhizosphere are significantly different from the bulk soil, positively affecting plant growth and yield production (Hinsinger et al., 2009). Root products are an important source of food and can also act as signals affecting the growth and activity of soil microbes and their interactions with the host plant (Hartmann et al., 2009).

Soil properties, including the activity of soil microbes, are affected by soil tillage. The adverse effects of CT on soil biological properties have resulted in shifting the agronomical practices toward the use of RT and NT practices with crop rotation. Accordingly, soil properties, including the physical, chemical, and biological ones, are positively affected by such types of tillage practices (Yang et al., 2013). The use of different sources of organic matter has also been another important contribution to the improved properties of the soil including the biological ones (Jordán et al., 2010).

Among the most important effects of NT on soil properties is the increased rate of soil organic matter affecting different soil properties and productivity. For example, Franchini et al. (2007) indicated that use of NT increased soil microbial biomass by 100% in less than 5 years, positively affecting soil microbial diversity and activity.

Shi et al. (2015) investigated the effects of crop and tillage practices during winter affecting soil P properties. Using CT (moldboard plowing), NT, and P fertilization (0, 17.5, 35 kg P/ha) as the experimental treatments the authors evaluated the properties and activities of soil microbes and the related enzymatic activities in a long-term

18-year experiment (from 1992) with the rotation of corn and soybeans in the province of Quebec, Canada.

The soil samples were taken from two different depths (0–10 and 10–20 cm) following the corn and soybean growing seasons. Although soil microbial biomass increased during the winter, the related enzymatic activities decreased compared with the autumn season. The total rate of phospholipid fatty acids of bacteria and mycorrhizal fungi increased following P fertilization after corn. The structures of soil microbial community following corn and soybeans were different. The authors accordingly indicated that soil microbial communities are significantly affected by winter, crop type and rotation, tillage practices, and chemical P fertilization.

The activity of soil microbes during the winter is mainly due to the presence of liquid water in the soil at the temperature of near zero. The presence of snow on the soil surface can also result in soil microbial activities by keeping the soil warm enough. However, temperatures less than zero alter the community and activity of soil microbes, as the more resistant soil microbes will be active. Such details indicate the importance of winter as an important part of the year, affecting the soil microbial community and activities and hence must be considered in ecological research (Shi et al., 2015).

Soil microbes can substantially affect the P behavior and P organic mineralization by producing different enzymes including phosphatases. Crop residue can also be mineralized by such activities affecting the use of soil P by crop plants and hence the microbial activities during the winter. The other important role for P is its effects on the activities of soil microbes, especially under Arctic conditions (Shi et al., 2013).

Conclusion and Future Perspectives

The soybean, the most important legume plant, was investigated in this chapter as it is affected by tillage practices under stress. Selecting the right tillage method for plant growth and yield production is among the most significant parameters affecting the productivity of agricultural fields. However, it is also important to utilize the right tillage method under stress, so that the adverse effects of stress are alleviated. Different tillage practices are used in different parts of the world including CT, RT, or MT, and NT. With respect to the benefits of RT, MT, and NT, including the economic and environmental ones, more agricultural fields are now cultivated using such methods, compared with CT. Under different soil stresses, including compaction, drought, acidity, and suboptimal root zone temperature, the right tillage method must be selected so that the adverse effects of stress can be alleviated. RT, MT, and NT can improve the properties of soil, including the physical, chemical, and biological ones, by enhancing the rate of soil organic matter, increasing the rate of soil moisture, adjusting the soil temperature, and increasing the activity of soil microbes. The most important details related to how the selection of the right tillage practice can alleviate the adverse effects of stress on soybean growth and yield production has been reviewed, presented, and analyzed in this chapter. Future research may investigate the other related details, which can improve the efficiency of tillage practices and hence the growth of soybean plants under stress. The increased efficiency of the tillage method under stress may result by combining it with other suitable parameters such as beneficial soil microbes and tolerant crop species.

References

Abdollahi, L., Hansen, E.M., Rickson, R.J., Munkholm, L.J., 2015. Overall assessment of soil quality on humid sandy loams: effects of location, rotation and tillage. Soil and Tillage Research 145, 29–36.

Alvarez, R., Steinbach, H., 2009. A review of the effects of tillage systems on some soil physical properties, water content, nitrate availability and crops yield in the Argentine Pampas. Soil and Tillage Research 104, 1–15.

Al-Kaisi, M., Kwaw-Mensah, D., 2007. Effect of tillage and nitrogen rate on corn yield and nitrogen and phosphorus uptake in a corn-soybean rotation. Agronomy Journal 99, 1548–1558.

Alva, A., Asher, C., Edwards, D., 1986. The role of calcium in alleviating aluminum toxicity. Australian Journal of Agricultural Research 37, 375–382.

Andales, A., Batchelor, W., Anderson, C., Farnham, D., Whigham, D., 2000. Incorporating tillage effects into a soybean model. Agricultural Systems 66, 69–98.

Anken, T., Weisskopf, P., Zihlmann, U., Forrer, H., Jansa, J., Perhacova, K., 2004. Long term tillage system effects under moist cool conditions in Switzerland. Soil and Tillage Research 78, 171–183.

Aziz, I., Mahmood, T., Islam, K., 2013. Effect of long term no-till and conventional tillage practices on soil quality. Soil and Tillage Research 131, 28–35.

Babujia, L., Hungria, M., Franchini, J., Brookes, P., 2010. Microbial biomass and activity at various soil depths in a Brazilian oxisol after two decades of no-tillage and conventional tillage. Soil Biology and Biochemistry 42, 2174–2181.

Bayer, C., Mielniczuk, J., Amado, T., Martin-Neto, L., Fernandes, S., 2000. Organic matter storage in a sandy clay loam acrisol affected by tillage and cropping system in southern Brazil. Soil and Tillage Research 54, 101–109.

Benitio, A., 2010. Carbon accumulation in soil. Ten year study of conservation tillage and crop rotation in a semi-arid areas of Castile-Leon, Spain Aurora Sombrero. Soil and Tillage Research 107, 64–70.

Blevins, R., Murdock, L., Thomas, G., 1978. Effect of lime application on no-tillage and conventionally tilled corn. Agronomy Journal 70, 322–326.

Caires, E.F., Barth, G., Garbuio, F.J., 2006. Lime application in the establishment of a no-till system for grain crop production in Southern Brazil. Soil and Tillage Research 89, 3–12.

Caires, E., Garbuio, F., Churka, S., Barth, G., Correa, J., 2008. Effects of soil acidity amelioration by surface liming on no-till corn, soybean, and wheat root growth and yield. European Journal of Agronomy 28, 57–64.

Calonego, J., Rosolem, C., 2010. Soybean root growth and yield in rotation with cover crops under chiselling and no-till. European Journal of Agronomy 33, 242–249.

Chen, Y., Palta, J., Clements, J., Buirchell, B., Siddique, K., Rengel, Z., 2014. Root architecture alteration of narrow-leafed lupin and wheat in response to soil compaction. Field Crops Research 165, 61–70.

Compant, S., Clément, C., Sessitsch, A., 2010. Plant growth-promoting bacteria in the rhizo-and endosphere of plants: their role, colonization, mechanisms involved and prospects for utilization. Soil Biology and Biochemistry 42, 669–678.

Conyers, M., Heenan, D., McGhie, W., Poile, G., 2003. Amelioration of acidity with time by limestone under contrasting tillage. Soil and Tillage Research 72, 85–94.

DeLaune, P., Sij, J., 2012. Impact of tillage on runoff in long term no-till wheat systems. Soil and Tillage Research 124, 32–35.

da Veiga, M., Horn, R., Reinert, D., Reichert, J., 2007. Soil compressibility and penetrability of an oxisol from southern Brazil, as affected by long-term tillage systems. Soil and Tillage Research 92, 104–113.

de Moraes, M., Debiasi, H., Carlesso, R., Franchini, J., da Silva, V., da Luz, F., 2016. Soil physical quality on tillage and cropping systems after two decades in the subtropical region of Brazil. Soil and Tillage Research 155, 351–362.

Ding, X., Zhang, B., Zhang, X., Yang, X., Zhang, X., 2011. Effects of tillage and crop rotation on soil microbial residues in a rainfed agroecosystem of northeast China. Soil and Tillage Research 114, 43–49.

Elmore, R.W., 1987. Soybean cultivar response to tillage systems. Agronomy Journal 79, 114–119.

Faust, K., Raes, J., 2012. Microbial interactions: from networks to models. Nature Reviews Microbiology 10, 538–550.

Fecák, P., Šariková, D., Černý, I., 2010. Influence of tillage system and starting N fertilization on seed yield and quality of soybean Glycine max (L.) Merrill. Plant, Soil and Environment 56, 105–110.

Feng, Y., Motta, A., Reeves, D., Burmester, C., van Santen, E., Osborne, J., 2003. Soil microbial communities under conventional-till and no-till continuous cotton systems. Soil Biology and Biochemistry 35, 1693–1703.

Fernández, F., Brouder, S., Beyrouty, C., Volenec, J., Hoyum, R., 2008. Assessment of plant-available potassium for no-till, rainfed soybean. Soil Science Society of America Journal 72, 1085–1095.

Franchini, J., Crispino, C., Souza, R., Torres, E., Hungria, M., 2007. Microbiological parameters as indicators of soil quality under various soil management and crop rotation systems in southern Brazil. Soil and Tillage Research 92, 18–29.

Franchini, J., Debiasi, H., Balbinot Junior, A., Tonon, B., Farias, J., Oliveira, M., Torres, E., 2012. Evolution of crop yields in different tillage and cropping systems over two decades in southern Brazil. Field Crops Research 137, 178–185.

Frey, D., Elliot, E., Paustian, K., 1999. Bacterial and fungal abundance and biomass in conventional and no-tillage agroecosystems along two climatic gradients. Soil Biology and Biochemistry 31, 573–585.

Gil, S.V., Meriles, J., Conforto, C., Basanta, M., Radl, V., Hagn, A., et al., 2011. Response of soil microbial communities to different management practices in surface soils of a soybean agroecosystem in Argentina. European Journal of Soil Biology 47, 55–60.

Gesch, R., Archer, D., Spokas, K., 2012. Can using polymer-coated seed reduce the risk of poor soybean emergence in no-tillage soil? Field Crops Research 125, 109–116.

Hammerbeck, A., Stetson, S., Osborne, S., Schumacher, T., Pikul, J., 2012. Corn residue removal impact on soil aggregates in a no-till corn/soybean rotation. Soil Science Society of America Journal 76, 1390–1398.

Hansen, N., Allen, B., Baumhardt, R., Lyon, D., 2012. Research achievements and adoption of no-till, dryland cropping in the semi-arid US Great Plains. Field Crops Research 132, 196–203.

Hartmann, A., Schmid, M., Van Tuinen, D., Berg, G., 2009. Plant-driven selection of microbes. Plant and Soil 321, 235–257.

Hatfield, J., Sauer, T., Prueger, J., 2001. Managing soils to achieve greater water use efficiency. Agronomy Journal 93, 271–280.

Hazarika, S., Parkinson, R., Bol, R., Dixon, L., Russell, P., Donovan, S., Allen, D., 2009. Effect of tillage system and straw management on organic matter dynamics. Agronomy for Sustainable Development 29, 525–533.

He, Z., Mao, J., Honeycutt, C., Ohno, T., Hunt, J., Cade-Menun, B., 2009. Characterization of plant-derived water extractable organic matter by multiple spectroscopic techniques. Biology and Fertility of Soils 45, 609–616.

Hinsinger, P., Bengough, A.G., Vetterlein, D., Young, I.M., 2009. Rhizosphere: biophysics, biogeochemistry and ecological relevance. Plant and Soil 321, 117–152.

Ibekwe, A.M., Kennedy, A.C., Halvorson, J.J., Yang, C.H., 2007. Characterization of developing microbial communities in Mount St. Helens pyroclastic substrate. Soil Biology and Biochemistry 39, 2496–2507.

Jina, H., Hongwena, L., Rasaily, R.G., Qingjiea, W., Guohuaa, C., Yanboa, S., Xiaodonga, Q., Lijinc, L., 2011. Soil properties and crop yields after 11 years of no tillage farming in wheat–maize cropping system in North China Plain. Soil and Tillage Research 113, 48–54.

Jordán, A., Zavala, L., Gil, J., 2010. Effects of mulching on soil physical properties and runoff under semi-arid conditions in southern Spain. Catena 81, 77–85.

Kennedy, A.C., Papendick, R.I., 1995. Microbial characteristics of soil quality. Journal of Soil and Water Conservation 50, 243–248.

Kihara, J., Martius, C., Bationo, A., Vlek, P., 2011. Effects of tillage and crop residue application on soybean nitrogen fixation in a tropical ferralsol. Agriculture 1, 22–37.

Lal, R., 2007. Evolution of the plow over 10,000 years and the rationale for no-till farming. Soil and Tillage Research 93, 1–12.

Li, W., Zhan, C., Jiang, H., Xin, G., Yang, Z., 2006. Changes in soil microbial community associated with invasion of the exotic weed, *Mikania micrantha* H.B.K. Plant and Soil 281, 309–324.

Malik, R., Yadav, A., Singh, S., 2004. Resource conservation technologies for maintaining health of natural resources in rice-wheat cropping system. Proceedings Workshop on Sustaining Agriculture: Problems and Prospects. Punjab Agricultural University, Ludhiana, India, pp. 9–11. November.

Marino, D., Frendo, P., Ladrera, R., Zabalza, A., Puppo, A., Arrese-Igor, C., González, E.M., 2007. Nitrogen fixation control under drought stress. Localized or systemic? Plant Physiology 143, 1968–1974.

Marsh, B., Grove, J., 1992. Surface and subsurface soil acidity: soybean root response to sulfate-bearing spent lime. Soil Science Society of America Journal 56, 1837–1842.

Mazzoncini, M., Di Bene, C., Coli, A., Antichi, D., Petri, M., Bonari, E., 2008. Rainfed wheat and soybean productivity in a long-term tillage experiment in central Italy. Agronomy Journal 100, 1418–1429.

Meese, B., Carter, P., Oplinger, E., Pendleton, J., 1991. Corn/soybean rotation effect as influenced by tillage, nitrogen, and hybrid/cultivar. Journal of Production Agriculture 4, 74–80.

Miransari, M., Smith, D., 2007. Overcoming the stressful effects of salinity and acidity on soybean nodulation and yields using signal molecule genistein under field conditions. Journal of Plant Nutrition 30, 1967–1992.

Miransari, M., Bahrami, H.A., Rejali, F., Malakouti, M., Torabi, H., 2007. Using arbuscular mycorrhiza to reduce the stressful effects of soil compaction on corn (*Zea mays* L.) growth. Soil Biology and Biochemistry 39, 2014–2026.

Miransari, M., Bahrami, H., Rejali, F., Malakouti, M., 2008. Using arbuscular mycorrhiza to alleviate the stress of soil compaction on wheat (*Triticum aestivum* L.) growth. Soil Biology and Biochemistry 40, 1197–1206.

Miransari, M., Bahrami, H., Rejali, F., Malakouti, M., 2009a. Effects of soil compaction and arbuscular mycorrhiza on corn (*Zea mays* L.) nutrient uptake. Soil and Tillage Research 103, 282–290.

Miransari, M., Bahrami, H., Rejali, F., Malakouti, M., 2009b. Effects of arbuscular mycorrhiza, soil sterilization, and soil compaction on wheat (*Triticum aestivum* L.) nutrients uptake. Soil and Tillage Research 104, 48–55.

Miransari, M., 2011. Soil microbes and plant fertilization. Applied Microbiology and Biotechnology 92, 875–885.

Miransari, M., 2013a. Corn (*Zea mays* L.) growth as affected by soil compaction and arbuscular mycorrhizal fungi. Journal of Plant Nutrition 36, 1853–1867.

Miransari, M., 2013b. Soil microbes and the availability of soil nutrients. Acta Physiologiae Plantarum 35, 3075–3084.

Miransari, M., Riahi, H., Eftekhar, F., Minaie, A., Smith, D., 2013. Improving soybean (*Glycine max* L.) N_2 fixation under stress. Journal of Plant Growth Regulation 32, 909–921.

Miransari, M., 2014. Plant growth promoting rhizobacteria. Journal of Plant Nutrition 37, 2227–2235.

Nawaz, M., Bourrie, G., Trolard, F., 2013. Soil compaction impact and modelling. A review. Agronomy for Sustainable Development 33, 291–309.

Oliveira, E., Pavan, M., 1996. Control of soil acidity in no-tillage system for soybean production. Soil and Tillage Research 38, 47–57.

Pedersen, P., Lauer, J., 2004. Soybean growth and development response to rotation sequence and tillage system. Agronomy Journal 96, 1005–1012.

Peralta, A., Wander, M., 2008. Soil organic matter dynamics under soybean exposed to elevated [CO_2]. Plant and Soil 303, 69–81.

Rhoton, F., 2000. Influence of time on soil response to no-till practices. Soil Science Society of America Journal 64, 700–709.

Soane, B., van Ouwerkerk, C., 2013. Soil Compaction in Crop Production. vol. 11. Elsevier.

Seufert, V., Ramankutty, N., Foley, J., 2012. Comparing the yields of organic and conventional agriculture. Nature 485, 229–232.

Shen, J., Lixing, Y., Zhang, J., Li, H., Bai, Z., Chen, X., Zhang, W., Zhang, F., 2011. Phosphorus dynamics: from soil to plant. Plant Physiology 156, 997–1005.

Shi, Y., Lalande, R., Hamel, C., Ziadi, N., Gagnon, B., Hu, Z., 2013. Seasonal variation of microbial biomass, activity, and community structure in soil under different tillage and phosphorus management practices. Biology and Fertility of Soils 49, 803–818.

Shi, Y., Lalande, R., Hamel, C., Ziadi, N., 2015. Winter effect on soil microorganisms under different tillage and phosphorus management practices in Eastern Canada. Canadian Journal of Microbiology 61, 315–326.

Silvente, S., Sobolev, A., Lara, M., 2012. Metabolite adjustments in drought tolerant and sensitive soybean genotypes in response to water stress. PLoS One 7, e38554.

Singer, J., Logsdon, S., Meek, D., 2008. Soybean growth and seed yield response to tillage and compost. Agronomy Journal 100, 1039–1046.

Six, J., Conant, R., Paul, E., Paustian, K., 2002. Stabilization mechanisms of soil organic matter: implications for C-saturation of soils. Plant and Soil 241, 155–176.

Teasdale, J., Coffman, C., Mangum, R., 2007. Potential long-term benefits of no-tillage and organic cropping systems for grain production and soil improvement. Agronomy Journal 99, 1297–1305.

Thomas, G., Dalal, R., Standley, J., 2007. No-till effects on organic matter, pH, cation exchange capacity and nutrient distribution in a Luvisol in the semi-arid subtropics. Soil and Tillage Research 94, 295–304.

Vetsch, J., Randall, G., 2002. Corn production as affected by tillage system and starter fertilizer. Agronomy Journal 94, 532–540.

Whalen, J., Prasher, S., Benslim, H., 2007. Monitoring corn and soybean agroecosystems after establishing no-tillage practices in Québec, Canada. Canadian Journal of Plant Science 87, 841–849.

Wang, Y., Shao, M., 2013. Spatial variability of soil physical properties in a region of the Loess Plateau of Pr China subject to wind and water erosion. Land Degradation and Development 24, 296–304.

West, T., Griffith, D., Steinhardt, G., Kladivko, E., Parsons, S., 1996. Effect of tillage and rotation on agronomic performances of corn and soybean: twenty-year study on dark silty loam soil. Journal of Production Agriculture 9, 241–248.

West, T., Post, W., 2002. Soil organic carbon sequestration rates by tillage and crop rotation. Soil Science Society of America Journal 66, 1930–1946.

Wilhelm, W., Wortmann, C., 2004. Tillage and rotation interactions for corn and soybean grain yield as affected by precipitation and air temperature. Agronomy Journal 96, 425–432.

Xue, L., Wang, L., Anjum, S., Saleem, M., Bao, M., Saeed, A., Bilal, M., 2013. Gas exchange and morpho-physiological response of soybean to straw mulching under drought conditions. African Journal of Biotechnology 12, 2360–2365.

Yang, Q., Wang, X., Shen, Y., 2013. Comparison of soil microbial community catabolic diversity between rhizosphere and bulk soil induced by tillage or residue retention. Journal of Soil Science and Plant Nutrition 13, 187–199.

Yin, X., Al-Kaisi, M., 2004. Periodic response of soybean yields and economic returns to long-term no-tillage. Agronomy Journal 96, 723–733.

Yu, H., Li, T., Zhang, X., 2010. Nutrient budget and soil nutrient status in greenhouse system. Agricultural Sciences in China 9, 871–879.

Zhu, B., Cheng, W., 2011. Rhizosphere priming effect increases the temperature sensitivity of soil organic matter decomposition. Global Change Biology 17, 2172–2183.

Soybean Production and Environmental Stresses

M. Hasanuzzaman[1], K. Nahar[1,2], A. Rahman[1,2], J.A. Mahmud[1,2], M.S. Hossain[2], M. Fujita[2]
[1]Sher-e-Bangla Agricultural University, Dhaka, Bangladesh; [2]Kagawa University, Miki-cho, Japan

Introduction

The soybean [*Glycine max* (L.) Merr.] is one of the nature's most versatile plants and an abundant source of protein and oil in both temperate and tropical environments. Since its domestication, the soybean has been considered as a major food crop in Eastern Asia. There are several uses of soybean such as oil and food for human consumption, feed for animals, and biofertilizer for crops. Increasing crop production is now the highest agricultural priority worldwide to feed the vast population. Although this crop is suitable for growing in diverse environmental conditions, it often suffers from various environmental or abiotic stresses, such as salinity, drought, temperature extremes, heavy metal toxicity, high light intensity, nutrient deficiency, ultraviolet (UV) radiation, ozone, etc., which cause substantial losses in yield and quality of a crop (Hasanuzzaman et al., 2012).

Plant growth and distribution are limited by the environment. If any environmental factor becomes diverse from the ideal, then it is considered as a limiting factor for plant growth. Either individually or in combination, they cause morphological, physiological, biochemical, or molecular changes within plant cells that adversely affect seed germination, growth and productivity, and ultimately yield. If the stress becomes very high and/or continues for an extended period, it may lead to an intolerable metabolic load on cells, reducing growth, and in severe cases, result in plant death (Hasanuzzaman et al., 2012). All stresses induce a variety of vital physiological and biochemical changes in plants including stomatal conductance, membrane electron transport, CO_2 diffusion, carboxylation efficiency, water use efficiency, respiration, transpiration, photosynthesis, and membrane functions. Disruption of these key functions limits growth and developmental processes and leads to a reduction in final crop yield.

The United Nations Food and Agriculture Organization (FAO) reported that more than 1 billion people are now suffering from malnutrition. In such cases, we have to go for food diversification by incorporating nutritious crops like soybeans. The potential yield of soybeans may be achieved by minimizing the damages incurred by environmental stress. For this reason, studies on plant responses and adaptation to environmental stress are necessary. In this chapter, we shed light on the soybean responses to different environmental stresses and the possible strategies to mitigate stress-induced damages.

Environmental Stresses in Soybean Production. http://dx.doi.org/10.1016/B978-0-12-801535-3.00004-8

Soybean: A Crop With Multifarious Uses

The soybean [*Glycine max* (L.) Merr.] is one of the principal food plants for humans (Kasmakoglu, 2004). It is an important legume crop worldwide and one of nature's most versatile plants, and it is considered as an abundant supply of protein and oil both in temperate and tropical environments. Since its domestication, the soybean has been a staple food in Eastern Asia. There are several uses of soybeans such as oil for human consumption, feed for animals, and food for human consumption (Figs. 4.1 and 4.2). In many areas it is one of the major oil crops. According to 2014 statistics, the United States, Brazil, and Argentina are the three largest soybean-producing countries in the world (FAOSTAT, 2014). The soybean-production area is second only to that of corn in US agriculture (Villamil et al., 2012). Among the food plants, the soybean is unique in that the traditional foods in Asia made from the soybean (eg, tofu, miso, and soy sauce) bear no resemblance to or association with the crop growing in the field. In China, Japan, and the Korean peninsula the soybean has been consumed since ancient times as a major food. The popularity of tofu (bean curd) in China took place during the latter half of the Song Dynasty (960–1279 CE) (Shinoda, 1971). The origin of soybean cultivation in China was the wild soybean (*Glycine soja*), the kindred ancestor of the current cultivated soybean (*G. max*). Miso is a fermented soybean paste that originated in China around the first century and still has the same name in many countries (Hymowitz, 2008).

Figure 4.1 Multifarious use of soybean plants and seeds.

The oil and protein contents in dry soybeans are 18–23% and 38–44%, respectively, which vary depending upon the genotype, growing environment, cultural practices, and processing techniques. It also contains 34% carbohydrate and 5% ash (Scott and Aldrich, 1983). The soybean contains both water-soluble and oil-soluble vitamins. The water-soluble vitamins are not lost during oil extraction. A kilogram of soy flour contains approximately 3.25, 3.11, 16.9, and 29.7 mg of vitamin B1 (thiamin), vitamin B2 (riboflavin), vitamin B5 (pantothenic acid), and vitamin B6 (niacin), respectively (Kumar et al., 2008a,b). Soybean meal accounts for 60–70% of the value of the soybean, with the balance from oil. The soybean supplies one-fourth of the world's fats and oils, about two-thirds of the world's protein concentrated animal feeds, and three-fourths of the world trade in high-protein meals (Keyser and Li, 1992). In many countries, such as Bangladesh, the soybean is mainly used as a pulse crop due to the lack of extended production and oil extraction facilities. The oil is converted to margarine, mayonnaise, shortening, salad oils, and salad dressings. Soya meal is used primarily as a source of high-protein feeds for poultry, fish, and cattle. For example, in Bangladesh, the majority of the soybean seeds are used in poultry farms. The soybean protein is also used in the form of protein concentrates and isolates and texturized protein for human consumption (Hymowitz and Newell, 1981).

Soybeans can be grown in a wide range of climate and soils and hence can be fit in many cropping seasons. Therefore, it is suitable as catch crops and restorative crops as well. Soybeans become an integral part of soil fertility management in agricultural

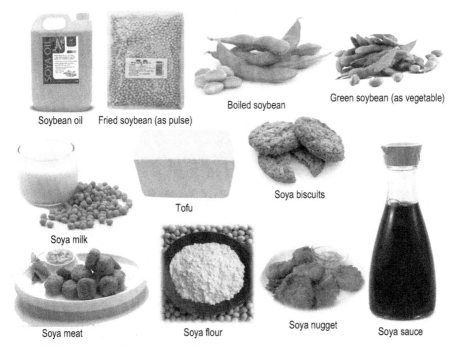

Soybean oil   Fried soybean (as pulse)

Boiled soybean

Green soybean (as vegetable)

Soya milk

Tofu

Soya biscuits

Soya meat   Soya flour   Soya nugget   Soya sauce

Figure 4.2 Different food products made from soybeans.

systems. Apart from food crops, soybeans have multifarious uses. It can fix atmospheric nitrogen (N) and thus can be used as green manuring or restorative crops. The symbiotic N-fixing ability of soybean makes it well suited for low N soils throughout the world and contributes to sustainable agriculture. Since the fertility and productivity of soils have been declining day by day the expansion of soybeans in those areas may sustain the productivity. However, improving N-fixation traits, such as nodulation mechanisms, should be emphasized to avail the ultimate benefit.

The soybean can be cultivated as pasture, and the crop can be used as hay and silage (Heuzé et al., 2015). In the United States before the 1950s, it was recommended to combine forage soybeans with maize or sorghum for annual pastures (Morse and Cartter, 1952). However, some late maturing varieties may be grazed from the flowering stage to near maturity (Blount et al., 2013). It may also be harvested from the flowering stage to near maturity for using as high-quality hay (Blount et al., 2013). Since forage soybean stems are coarse and fibrous and feeding forage soybeans alone may cause digestive troubles, an early harvest is sometimes recommended (Blount et al., 2013). Soybeans may be grown as a silage crop in pure culture or intercropped with maize or sorghum (Blount et al., 2013). A soybean silage is seldom ensiled alone because of its bitter taste and high concentration of ammonia and butyric acid (Gobetti et al., 2011). Soybean and sorghum ensiled together (40:60) with molasses give a good, preserved silage (Lima et al., 2011). Sometimes the combination of inoculants (ie, *Lactobacillus brevis*) and molasses were reported to have a synergistic effect on soybean silage quality (Tobia et al., 2008; Rigueira et al., 2008).

The soybean is not only a food crop but also a potential source of green bioenergy. Researchers elucidated that both corn grain ethanol and soybean biodiesel produce more energy compared with the energy to grow the crops and convert into biofuels (Adie and Krisnawati, 2015). Although soybeans contain less oil than other oil-yielding crops, its biodiesel returns 93% more energy than is used to produce it, while corn grain ethanol currently provides only 25% more energy. A study of comparative advantages to the use of soybeans as a source of biodiesel showed a very strong indication that the soybean biodiesel has a favorable net energy ratio, more than 2.08 (Pradhan et al., 2008). Moreover, since soybeans can be produced without or nearly zero N, it makes soybeans advantageous for the production of biodiesel (Pimentel and Patzek, 2005). However, soybean oil and its derivatives do not generally have the volatility demanded by spark-ignited engines, so most attempts to use these fuels have focused on diesel engines.

Environmental Stress and Plant Responses

Plants are always confronted with various environmental adversities, termed as abiotic stresses, that seriously reduce their productivity. Plant responses to these stresses are complex and involve uncountable physiological, molecular, and cellular adaptations. The major abiotic stresses include salinity, drought, extreme temperature, flooding, toxic metal/metalloids, ozone, UV radiation, high light, etc. (Fig. 4.3).

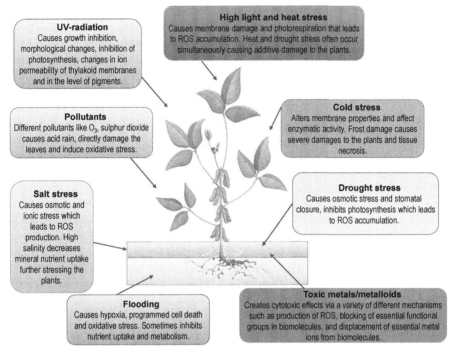

UV-radiation
Causes growth inhibition, morphological changes, inhibition of photosynthesis, changes in ion permeability of thylakoid membranes and in the level of pigments.

High light and heat stress
Causes membrane damage and photorespiration that leads to ROS accumulation. Heat and drought stress often occur simultaneously causing additive damage to the plants.

Pollutants
Different pollutants like O_3, sulphur dioxide causes acid rain, directly damage the leaves and induce oxidative stress.

Cold stress
Alters membrane properties and affect enzymatic activity. Frost damage causes severe damages to the plants and tissue necrosis.

Salt stress
Causes osmotic and ionic stress which leads to ROS production. High salinity decreases mineral nutrient uptake further stressing the plants.

Drought stress
Causes osmotic stress and stomatal closure, inhibits photosynthesis which leads to ROS accumulation.

Flooding
Causes hypoxia, programmed cell death and oxidative stress. Sometimes inhibits nutrient uptake and metabolism.

Toxic metals/metalloids
Creates cytotoxic effects via a variety of different mechanisms such as production of ROS, blocking of essential functional groups in biomolecules, and displacement of essential metal ions from biomolecules.

Figure 4.3 Types and nature of environmental stresses in plants.

Being a sessile organism, plants cannot avoid adverse environmental conditions and thus try to adapt under stress conditions. Due to climate change the environmental adversities are getting worse day by day. Salinity is a major constraint on crop plant productivity (Munns and Tester, 2008). More than 800 million hectares of land throughout the world are salt affected, which accounts for 6% of the world total land area (Munns and Tester, 2008). Plants growing under saline conditions are affected by osmotic stress as well as ionic toxicity (Munns, 2002). Drought is another complex and devastating stress, and its frequency is expected to increase as a consequence of climate change (Hasanuzzaman et al., 2014). Water shortage is expected to lead to global crop production losses of up to 30% by 2025, compared to current yields, according to the "Water Initiative" report of "The World Economic Forum (2009)" at Davos (Zhang, 2011). Due to global warming, plants are intimidated by adverse effects of high temperature (HT) stresses. Protein denaturation, inactivation of enzymes, production of reactive oxygen species (ROS), and disruption of membrane structure are some primary damage effects of HT, which are also responsible for damage to the ultrastructural cellular components. These anomalies hamper plants' growth and development as a whole (Howarth, 2005). Although higher plants develop their own defense strategies to struggle against HT stress, those are often not enough and thus result in substantial damages (Nahar et al., 2015). Low temperature (LT) or cold stress is another major environmental factor that often affects plant growth and crop productivity, again leading to substantial crop losses (Sanghera et al., 2011). Plants differ in their tolerance to chilling (0–15°C) and freezing (<0°C) temperatures. Both chilling and freezing stresses are together termed cold stress. The damage due to cold stress can range from chilling injury and freezing injury to suffocation and heaving. In general, plants from temperate climatic regions are considered to be tolerant to chilling to variable degrees, and these can often increase their freezing tolerance by being exposed to cold, but nonfreezing temperatures; this process is known as cold acclimation. However, plants of tropical and subtropical origins are sensitive to chilling stress and lack this mechanism of cold acclimation (Sanghera et al., 2011). Low temperature may affect several aspects of crop growth; viz, survival, cell division, photosynthesis, water transport, growth, and finally crop yield. Chilling stress results from temperatures cool enough to produce injury without forming ice crystals in plant tissues, whereas freezing stress results in ice formation within plant tissues. Due to rapid industrialization the concentrations of toxic metals and metalloids have been increasing day by day. After an excessive uptake by plants, they may participate in some physiological and biochemical reactions, which destroy the normal growth of the plant by disturbing absorption, translocation, or synthesis processes. More seriously, they may combine with some huge molecules like nucleic acid, protein, and enzyme or may substitute special functional elements in protein or enzyme, so as to induce a series of turbulence of metabolism. Therefore, the growth and procreation of plants is prohibited, and the plant dies (Wei and Zhou, 2008). In many tropical and subtropical regions, flooding may develop due to several direct (improper irrigation practices) and indirect (global warming) anthropogenic and natural consequences (meteorological) leading to altered plant metabolism, architecture, and ecogeographical distribution depending upon a plant's responses. Flooding, either complete or partial waterlogging, induces the progressive reduction in soil O_2 concentration and in redox potential, which contribute to the appearance of several reduced compounds of either chemical or biochemical origin. Plants are exposed to UV-B radiation intrinsic to sunlight. As sessile photosynthetic organisms, plants necessarily have to deal with UV-B radiation and its impact. However, UV-B is not only a stress factor but also triggers specific and protective photomorphogenic responses mediated by a molecularly unknown UV-B photoreceptor (Gruber et al., 2010).

In nature, the stress-induced damages vary depending on the type of stress. In general, the stress signal is first perceived by the receptors present on the membrane of the plant cells. Following this the signal information is transduced downstream, resulting in the activation of various stress-responsive genes. The products of these stress genes ultimately lead to stress-tolerance response or plant adaptation and help the plant to survive and surpass unfavorable conditions. The response could also result in growth inhibition or cell death, which will depend upon how many and which kinds of genes are up- or downregulated in response to the stress. The various stress-responsive genes can be broadly categorized as early and late induced genes (Tuteja et al., 2011).

Environmental Stress-Induced Oxidative Stress

One of the most common consequences of abiotic stress is the production of ROS, which leads to oxidative stress, resulting in direct damage to plant cells through oxidation of biological components (nucleic acids, proteins, and lipids) and can instigate chain reactions resulting in accumulation of more ROS and initiation of programmed cell death (Apel and Hirt, 2004; Hasanuzzaman et al., 2012). Reactive-oxygen species include some free radicals, such as superoxide ($O_2\cdot^-$), hydroxyl radical ($\cdot OH$), alkoxyl ($RO\cdot$), and peroxyl ($ROO\cdot$), and nonradical products like hydrogen peroxide (H_2O_2), singlet oxygen ($^1O_2$), etc. (Gill and Tuteja, 2010; Sandalio et al., 2013). The generation of ROS is an inevitable part and byproduct in different metabolic processes. Under normal condition (nonstress), their overproduction is controlled by a well-equipped antioxidant defense system, but under an adverse environment or stressful condition, electron transport chains (ETC) are deregulated or overflowed or even disrupted, which cause overwhelmed production of ROS leading to oxidative stress. The generation of ROS happens due to both physical and chemical activation (Fig. 4.4). In the case of physical activation, energy is transferred from photoactivated pigment such as excited chlorophyll ($chl\cdot$). The latter absorbs adequate energy, resulting in the spin of one electron (inverted) toward generation of $^1O_2$ during photosystem II (PS II). Chemical activation occurs by the univalent reduction of molecular oxygen (Fig. 4.4; Gill et al., 2015). In the processing of the reduction of dioxygen or triplet oxygen to water, four electrons and four protons are engaged, which leaves three major ROS, viz, $O_2\cdot^-$, H_2O_2, and $\cdot OH$ (Fig. 4.4). The superoxide radical is dismutated either spontaneously or by the action of superoxide dismutase (SOD) to produce H_2O_2. Several cell organelles such as chloroplast, mitochondrion, plasma membrane, apoplast, and nucleolus are the sources of ROS production. However, since ETC is present in chloroplasts and mitochondria, they are considered as the most prominent sources of ROS.

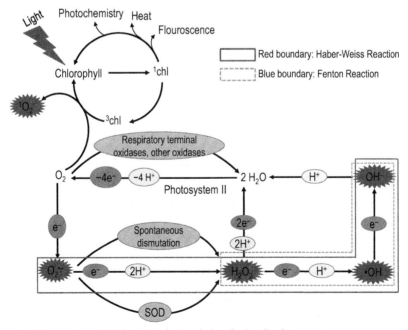

Figure 4.4 Formation of ROS by physical and chemical activation.
Adapted from Gill, S.S., Anjum, N.A., Gill, R., Yadav, S., Hasanuzzaman, M., Fujita, M., Mishra, P., Sabat, S.C., Tuteja, N., 2015. Superoxide dismutase–mentor of abiotic stress tolerance in crop plants. Environmental Science and Pollution Research 22, 10375–10394, with permission from Springer.

Peroxisome is also considered as a powerful place of ROS generation in cells, and it has been widely studied (Gill and Tuteja, 2010; Sandalio et al., 2013).

Although there are a number of routes to generate ROS in plant cells and most cellular compartments have the potential to become a source of ROS (Fig. 4.5), the chloroplast is considered as the key source of ROS in plants. During photosynthesis, inadequate energy dissipation often forms a chl triplet state (^{3}chl) that is involved in the transfer of its excitation energy onto O_2 to make $^1O_2$ (Fig. 4.5; Gill et al., 2015). In case of impairment of carbon dioxide (CO_2) fixation, the oxygenase activity of ribulose-1,5-bisphosphate carboxylase/oxygenase (RuBisCO) in the chloroplast is impaired, resulting in the formation of glycolate, which moves to peroxisomes (Takashi and Murata, 2008; Fig. 4.5). Electron transport chain reactions in light harvest complex or photosystem (PS: PS I and PS II) may generate ROS. Under a normal environment, electron flow from excited PS centers is transferred to $NADP^+$ that is reduced to NADPH. This NADPH enters the Calvin/C_3 cycle, where it reduces the final electron acceptor (CO_2), and through the Calvin cycle, NADP is recycled (Vašková et al., 2012; Fig. 4.5). But under stress conditions, NADP recycling can be impaired and ETC can be overloaded. Peroxisomes' metabolic processes are responsible for the production of the majority of intracellular H_2O_2 (Noctor et al., 2002a,b). Peroxisomes have oxidative metabolisms, and the generation of H_2O_2 involves

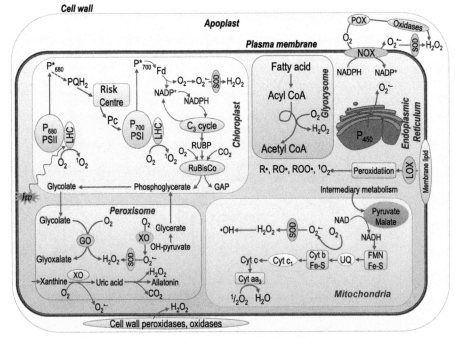

Figure 4.5 Possible mechanisms of ROS formation in different sites of plant cells.
Adapted from Gill, S.S., Anjum, N.A., Gill, R., Yadav, S., Hasanuzzaman, M., Fujita, M., Mishra, P., Sabat, S.C., Tuteja, N., 2015. Superoxide dismutase–mentor of abiotic stress tolerance in crop plants. Environmental Science and Pollution Research 22, 10375–10394, with permission from Springer.

glycolate oxidation catalyzed by glycolate oxidase, the β-oxidation of fatty acids, catabolism of lipids, enzymatic reaction of flavin oxidases, and the disproportionation of $O_2 \cdot^-$ (del Río et al., 2009). This process during photorespiration in peroxisomes is the considerable source of H_2O_2 (Noctor et al., 2002b). The $O_2 \cdot^-$ in peroxisomes can be generated mainly in two ways. Oxidation of xanthine or hypoxanthine to uric acid by the activity of xanthine oxidase generates $O_2 \cdot^-$, and this reaction occurs in organelle matrix (Sharma et al., 2012; Fig. 4.5). In glyoxysomes, ROS can be produced due to peroxisomal β-oxidation enzyme-induced hydrolysis of fatty acids to acetyl-CoA (Olsen and Harada, 1995; Fig. 4.5). Mitochondrial ETCs are considered a major site of ROS production. Mitochondrial ETC and ATP syntheses are coupled strongly and abiotic stress-induced imbalance between these two may lead to overreduction of electron carriers. For this reason, ROS may generate (Noctor et al., 2007; Blokhina and Fagerstedt, 2010). Reduction of O_2 to $O_2 \cdot^-$ occurs in the flavoprotein region of an NADH dehydrogenase segment (Fig. 4.5).

A number of enzymes present in the mitochondrial matrix are responsible to produce ROS. The NADPH-dependent electron transport process in the endoplasmic reticulum is involved with Cyt P_{450} and produces $O_2 \cdot^-$ (Mittler, 2002). Peroxidase and oxidases associated with the cell wall results in H_2O_2 generation. Environmental stresses such as salinity, drought, flood or waterlogging stress, HT, LT, metal toxicity, UV-B radiation, O_3, high light, and mechanical damage led to excessive ROS generation. In the absence of a well-organized and sufficient antioxidant system within the plant cell, the overproduction of ROS causes oxidative stress. The molecular mechanism of different ROS production under different abiotic stresses might be almost similar, although the magnitude of ROS production depends on the plant species: their stress-tolerant capability, growth stages and stress type, stress extremity, duration, and combination or complexity of stresses (Gill and Tuteja, 2010; Hasanuzzaman et al., 2012; Sharma et al., 2012). However, oxidative stress induced by ROS exerts extreme damage on the basic biomolecules such as lipids, proteins, and DNA (Gill and Tuteja, 2010; Hasanuzzaman et al., 2012).

Soybean Responses to Environmental Stresses

Soybean plants are greatly affected by different adverse environmental stresses. However, the effect of stressors greatly depends on the genotypes, nature, and duration of stress.

Salinity

As a moderately salt-sensitive crop, the productivity of soybeans is significantly affected by salt stress (Munns and Tester, 2008; Phang et al., 2008). The presence of a higher concentration of salt in the plant growth medium negatively affects growth (Fig. 4.6), nodulation, agronomic traits, seed quality, and the yield of soybeans. Several studies indicate that higher salt in the growth medium causes ionic and water imbalance and osmotic and oxidative stress, which ultimately lead to growth inhibition and yield reduction (Sheeren and Ansari, 2001; Egbichi et al., 2013; Simaei et al., 2012; Gu-wen et al., 2014).

Figure 4.6 Salt-induced growth inhibition and chlorosis in soybean cv. BARI soybean 5.

Under salt stress condition, ROS production and lipid peroxidation increased in soybean seedlings, which affect antioxidant machinery and consequently cause oxidative damage (Egbichi et al., 2013). The salt stress increased ionic toxicity by decreasing potassium (K) uptake and increasing sodium (Na) uptake and the Na/K ratio in soybean seedlings (Simaei et al., 2012). The soybean seedlings exposed to higher concentrations of salt caused chlorosis by decreasing chl *a*, chl *b*, chl (*a+b*), and carotenoid (Car) contents (Sheteawi, 2007).

The damages caused by higher salinity in soybeans start from the germination and exist until the death of the plant (Phang et al., 2008). The germination of soybean seeds were delayed due to low salinity (0.05% and 0.1%) and significantly decreased the germination rate by higher salinity (Simaei et al., 2012). Compared with the germination stage, the seedling stage of soybeans is considered as much more sensitive to salt stress. The growth of soybean seedlings was completely inhibited when Na^+ content in plant tissue was 6.1 mg/g FW, where 40% germination is possible even at 9.1 mg/g FW of Na^+ in embryonic axis (Hosseini et al., 2002; Phang et al., 2008). The plant growth in terms of root fresh and dry weight, shoot fresh and dry weight, total fresh and dry weight, plant height, root length, leaf number, and leaf area of soybean seedlings were significantly decreased under salt stress conditions (Hamayun et al., 2010; Dolatabadian et al., 2011; Egbichi et al., 2014; Kang et al., 2014; Klein et al., 2015). Salt stress reduced nodule dry weight, nodule number, nodule length, hemoglobin content, nodule nitrogenase activity, and nodule N fixation efficiency by attenuating the aerobic respiration of N-fixing bacteria in soybean seedlings (Delgado et al., 1994; Egbichi et al., 2014; Klein et al., 2015). The seed yield and quality of soybean seeds were also affected by salt stress. The yield and yield attributes of soybeans in terms of number of branches per plant, number of pods per plant, 1000 seed weight, and seed yield considerably decreased under salt stress conditions. The seed quality of soybeans also deteriorates in terms of oil content, protein content, mineral content, soluble carbohydrate, and amino acid under salt stress conditions (Sheteawi, 2007; Rezaei et al., 2012; Weisany et al., 2014; El-Sabagh et al., 2015).

Drought

Drought stress is considered one of the major limitations of crop growth, development, and production. This stress induces various biochemical and physiological responses in plants. Several metabolic processes of plants were negatively interfered by water shortage conditions in plant. Drought stress reduces germination percentage, as well as the ratio of chl a and b and Car, and increases the production of ROS that causes lipid peroxidation, protein degradation, and nucleic acid damages. Water deficiency also damages the photosynthetic apparatus, essentially by disrupting all major components of photosynthesis including the thylakoid electron transport, the carbon reduction cycle, and the stomatal control of the CO_2 supply, together with an increased accumulation of carbohydrates and the disturbance of water balance (Fanaei et al., 2015). The soybean is one of the major oil seed crops considered as a potential economic crop to contribute in the agricultural economy. Extensive research has been conducted on soybeans under drought stress to understand its responses (Souza et al., 2013). Soybean seed germination was greatly hampered by drought stress, especially in arid and semiarid regions (Neto et al., 2004). Salimi (2015) investigated the germination percentage of three soybean cultivars at different levels of drought induced by PEG 4000. He found that the germination percentage reduced significantly with increased drought levels (0, −3, −6, and −9 bar) and the M7 cultivar performed better than the other two cultivars (cv. L17 and Hamilton) at drought stress. All physiological and biochemical processes in plants depend on water functioning as a solvent, through which gases, minerals, and other solutes enter into cells and move within plants, and also a thermal regulator of plants (Taiz and Zeiger, 2009). Depending on growth and development stages, the water demands of plants vary considerably. Soybean plants require the highest amount of water at flowering-grain filling stages, when the water requirement is 7–8 mm/day for successful soybean production (Embrapa, 2011). As the requirement of water varies with different growth phases, plants show different responses under water shortage conditions at different growth phases. The water deficit at the reproductive phase was considered as the most sensitive and caused significant reduction in yield components and yield of soybeans (Desclaux et al., 2000). The decreases in the pod number per plant (by 18.22%), grain number per plant (by 14.81%), grain number per pod (by 3.41%), 100 seed weight (by 13.16%), biological yield per plant (by 16.06%), and grain yield per plant (by 21.88%) were observed under drought stress compared to control. The yield reduction was caused by oxidative stress indicated by malondialdehyde (MDA) level and lower tissue water status that causes the inactivation of enzymes (Anjum et al., 2011). Decrease in leaf area and acceleration of the senescence, abscission of leaves (Santos and Carlesso, 1988; Catuchi et al., 2011), decreased stomatal pore size, and the reduced numbers of stomata (Widiati et al., 2014) are the most prominent responses of plants under water deficit condition. However, stomatal length increased with the decrease in moisture content from 100% field capacity to 40–60% field capacity (Widiati et al., 2014).

Photosynthetic pigments such as chl a and chl b play vital role in photosynthesis. During stress conditions, net photosynthesis is greatly reduced as chl a, chl b, and chl $(a+b)$ contents decreased in soybean plants (Anjum et al., 2013). Some osmo-regulatory elements such as proline (Pro) and sugars are produced in plants under a wide range of abiotic stresses including drought, salinity, and HT stress to maintain metabolic activity and to improve osmotic adjustment. Thus, these osmoregulators increase the chance of plant survival under abiotic stresses, maintaining cell homeostasis (Abbaspour, 2012). The increased Pro and sugar contents in soybeans under water-deficient conditions were reported by Abu-Muriefah (2015). Drought stress caused a significant reduction in N fixation, leaf expansive growth, biomass accumulation, and increasing leaf resistance in two contrasting soybean cultivars where Biloxi was a drought-sensitive variety and Cook was relatively tolerant to drought (Agboma et al., 1997). Like different abiotic stresses, overproduction of ROS is very common under drought stress (Agarwal et al., 2005). Activities of SOD, catalase (CAT), and glutathione peroxidase increased to scavenge ROS, as investigated by Anjum et al. (2011) and Masoumi et al. (2010). Nutrient availability in the soil and nutrient uptake are hampered when plants are exposed to drought because drought causes stomatal closure, which reduces transpiration and imbalance in active transport and membrane permeability, resulting in a reduced absorption power in the roots (Hu et al., 2007). A significant reduction of N, phosphorus (P), and K uptake in soybeans was also caused by water-deficit conditions. Thus amounts of N, P, and K found were lower on shoots and roots compared to a control (Tanguilig et al., 1987; Souza et al., 2013).

Toxic Metals/Metalloids

The increasing human population and the rapid industrialization in the world are creating serious environmental problems, including the production and release of considerable amounts of toxic metals into the environment (Sarma, 2011). Among the toxic metals, some are considered as heavy metals, and metal toxicity is one of the major abiotic stresses leading to hazardous health effects in animals and plants. Because of their high reactivity, they can directly influence growth, senescence, and energy synthesis processes (Maksymiec, 2007). The excessive uptake of toxic metals by plants may occur, and these elements may participate in some physiological and biochemical reactions that can harm the normal growth of the plant by disturbing absorption, translocation, or synthesis processes (Hasanuzzaman and Fujita, 2012a,b; Table 4.1). Several research works proved that soybean plants suffered a lot by toxic metal/metalloids in the whole life cycle (from germination to maturity).

The germination of soybeans is greatly inhibited by the different metals/metalloids stresses. According to Luan et al. (2008), germination percentage was decreased under different levels of cadmium (Cd), lead (Pb), and arsenic (As) stresses, and the inhibition of germination increased with the increase of metal concentration; for the dose 1600 mg As/kg soil, no germination was recorded. A similar result was obtained from mercury (Hg) stress by Palanisamy et al. (2012). They found 93, 73, 70, 54, 46,

Metal or metalloids; dose; duration	Effects	References
Cd, Ni; 0.05 mg Cd/L, 1.0 mg Ni/L; whole lifetime with nutrient	Markedly reduced plant biomass and seed production. Significant amount of metal entered into seeds.	Malan and Farrant (1998)
Al; 150 µM; 20 days	Caused marked reduction in root length, shoot height, dry weight, chl content, (SPAD value), and photosynthetic rate.	Shamsi et al. (2007)
Cd, Pb, and As; 100, 200, 400, 800, and 1600 mg/kg soil; 7 days	Percent germination, root and shoot growth of soybean decreased.	Luan et al. (2008)
Cr; 5, 10, 25, 50, 100, and 200 mg/L; 7 days	Reduced growth, dry weight, and vigor index.	Ganesh et al. (2009)
HgCl₂; 1.0, 2.0, 3.0, 4.0, and 5.0 ppm; seeds were germinated at 25 mL metal solution	Germination percentage, root length, shoot length, fresh weight, and dry weight of seedlings were decreased.	Palanisamy et al. (2012)
Cd, Cd(NO₃)₂; 16 mg/kg sand; 10 weeks	Adversely affected growth, nodulation, and N fixation.	Sheirdil et al. (2012)
Cd; 200 mg mg/L; seed exposure for 96 h	Increased MDA content, GST activity, and Cd accumulation in seedlings. Reduction of GSH/hGSH ratio and concomitant increase of GST activity.	Yang et al. (2012)
Cd, CdCl₂; 223 µM; 24 h	Leads to oxidative stress, inhibition of respiration, and photosynthesis, increased rate of mutation and, as a consequence, growth and yield decreased. Increased production of ethylene.	Chmielowska-Bąk et al. (2014)
Cd, CdCl₂; 200 µM; 72 h	Inhibited growth (shoot length, root length, number of leaves per plant, and mean leaf area per plant) and reduced photosynthetic pigment content. Increased lipid peroxidation, H₂O₂ levels, and the activities of SOD, POX, CAT, APX, and altered GSH level.	Hashem (2014)
Co and Pb; 150 µM for each; 2 weeks	Significantly reduced seed germination percentage, seedling growth, leaf area, root development, and biomass production.	Imtiyaz et al. (2014a)
Co and Pb; 150 µM for each; 2 weeks	Reduction of chl, Car, carbohydrate, and protein content and addition in Pro content.	Imtiyaz et al. (2014b)
Cd, CdSO₄.4H₂O; 40 µM; whole life cycle	Reduced nodulation in root in all growth stages and induced earlier nodular senescence.	Thakur and Singh (2014)
Cr; 50, 100, 200, 300 µg/L; 15 d	Decreased germination percentage, root length, shoot length, fresh weight, dry weight, photosynthetic pigments, and protein content and increased Pro content, Cr uptake, and accumulation. In antioxidant system, CAT and polyphenol oxidase decreased, whereas POX and SOD were increased.	Sundarmoorthy et al. (2015)

and 31% germination under the control, 1.0, 2.0, 3.0, 4.0, and 5.0 mg/kg Hg stresses, respectively. The percentage of germination decreased with the increase of chromium (Cr) concentrations (from 50 to 200 mg/L). No germination was recorded beyond 300 mg/L concentration of Cr (Sundarmoorthy et al., 2015). Similarly, 150 μM of both cobalt (Co) and Pb stresses reduced the seed germination percentage significantly (Imtiyaz et al., 2014a).

Studies on soybean growth revealed that toxic metals/metalloids hamper the normal growth of plants. Aluminum (Al) stress (150 μM with pH 4.0) for 20 days caused marked reduction in root length, shoot height, and dry weight (Shamsi et al., 2007). According to Luan et al. (2008), root and shoot growth of the soybean were adversely affected when exposed to Cd, Pb, and As stresses (100, 200, 400, 800, and 1600 mg/kg soil). Compared with the control treatment, the percent of growth of the roots and shoots decreased with the increased concentrations of Cd, Pb, and As in the soil, and the rate of decrease results from As exposure was the highest. The presence of Cr in the external environment leads to changes in the growth and development pattern of the seedlings. There was a reduction in growth, dry weight, and vigor index in all tested genotypes of soybeans at increasing concentrations of Cr. The reduction in seedling growth under Cr stress might be due to the poor root growth, which inhibits the transportation of water and nutrients to the shoot of the plants (Ganesh et al., 2009). In another study, seedlings of soybeans were exposed to Cd (CdCl$_2$, 200 μM for 72 h) which significantly inhibited shoot length, root length, number of leaves per plant, and mean leaf area per plant (Hashem, 2014). Significant growth reductions were noticed under Hg (Palanisamy et al., 2012), Cd (Yang et al., 2012; Chmielowska-Bąk et al., 2014; Hashem, 2014), Co and Pb (Imtiyaz et al., 2014a), and Cr (Sundarmoorthy et al., 2015) stresses. The reduction of growth under metal stress might occur due to the inhibition of photosynthesis. A marked reduction of chl content (SPAD value) and photosynthetic rate were found under Al stress (Shamsi et al., 2007). Similarly, damages of different photosynthetic pigments were reported under 223 μM Cd (Chmielowska-Bąk et al., 2014; Hashem, 2014), Co, and Pb stress (Imtiyaz et al., 2014b). The toxicity symptoms of soybeans were characterized by generalized interveinal chlorosis in young leaves with the appearance of dark brown spots, progressing to the drying of the leaves that were observed under Cd, copper (Cu), iron (Fe), manganese (Mn), Pb, and zinc (Zn) polluted soil (Silva et al., 2014). On the other hand, root nodulation was severely affected, and their number reduced drastically in Cd-treated soybean plants (Thakur and Singh, 2014), which adversely affected the N fixation of plants (Sheirdil et al., 2012). In relation to inhibited photosynthesis, Cd leads to oxidative stress, inhibition of respiration, increased production of ethylene, and rate of mutation; as a consequence, stunted growth and yield decrease were noticed (Chmielowska-Bąk et al., 2014). Hashem (2014) reported that 200 μM Cd stress for 24 h caused lipid peroxidation, increased H$_2$O$_2$ levels and the activities of SOD, peroxidase (POX), CAT, and ascorbate peroxidase (APX), and altered glutathione (GSH) content in soybeans. On the other hand, Yang et al. (2012) found that Cd stress increased MDA and glutathione *S*-transferases (GST) activities in seedlings and depressed the ratio of GSH and homoglutathione (hGSH). Reduction of GSH/hGSH ratio and concomitant increase of GST activity suggested a possible participation of GSH into GSH-Cd conjugate synthesis. Under

Cr stress (50, 100, 200, 300 µg/L; 15 days), the reduction of protein content and the increase of Pro content were observed by Sundarmoorthy et al. (2015). Similar results were reported by Imtiyaz et al. (2014b) under Co and Pb stresses. Sundarmoorthy et al. (2015) also reported that activity of antioxidant enzymes such as CAT and polyphenol oxidase decreased, whereas POX and SOD activities increased. Mittona et al. (2015) showed that dichlorodiphenyltrichloroethane (DDT) had no effect on soybean photosynthetic pigments but different responses in the protein content, antioxidant capacity, GST activity, and thiol groups on roots, stems, and leaves, indicating that DDT affected soybean plants. Accumulation of metal in the root, shoot, or seed of soybeans was reported under Cd (Yang et al., 2012) and Cr stresses (Sundarmoorthy et al., 2015), which may lead to an imbalance in the nutrient uptake of plants. Seed yield and quality of seeds were hampered by the different toxic metals. Malan and Farrant (1998) reported that Cd and nickel (Ni) (0.05 mg/L) markedly reduced plant biomass and seed production. Significant amount of metal also entered into seeds, which deteriorated the quality. Similar yield reduction was obtained by Chmielowska-Bąk et al. (2014) under Cd stress. Khan et al. (2013) showed that the exposure of increasing the concentration of heavy metals (Cd and Hg) significantly reduced the oil content when applied separately, while the interactive effect of heavy metal showed a smaller decrease in oil content and showed an antagonistic impact of heavy metal on oil content. The study also revealed considerable changes in major and minor fatty acids of the soybean seeds due to heavy metal exposure. There was a noteworthy decrease in the amounts of fatty acids such as oleic acid and linoleic acid, while the fatty acids such as palmitic acid, stearic acid, and linolenic acid were markedly increased as a result of the increasing concentration of heavy metals. The results suggested that the heavy metal exposures adversely affected the seed oil content and changed the fatty acid composition of soybean oil.

Extreme Temperature

The effect of extreme temperatures, HT, or LT on soybeans may vary depending upon growing season and region or soybean cultivar. Different growth stages show various damage effects under HT or LT stress. Germination and healthy seedling establishment of soybeans are dependable on temperature. The effect of temperature on germination was examined in different studies. Liu et al. (2008) reported that suitable temperature range for the emergence of soybeans is 15–22°C. In other study, seeds of soybeans were exposed to different HTs (50, 60, and 70°C for 10 h). Seeds could withstand up to 70°C. But compared to a control, the gradual increase of temperature reduced the germination percentage, moisture content, and seedling vigor of soybeans (Anto and Jayaram, 2010). The germination percentage of soybean seeds was 62% under the HT stress of 39°C, compared to the control temperature (29°C). The length of seedlings also reduced significantly under that temperature stress (Ndunguru and Sununerfield, 1975). At different temperature levels, such as 24.5, 28.5, 33.0, 36.5, and 40.0°C, germination performance was examined. The highest germination percentage was recorded under 33.0–36.5°C.

Without this temperature range the highest time was required for germination, or the least number of seeds were germinated, or seedlings showed abnormal appearances (Edwards, 1934). Increasing temperature due to delayed sowing decreased plant height and leaf area, biomass of leaves, stem, pods, roots, and total biomass (Kumar et al., 2008a,b). High temperatures (10, 25, 50, and 110°C) reduced the germination and disrupted phenological events of soybeans at different growth stages (Sapra and Anaek, 1991). High temperatures increased the thicknesses of palisade and spongy parenchyma layers and the lower epidermis in soybean leaves (Djanaguiraman et al., 2011b). Sinclair and Weisz (1985) reported that soil temperatures above 34°C negatively affected the N fixation of soybean root. Nitrogen fixation activity decreased in soybean due to HT exposure of 40°C. After 2 hours of HT, the fixation rate decreased by 15%, which decreased by 70% after 6 hours of HT exposure (Keerio et al., 2001).

The response of different soybean genotypes was evaluated under 30/22°C and 38/30°C. High temperatures hampered net photosynthesis, total chl, phenolic and wax contents, and vegetative growth (Koti et al., 2007). The high temperature of 38/28°C distorted the plasma membrane, chloroplast, thylakoid and mitochondrial membranes, cristae, and matrix, compared to the normal growth temperature (28/18°C). High temperatures also decreased leaf photosynthetic rate and stomatal conductance by 20.2% and 12.8%, respectively (Djanaguiraman et al., 2011b). In another study, it was reported that HT (45 and 48°C) decreased efficiency of identical maximum PSII photochemistry (Fv/Fm ratio) in soybean seedlings (Li et al., 2009). Decrease of total chl content (by 17.8%), chl a content (by 7.0%), chl a/b ratio (by 2.5%), sucrose content (by 9.0%), and stomatal conductance (by 16.2%) were reported under HT of (38/28°C for 14 d), compared to normal growth temperature (28/18°C). High temperature also decreased PSII quantum efficiency and photosynthetic rate (Djanaguiraman et al., 2011a). Djanaguiraman et al. (2013) again reported that leaf chl content, photosynthetic rate, photochemical quenching, and electron transport rate decreased due to HT (39/20°C) exposure, compared to 30/20°C. Decreased quantum yield of PS II (35%) and photochemical quenching (25%) by high day temperature imply that the maximum inhibition of electron transport occurred at a quantum yield of PS II, rather than photochemical quenching (Djanaguiraman et al., 2013). Chloroplast and mitochondria are major sites of ROS production. Chloroplasts, mitochondria, and cell plasma membrane damage in HT-affected soybeans indicated oxidative damage. High temperature caused swollen chloroplasts, dilated thylakoids, and loss of integrity of stomal and chloroplast-envelope membranes. High temperature also resulted in a discontinuous mitochondrial membrane (Djanaguiraman et al., 2011b). In another study, Djanaguiraman et al. (2011a) reported that HT (38/28°C day/night temperature, compared to control 28/18°C) increased H_2O_2 content. High temperature also significantly increased O_2·⁻ (by 85.6%) and MDA contents (by 174%) (Djanaguiraman et al., 2011a). Developing seeds of soybeans showed higher accumulation of ROS, including H_2O_2, which increased lipid peroxidation under HT (40°C/30°C with 100%/70% relative humidity under light/dark), compared to the control (Wang et al., 2012).

Low temperature caused poor germination rate, reduced seedling emergence, decreased seedling vigor, and resulted in severe yield loss (Cheng et al., 2010).

Soybean plants showed physiological and biochemical dysfunctions when exposed to temperature below 10°C. The soybean is a chilling-sensitive plant, and its seeds are the most vulnerable to LTs during the first hours of the imbibition phase (Bramlage et al., 1978). The seeds of soybeans were exposed to a chilling temperature (4°C) and then were transferred to 22°C for germination. Soybean seeds exposed to low temperature imbibition increased alcohol dehydrogenase I and *RAB21* might contribute in decreasing the effect of anoxia (resulting from water uptake during imbibition). LT increased stress-related proteins such as *LEA* and *GST24*, which is related to low temperature stress response. Expression of some crucial enzymes (malate dehydrogenase and phosphoenolpyruvate carboxylase) involved in the TCA cycle was improved that might be beneficial during germination (Cheng et al., 2010). The influence of low temperature on *G. max* on energy transduction via mitochondrial respiration and activity of dehydrogenases enzymes was investigated during imbibition and germination. Mitochondria were isolated from embryonic axes of seeds and exposed to 10 and 23°C (control) by submergence in H_2O for 6 hours and kept an additional 42 h in a moist environment. Activities of glutamate dehydrogenase, alcohol dehydrogenase, glucose-6-phosphate dehydrogenase, and NADP-isocitrate dehydrogenase decreased during first 6 hours. But after germination at 23°C (48 h), their activities increased. Activities of these enzymes may limit mitochondrial respiration at LT in soybean tissues (Duke et al., 1977). Janas et al. (2000) studied the root growth under LT, and root growth was inhibited under LT. Compared to the relative growth rate at 25°C, the LT of 10°C resulted in 10-fold decrease of root growth, which was 1 mm/day. Infection and/or early nodule development were very sensitive to low root zone temperature, which determine onset of N fixation and subsequent N accumulation and plant growth. Soybean plants were grown under different root zone temperatures (15, 17, 19, 21, 23, and 25°C). Between 25 and 17°C, the onset of N_2 fixation was linearly delayed by 2.5 days for each degree decrease in temperature. Below 17°C, the decrease in each degree Celsius temperature appears to delay the onset of N_2 fixation by about 7.5 days (Zhang et al., 1995). Low temperatures limit photosynthesis and growth. The decrease in photosynthetic capacity in soybeans was related to chilling-associated oxidative damage to chloroplast components. Soybean plants exposed to 7°C (24 h) damaged PSII decreased the ratio of variable to maximum fluorescence (Fv/Fm) and caused oxidative damage to thylakoid proteins and lipids (Tambussi et al., 2004).

The survival and succession of seed crop plants are ensured by the successful reproduction, and the reproductive phase is considered as the most sensitive among the growth stages. High temperature decreases photosynthesis and causes abscission and abortion of flowers, young pods, and developing seeds (Prasad et al., 2002, 2006, 2008). Pollen viability and stigma receptivity were obstacles under HT, which decreased seed set (Prasad et al., 2002, 2006; Snider et al., 2009). Lindsey and Thomson (2012) mentioned 25–29°C as the optimum temperature for pod setting, which was severely affected above 37°C. A moderate increase in daytime temperature (18–26°C) during seed filling was considered as favorable for desired soybean yield (Sionit et al., 1987). Pollen germination percentage, pollen tube length, pollen number anther^{-1}, and flower length decreased under HT of 38/30°C, compared to the control temperature (Koti et al., 2005). Hatfield et al. (2008) studied different soybean cultivars under HT. The cardinal, optimum, and failure point temperatures were determined as 13.2°C, 30.2°C, and 47.2°C for pollen germination. Cardinal, optimum, and failure point temperatures were 12.1°C, 36.1°C, and 47.0°C for pollen tube growth. Djanaguiraman et al. (2011a) studied soybean plants under HT (38/28°C, 14 days). Compared to the control (28/18°C), HT stress increased flower abscission, which was the reason for decreased pod set percentage. Pollen production decreased by 34%, pollen germination by 56%, and pollen tube elongation decreased by 33% under HT of 38/30°C, compared to 30/22°C (Salem et al., 2007). The high temperature (37/27°C, day/night) reduced pod setting, which was due to lower pollen viability (Kitano et al., 2006). Boote et al. (1998) motioned the cardinal or base temperature of 6°C and an optimum temperature of 26°C for anthesis. The postanthesis phase optimum temperature was mentioned as 23°C in different research findings (Pan, 1996; Thomas, 2001; Boote et al., 2005). Reduced photosynthate supply under HT stress during flowering and pod set growth stages may result in abortion of flowers and young pods and seeds (Egli and Bruening, 2006). The high temperature reduced saturated phospholipids and phosphatidic acid, which resulted in decreased pollen viability and germination. Pollen viability and pollen germination decreased with increasing nighttime temperature (from 23 to 29°C pollen viability decreased by 19, 29, and 39%, and pollen germination decreased 11, 20, and 28%, respectively, in 23, 26, and 29°C, compared to optimum temperature [20°C]) (Djanaguiraman et al., 2013). Liu et al. (2008) mentioned that the suitable temperature for the flowering and maturity of soybeans are 20–25°C and 15–22°C, respectively.

High temperatures at the flowering stage caused flower abscission, reduced pollen viability, reduced pod set percentage, and individual seed weight, rather than the optimum temperature, which reduced yield of soybeans (Koti et al., 2005). The thermosensitivity of pod set is higher than the accumulation of assimilate in seed. Thus pod set decreased at nighttime temperatures >23°C, whereas seed weight decreased at nighttime temperatures >26°C. However, high daytime (39/20°C) or nighttime temperature (30/29°C) reduced the pod set and seed weight, compared to optimum temperature (30/20°C) (Djanaguiraman et al., 2013). While studying three soybean varieties, viz, AGS190 (large seeded), Willis (medium seeded), and Dieng (small seeded), irrespective of varieties, the reproductive stage showed higher sensitivity to HT to cause yield loss resulting from the reduction in number of pods plant[-1], number of seeds pod[-1], and 100-seed weight. Compared to 30°C, the temperature 35°C had more damage effects (Puteh et al., 2013). A temperature increase from 29/20 to 34/20°C at the seed filling stage decreased seed yield, which was reported by Dornbos and Mullen (1991). Soybean plants were subjected to HT stress in different growth stages. During flowering and pod set, seed filling and maturation, and the entire reproductive period, the HT stresses were 30/20, 30/30, 35/20, and 35/30°C, respectively. A significant reduction of seed weight plant[-1] was observed during flowering and pod set stage. The highest yield reduction was 27% due to exposure at 35°C, 10 h/day from flowering to seed maturity (Gibson and Mullen, 1996). According to Dornobos and Mullen (1991), rise in the temperature from 29/20 to 34/20°C during the seed filling stage significantly decreased soybean seed yield. In other research, the temperatures above 30/25°C (day/night) during flowering and pod development adversely affected soybean physiology and reduced weight of seeds (Egli and Ward-law, 1980). Soybean plants are often affected by HT stress due to a delayed growing season. Number of pods plant^{-1}, number of seeds pod^{-1}, dry matter accumulation, stover yield, and grain yield significantly reduced due to HT (induced by delayed growing season) (Kumar et al., 2008a,b). Djanaguiraman and Prasad (2010) studied the effect of HT on yield components of soybeans. The number of filled pods plant^{-1} reduced by 50.9%, seed set reduced by 18.6%, number of filled seeds plant^{-1} by 30.5%, and seed size reduced by 64.5% under HT, compared to the control. The reduction in yield components under HT resulted in a severe reduction of seed yield of soybeans by 71.50% and harvest index by 78.2% in soybeans, compared to the control (Djanaguiraman and Prasad, 2010). Baker et al. (1989) and Boote et al. (2005) concluded that seed harvest index reduced above the temperature range 23~27°C depending upon the cultivar of soybean. A mean temperature of 39°C or higher was stated as lethal for soybean yield production (Pan, 1996; Thomas, 2001). In their findings, Ferris et al. (1999) reported that HT at maturity period decreased the yield by 29%.

Low temperature adversely affects the yield attributes and yield of soybeans. Soybean seed weight and its biochemical parameters vary depending upon temperature and cultivar. Soybean plants were subjected to one of three night/day temperature regimes (eg, 13/23, 18/28, or 23/33°C). Lower temperatures decreased mean seed weight by 5%, and it decreased by 16% under higher temperatures, whereas mean germ weight decreased by 15% under HT. Low temperature decreased mean protein concentration (by 7%), increased cotyledon mean genistein by 130%, and daidzein by more than 200%. High temperature treatment significantly reduced the total isoflavone contents. The mean isoflavone content decreased by 18% under HT, whereas the cold temperature had no effect (except the increase in Imari cultivar). Lower temperatures increased the glycitein content in cultivar Jack and the daidzein and genistein contents in Imari (Rasolohery et al., 2008). Lower temperature reduced growth duration (due to delayed sowing), declined crop biomass, harvest index, seed number, and individual seed weight of soybeans. The low-temperature effect was severe in two growth stages from R5 (seed was 3 mm long in a pod at one of the four uppermost nodes on the main stem with a fully developed leaf) to R7 (one normal pod on the main stem that has reached its mature pod color), which accounted for 77% variation in yield, 56% variation in seed number, and 62% variation in seed weight (Calviñoa et al., 2003).

Waterlogging

Excess rainfall, tides, floods, and lack of proper drainage facilities are the reasons causing waterlogging stress in plants. The primary effect of waterlogging in crop plants is anoxia or oxygen deprivation. Plants need oxygen for growth and physiological processes such as cell division, respiration, uptake, and transportation of nutrients. Chlorosis, necrosis, defoliation, growth reduction, reduced N fixation, yield loss, and plant death are effects of waterlogging stress in soybean plants, which are often occurring at various vegetative and reproductive stages (Scott et al., 1989; Linkemer et al., 1998;

Vantoai et al., 1994; Bacanamwo and Purcell, 1999; Fig. 4.7). The growth of soybean plants is suppressed under anoxic condition. Root system of soybeans usually becomes shallow under waterlogged condition because of decreasing the tap root length, whereas the number of adventitious roots increases (Sallam and Scott, 1987; Scott et al., 1989). The formation of adventitious roots under waterlogging condition is also evident (Fig. 4.7).

Nitrogen fixation and nitrate reduction are important for N accumulation in soybeans. The effect of waterlogging on activities of nodule nitrogenase and leaf nitrate reductase was studied in two field-grown soybean lines, such as in line PI200492 and Chung Hsing No. 1. Waterlogging stress was imposed on the anthesis and seed filling stage. Under both growth stages the activities of nodule nitrogenase and leaf nitrate reductase decreased. But the line PI200492 fixed more N and was involved in reducing more nitrate, compared to the line Chung Hsing No. 1 (Sung, 1993). Growth, C, and N metabolisms at seedling and flowering stages of soybean plants were examined under different durations of waterlogging stress. Waterlogging decreased growth, biomass, leaf area, leaf pigment content, and photosynthesis rate. The significant increase of lipid peroxidation or the leaf MDA content indicated the oxidative damage of soybean plants under waterlogging stress. Waterlogging increased soluble sugar content and activities of glutamine synthase and sucrose synthase. But soluble protein content was decreased under waterlogging stress. The plants in their flowering stage showed

Control   Waterlogging

Figure 4.7 Growth inhibition, chlorosis, and formation of adventitious roots in soybeans cv. Shohag grown under waterlogging condition for 30 days.
Taufika Islam Anee.

higher sensitivity, compared to the seedling stage. The shorter the duration of stress, the shorter was the adverse effect of waterlogging stress, and vice versa (Zhou et al., 2012). The reduction of nodule number and weight corroborated the reduction of shoot and root dry weight, leaf area, seed dry weight, and harvest index in soybean plants under waterlogging conditions. The reduction of shoot dry weight was up to 75–77%, root dry weight was up to 64–75%, nodule number decreased by about 68–73%, shoot N content of waterlogged plants was 31–49% of the control, and seed dry weight decreased up to 69% depending upon genotype (Youn et al., 2008). The early vegetative and early reproductive stage was considered as the most sensitive stage for soybean under the waterlogging stage. Both greenhouse and field studies were conducted to evaluate the performance of soybean plants. The growth rate of soybean plants decreased noticeably under waterlogging stress. Decreased pod production resulting from fewer pods per reproductive node was the reason for decreased yield by 37% (Linkemer et al., 1998). Scott et al. (1989) reported that only 2 days of waterlogging can cause 18% yield loss at the late vegetative stage. In contrast, waterlogging stress at early reproductive stages may cause yield loss above 26%.

Ultraviolet Radiation

Due to the depletion of the stratospheric ozone layer, the entrance of UV radiation on the earth surface has increased markedly (McKenzie et al., 1999). Plant photosynthesis (Sunita and Guruprasad, 2012) and the accumulation and translocation of photosynthetic assimilates (Bassman, 2004) are altered by UV radiation exposure, which changes morphology, reproductive development, and yield potential in a detrimental way (Shena et al., 2010; Liu et al., 2013). Considering the importance, the effect of enhanced UV-B radiation on soybean growth and yield has been studied by researchers. Barnes et al. (1990) suggested enhanced UV-B radiation makes the soybean plant dwarf by shortening the internode length. The content of macroelements in the root, stem, and leaf of soybean seedlings decreased due to UV-B radiation exposure. But micronutrient contents of different parts of soybean plants were differentially affected under UV-B stress. However, the altered distribution of mineral elements in the root, stem, and leaf led to the decrease in the accumulation of dry matter and soybean growth (Peng and Zhou, 2010). Feng et al. (2001a) reported enhanced UV-B radiation-induced changes in flowering time in some soybean cultivars. They also noticed decreased ratio of chl a/b, decreased total leaf number, and total leaf area in different soybean cultivars under UV-B stress. In their other study, they reported that total biomass and seed yield of 10 different soybean cultivars decreased averaged by 24.2% and 23.3%, respectively, due to UV-B exposure (Feng et al., 2001b). UV-B imposition produced smaller flowers with shorter petals and staminal columns. Flowers also produced a lower number of pollen with shriveled shapes, poor pollen germination, and shorter pollen tube lengths (Koti et al., 2005). Baroniya et al. (2011) studied eight cultivars of soybeans to determine the effect of solar UV-B. Vegetative growth parameters including plant height, leaf area, and number of nodes markedly decreased. Yield attributes such as number of pods and seed weight also decreased significantly, which resulted in yield reduction (Baroniya et al., 2011). Three determinate soybean cultivars, Hai339 (H339), Heinong35 (HN35), and Kennong18 (KN18) were exposed to enhanced UV-B radiation to study the effects on yield components and seed growth characteristics. Plant height, dry weight of individual stem, seed size, and yield per plant of three soybean cultivars decreased on average by 15.5, 16.9, 12.3, and 43.7%, respectively (Liu et al., 2013). Reduced concentrations of isoflavones and phenolic compounds in soybean seeds were reported by Kim et al. (2011). After a week of UV-B radiation imposition decreased transpiration of soybean seedlings was accompanied by decrease of leaf relative water content (RWC) and an increase of Pro level. Lipid peroxidation and electrolyte leakage were evident under stress conditions. Exposure to UV-B radiation increased contents of anthocyanin and soluble phenols. Decrease of stomatal conductance increased intercellular CO_2 concentration, decreased chl content, and net photosynthetic rate were the direct reasons for reducing biomass accumulation and growth of soybean seedlings under UV radiation stress (Shena et al., 2010).

Possible Strategies for Environmental Stress Tolerance

Use of Exogenous Protectants

Several studies indicated that the use of exogenous phytoprotectants like osmoprotectants, plant hormones, antioxidants, signaling molecules, polyamines, nutrient elements, etc. ensures significant protection against salt stress in soybean plants (Table 4.2). Yin et al. (2015) reported that exogenous application of calcium (Ca; 6 mM $CaCl_2$) improves salt stress tolerance of germinating soybeans via enriching signal transduction, promoting protein biosynthesis, enriching antioxidant enzymes, and activating their activities. Exogenous nitric oxide (NO) increased shoot, root, and nodule weights of soybean seedlings under salt stress by reversing H_2O_2 production and increasing APX activity (Egbichi et al., 2014). External Zn improved shoot dry and fresh weight in salt-induced soybean seedlings by increasing P, K, Ca, Zn, and Fe contents and decreasing Na content (Weisany et al., 2014). Putrescine (Put) also improved the growth of soybean seedlings by stimulating antioxidant enzymes and decreasing lipid peroxidation and ROS production under salt stress conditions (Gu-wen et al., 2014). Klein et al. (2015) showed that exogenous caffeic acid improved the growth of soybean seedlings by decreasing lipid peroxidation and increasing chl content via regulation of NO signaling. Exogenous glycine betaine (GB) improved yield and yield attributes of soybeans by decreasing Na uptake under salt stress conditions (Rezaei et al., 2012).

Yield reduction of soybean due to drought stress has been reported in research. Thus it has become compulsory to develop a drought-tolerant soybean variety. Scientists are adopting various tools and techniques to improve drought tolerance in soybeans. Use of plant growth regulators and osmoregulators as phytoprotectants is an effective way to cope with drought stress (Agboma et al., 1997; Anjum et al., 2011, 2013; Abu-Muriefah, 2015; Table 4.2). Agboma et al. (1997) reported that GB plays an important role in improving plant performance under the conditions of drying soil. They found that the transpiration rate of GB-treated plants decreased to 85% compared to untreated plants, suggesting an antitranspirant property of GB. Glycine betaine slowed down the transpiration rate; as a result, the plant can access water for a longer period and increased photosynthesis and N fixation rates under drought stress. Lecube et al. (2014) suggested that pretreatment with $100\,\mu M$ indole acetic acid (IAA) enhanced drought tolerance by modulating the AsA-GSH cycle and they concluded that IAA serves as an important molecule for signaling the dehydration/drought response in soybeans. Anjum et al. (2011) found that foliar spray of $50\,\mu M$ methyl jasmonate (MeJa) enhanced yield and the yield-contributing character of soybeans under drought stress, as MeJa acts as signaling molecule and plays a role in modulating antioxidant defense system. Bîrsan et al. (2014) reported biologically active substances by phenolic nature such as salicylic acid (SA) spraying at a low concentration (0.01%) improves the adaptation of plants to water deficit by adjusting Pro level. Anjum et al. (2013) reported that foliar application of 0.5 mM benzoic acid (BZA) in soybeans markedly enhanced the net photosynthesis by 11.54% and chl a contents by 6.57%, whereas net photosynthesis and chl content reduced due to drought stress. Foliar spray of paclobutrazol (40 mg/L) at vegetative growth and just before blooming enhanced tolerance against drought through maintaining osmoregulators (Pro and sugars), changing phytohormone concentration, and modulating the antioxidant system (Abu-Muriefah, 2015).

Toxic metals/metalloids stress are considered as one of the major causes of yield reduction in soybeans. So to improve yield under stress conditions, plant scientists are trying to establish tolerant varieties and are exploring suitable stress ameliorants. Exogenous protectants have been found to be effective in mitigating toxic metal stress-induced damage in plants. These phytoprotectants can enhance plant growth and productivity. Silva et al. (2001) used magnesium (Mg) and Ca as an Al stress ameliorant. The use of $50\,\mu M$ Mg and 3 mM Ca in Al stress increased malate and citrate concentrations in the tap root of soybeans, which ameliorate rhizotoxicity by reducing the Al accumulation by root tip. Similarly, Yang and Juang (2015) used different concentrations of Ca and K for the alleviation of Cd stress. They found that Ca and K levels alleviated dramatic reduction in root elongation of soybean due to increased Cd. Yathavakilla and Caruso (2007) reported that selenium ($60\,\mu M$ Se(IV)) played a protective role against Hg stress ($45\,\mu M$ Hg(II)). Yeast (60 mL/L) and garlic extracts (30 mL/L) were able to minimize the harmful effects of Cd and improved the concentrations of photosynthetic pigments, total sugars, IAA, and gibberelic acid (GA_3) in leaves of Cd-polluted soybean plants. Foliar application with yeast or garlic extract was sufficient for reducing the harmful effect of Cd and improved the percentage of crude protein and of total lipids in seeds of Cd-polluted soybean plants (Abdo et al., 2012).

Extreme temperatures adversely affect soybean plants from germination to the reproductive stage and result in severe yield loss. Few studies are available that used exogenous phytoprotectants to prevent or protect extreme temperature damages in soybean plants (Table 4.2). Soybean genotypes were exposed to HT (45°C), which reduced the germination of pollen and hampered the reproductive growth. When different soybean genotypes were applied with GB (at 2 kg/ha) as foliar spray before flowering pollen germination was found to be 15% and 21% higher in GB-treated plants under HT. Depending upon response to GB application under HT, the tolerance of soybean genotypes was selected. Extremely responsive genotypes were Hutcheson,

Table 4.2 Protective effect of exogenous phytoprotectants under environmental stress in soybeans

Level of Stress	Name and Dose of Protectants	Protective Effect	References
11.1 dS/m salinity, through-out the life cycle	Osmoprotectant, (GB) (0, 2.5, 5, 7.5, and 10 kg/ha) foliar spray	Decreased Na uptake, increased endogenous GB, restored yield attributes and yield.	Rezaei et al. (2012)
50 and 100 mM NaCl, 76 days	Plant hormone (jasmonic acid), 1 μM	Increased chl content, carotenoid content, mineral content, recovered growth, and yield.	Sheteawi (2007)
100 mM NaCl, 14 days	Plant hormone (GA$_3$), 0.5, 1, 5 μM GA$_3$	Regulated phytohormone, increased chl content and growth.	Hamayun et al. (2010)
100 mM NaCl, 7 days	Plant hormone (SA), 100 μM	Increased flavonoid content, decreased Na uptake, Na/K ratio, LOX activity.	Simaei et al. (2012)
70 mM NaCl, 12 days	Plant hormone (caffeic acid), 100 μM	Enhanced NO biosynthesis, chl content, N fixation efficiency, and restored growth.	Klein et al. (2015)
12.5, 50 mM NaCl, 10 days	Antioxidant (AsA) 400 mg/L	Increased CAT and POX activity, restored growth.	Dehghan et al. (2011)
80 mM NaCl, 150 mM NaCl, 1 day; 100 mM NaCl, 7 days	Signaling molecule (NO), 10 μM diethylene tri-amine-NO; 100 μM SNP	Increased APX activity, antioxidant component, restored growth; increased flavinoid content, decreased Na uptake, Na/K ratio, and LOX activity.	Egbichi et al. (2013, 2014); Simaei et al. (2012)
100 mM NaCl, 15 days	Polyamine (putrescine), 10 mM	Stimulate antioxidant enzyme, decreased lipid peroxidation and ROS production, increased growth.	Gu-wen et al. (2014)
50 mM NaCl, 4 days	Nutrient element (Ca), 6 mM CaCl$_2$	Signal transduction, promotion of protein biosynthesis, increased germination.	Yin et al. (2015)
5, 8, 10 mM dS/m salinity, 35 days	Nutrient element (Zn), 10 μM zinc sulfate (ZnSO$_4$)	Maintained ionic balance, increased growth.	Weisany et al. (2014)
Drought stress, (different watering regime)	GB 3 kg/ha, foliar spray	Reduced transpiration, increased net photosynthesis, N$_2$ fixation, and yield.	Agboma et al. (1997)
Drought stress, 0–8% PEG, 3 days	IAA 100 μM, seed pre-treatment	Protection against oxidative stress, modulation of NO levels, enhanced HO-1 (Heme oxygenase) synthesis and activity.	Lecube et al. (2014)
Drought stress, 80% and 35% soil field capacity, until harvest	BZA 0.5 mM, foliar application	Improved growth, development, yield, and yield components, played a role in improving gas exchange and chl contents.	Anjum et al. (2011)
Drought stress, 35% TCW (total capacity of water in	SA 0.01%, foliar spray	Increased Pro content.	Bîrsan et al. (2014)

Stress	Treatment	Effect	Reference
Drought stress, 60% of the soil field capacity, until harvest	Paclobutrazol 40 mg/L, foliar application	Increased leaf water potential, IAA, and zeatin levels and the content of Pro and soluble sugars.	Abu-Muriefah (2015)
Drought stress, 20% of field capacity, until harvest	Melatonin 50 μM, 100 μM seed coating	Promotional roles in soybeans through enhancement of genes involved in cell division, photosynthesis, carbohydrate metabolism, fatty acid biosynthesis, and AsA metabolism.	Wei et al. (2014)
Drought stress, 0.3 MPa	Put 0.5 mM	Improvement of growth parameters (length and weight of seedlings, root, shoot, stem, hypocotyl, first internode).	Zabihi et al. (2014)
4.6 μM Al^{3+}, 72 h	50 μM Mg and 3 mM Ca, 72 h	Reduction of Al accumulation in root tips.	Silva et al. (2001)
45 μM Hg(II) every week as a solution	60 μM Se(IV), 4 times	Formed a high molecular weight entity containing Se and Hg in plants.	Yathavakilla and Caruso (2007)
50, 100, and 200 ppm Cd with irrigation water start from 30 days-old seedling	60 mL/L yeast extract and 30 mL/L garlic extract, three times in 2 weeks	Improved the concentrations of photosynthetic pigments, total sugars, IAA, and GA_3 in leaves as well as yield of soybeans.	Abdo et al. (2012)
Cd levels of 0, 4, 8, 12, and 16 mM, 7 days	0.5, 1, and 10 mM Ca and 0.3, 1.2, and 9.6 mM K, 7 days	Ca and K levels alleviated reduction in root elongation of soybeans due to increased Cd.	Yang and Juang (2015)
45°C	GB at 2 kg/ha as a foliar spray before flowering	Improved pollen germination and reproductive development.	Salem et al. (2005)
45°C, 2 h	PAs: Put (1 mM), Spd (1 mM), and Spm (1 mM) were applied as a pretreatment, 2 h	Improved membrane integrity, decreased electrolyte leakage and MDA, improved growth performance.	Amooaghaie and Moghym (2011)
38/28°C, 14 days	Foliar spray of 1-MCP, 1 μg/L	Decreased ethylene production rate and premature leaf senescence, increased seed set percent and seed size and yield.	Djanaguiraman and Prasad (2010)
38/28°C, 14 days at flowering stage	1-MCP, 1 μg/L	Decreased rate of ethylene and ROS ($O_2^{\cdot-}$ and H_2O_2) production, oxidative damage, enhanced antioxidant enzyme activities (SOD and CAT), delayed leaf senescence, reduced flower abscission, and increased pod set percentage.	Djanaguiraman et al. (2011a)

Continued

Table 4.2 Protective effect of exogenous phytoprotectants under environmental stress in soybeans—cont'd

Level of Stress	Name and Dose of Protectants	Protective Effect	References
10±0.5°C, 72 h	5-aminolevulinic acid (ALA), 0, 0.3, 0.6, and 0.9 mM	Enhanced plant height, shoot fresh and dry weight, chl content, photosynthesis, Pro content, as well as RWC, decreased electrolyte leakage, enhanced SOD, and CAT activities.	Manafi et al. (2015)
1°C, 10 weeks	Spm, 0.5, 1, and 2 mM, 20 min	Reduced lipid peroxidation and MDA accumulation, enhanced activities of antioxidant enzymes (CAT, POX, and SOD).	Chun-quan et al. (2015)
Waterlogging	SA, 2 mM	Promoted seedling growth, enhanced root activity, chl content, and Pro content. Reduced MDA content and reduced the cell permeability.	Jianguo et al. (2006)
Waterlogging	SA, 100, 200, and 400 ppm	Increased total protein content and decreased ROS (H_2O_2 and $O_2 \cdot^-$).	Mishra et al. (2013)
UV-B radiation (5.4 kJ/ m^2 day^{-1}), 7 days	Si, 1.7 mM	Reduced membrane damage/lipid peroxidation and osmolyte leakage, improving antioxidant system. Increased leaf RWC, photosynthesis, and growth.	Shena et al. (2010)
UV-B radiation (30 kJ/m^2)	SNP (NO donor), 0.4, 0.8, 1.2, and 1.6 mM	Prevented chl loss, increased activity of CAT and APX, decreased H_2O_2 and $O_2 \cdot^-$ accumulation, and ion leakage.	Santa-Cruz et al. (2010)
UV-B (7.1 W/m^2, 6 h/day), 3 and 6 days	SA, 0.5 mM	Mitigated the reduction of chl, Car, photosynthesis, stomatal conductance.	Zhang and Li (2012)

D 68–0102, and Stress land; the more responsive genotypes were DP 4690RR, DK 3964RR, Williams 82, and Stalwart III, and the less responsive genotypes were P 9594, Maverick, and DARE, and the not-responsive genotype was DG 5630RR (Salem et al., 2005). Exogenous $CaCl_2$ application decreased electrolyte leakage and lipid peroxidation from root and hypocotyl tissue sections under HT stress (45°C, 2 h). The advantageous effects of $CaCl_2$ were reflected from enhanced recovery of seedlings, increased hypocotyl, and root growth of seedlings. Again, application of EGTA (a chelator of Ca) increased stress injury and growth inhibition, which further confirmed the protective role of $CaCl_2$ to recover HT injury (Amooaghaie and Moghym, 2011). Moghym et al. (2010) studied the effects of exogenous polyamine (PA) in alleviating the adverse effects of HT stress in soybean. Temperature up to 40°C inhibits its seed germination. Exogenous PAs (Put; spermidine, Spd and spermine, Spm) stimulated heat-shock protection of soybean seedlings (Moghym et al., 2010). A high temperature of 45°C (for two hours) decreased primary root length (by 47%) and decreased hypocotyl length (by 50%). Exogenous PA pretreatment showed protective effects against the HT damage. Pretreatment with PAs [Put (1 mM), Spd (1 mM), and Spm (1 mM) at the normal temperature of 28°C for 2 h before subjecting the seedlings to heat-shock] alleviated the damage effects of HT. Application of PAs imparted membrane integrity and decreased electrolyte leakage and MDA content. Seedling growth was also improved by PA pretreatment. Among the PAs, the effect of Put was better in conferring thermotolerance, followed by Spm (Amooaghaie and Moghym, 2011). High-temperature (38/28°C) stress decreased photochemical efficiency (by 5.8%), photosynthetic rate (by 12.7%), sucrose content (by 21.5%), activities of SOD (by 13.3%), CAT (by 44.6%), and POD (by 42.9%) and increased $O_2 \cdot^-$ (by 63%) and H_2O_2 (by 70.4%) contents and membrane damage (by 54.7%), compared to 28/18°C temperature. Application of 1-MCP (1-methyl cyclopropene, an inhibitor of ethylene production) decreased ethylene production rate and premature leaf senescence. An antioxidant defense system was enhanced by 1-MCP application. High-temperature stress decreased seed set percentage, seed size, and seed yield plant^{-1}, compared with normal temperature. In contrast, foliar spray of 1-MCP increased the seed set percent, seed size, and yield (Djanaguiraman and Prasad, 2010). Another study also revealed a similar result. High-temperature stress (38/28°C day/night temperature) increased ethylene production rate, caused oxidative damage, decreased antioxidant enzyme activity, caused premature leaf senescence, increased flower abscission, and decreased pod set percentage. Application of 1-MCP reduced ethylene production, enhanced antioxidant enzyme activity, including SOD, POX, and CAT (by 14.4, 50.7, and 23.1%, respectively), decreased ROS production, and increased membrane stability, delayed leaf senescence, decreased flower abscission, and increased pod set percentage (Djanaguiraman et al., 2011a). Soybean seedlings treated with SA improved seedling emergence and vigor evaluated under HT of 41°C or 10°C. Germination and the length of the shoot were improved by maximum concentrations of 44.1 and 78.8 mg/L SA (Brunes et al., 2014).

Exogenous application of 5-aminolevulinic acid (ALA) at low concentrations (0, 0.3, 0.6, and 0.9 mM) protected soybean plants from LT damage (10 ± 0.5°C, 72 h). Alleviation of stress damage effects was evident from enhanced plant height, shoot fresh and dry weight, chl content, photosynthesis, stomatal conductivity, and Pro content, as well as RWC. The application of ALA also enhanced antioxidant defense (enhanced SOD and CAT activities), which prevented overaccumulation of ROS and decreased electrolyte leakage (Manafi et al., 2015). Exogenous application of the AsA precursor L-galactono-1,4-lactone (Gal, 50 mM) or of the α-tocopherol analog 6-hydroxy-2,5,7,8-tetramethylchroman-2-carboxylic acid (Trolox, 0.5 mM) increased the AsA content of the leaves but did not prevent the chilling-induced (7°C, 24 h) decrease of Fv/Fm (Tambussi et al., 2004). Markhart (1984) reported that ABA-induced changes in root hydraulic conductance alleviated the LT-induced water stress in soybeans. The increase of germination rates and inhibition of cell membrane permeability and increased antioxidant defense in germinating soybean seeds under LT (5°C) stress were attributed by H_2O_2 (10, 20, 50, 100, and 200 mM of H_2O_2 for 10 min) priming of soybean seeds (Li-jun et al., 2008). Vegetable soybeans were immersed in 0.5, 1, and 2 mM Spm for 20 min. Soybeans were stored at 1°C for 10 weeks (relative humidity of 85–90%). Low temperature had detrimental effect on soybean physiology. Spermine application reduced lipid peroxidation, enhanced activities of antioxidant enzymes (CAT, POX, and SOD), and reduced the sucrose loss; fructose and glucose contents were noticed in Spm-treated vegetable soybean grains (Chun-quan et al., 2015).

Waterlogging stress creates an imbalance in assimilate translocation and distribution among different plant parts including seeds, which reduced seed nutrient accumulation. Exogenous boron (B) application was effective to ameliorate the adverse effect of waterlogging (Table 4.2). Foliar application of B improved seed protein, oil, fatty acid, and N metabolism. In another study, Bellaloui et al. (2013a) reported that foliar B application improved the seed composition. Cell wall B content and seed N and C contents were improved by exogenous B application in water-stressed soybean plants, which suggests the role of B in regulation of N fixation. Exogenous SA application promoted seedling growth, enhanced root activity, chl content, and Pro content. Exogenous SA application ameliorated the cellular oxidative damage as it efficiently reduced MDA content and reduced the cell permeability in soybean plants subjected to waterlogging stress (Jianguo et al., 2006). Waterlogging disrupted the antioxidant defense system, amplified oxidative burst, and decreased the growth of soybean plants. Exogenous SA supplementation in waterlogged soybean plants increased total protein content and decreased the content of reactive oxygen species including H_2O_2 and $O_2^{\cdot -}$ modulating enzymatic (SOD, CAT, APX, and glutathione reductase, GR) activities and nonenzymatic (Car, AsA, nonprotein thiol, and Pro) components of the antioxidant system (Mishra et al., 2013). During waterlogging stress, the application of NH_4NO_3 showed differential responses in soybean plants. NH_4^+ restricted amino acid metabolism during waterlogging, whereas NO_3^- facilitated amino acid metabolism (Thomas and Sodek, 2006).

Few reports demonstrated the roles of exogenous protectants under UV radiation stress. Silicon (Si) has metabolic and physiological activity in higher plants to enhance the stress-tolerance capacity. Exogenous supplementation of Si in UV-B affected soybean seedlings and improved physiological performances. Anthocyanin and phenol levels decreased 91.5% and 10.0% after Si cotreatment with

UV-B imposition. Silicon application significantly reduced the membrane damage and electrolyte leakage but increased chl content and improved stomatal conductance and net photosynthetic rate, which significantly improved growth performance of UV-B-affected soybean plants (Shena et al., 2010). Exposure to UV-B (280–315 nm) disturbed dry matter partitioning, which significantly decreased soybean plant biomass (by 74.9–135.6%). Silicon application in UV-B radiation affected seedlings and improved the uptake of P and Mg by 11%, which favored the partitioning of dry mass to shoots and the allocation of tissue P and Ca to roots. The overall favorable effect of Si was reflected from an increase in dry mass (Shen et al., 2009). Three-week-old soybean seedlings were subjected to UV-B radiation, which caused oxidative burst and membrane damage. But these seedlings, when were sprayed with SNP (sodium nitroprusside, NO donor, 0.4, 0.8, 1.2, and 1.6 mM) 12 h prior to UV radiation exposure, performed better. Sodium nitroprusside pretreatment prevented chl loss, H_2O_2 and $O_2 \cdot^-$ accumulation, and ion leakage in UV-B-treated plants. The major reasons for decreased oxidative damage and improved plant performance were NO-induced improved antioxidant system (Santa-Cruz et al., 2010). Exposure of UV-B significantly reduced the contents of photosynthetic pigments, stomatal conductance, photosynthesis, shoot height, root length, and shoot and root fresh weight. Exogenous application of SA alleviated the reduction of chl, Car, stomatal conductance, and photosynthesis and improved the growth performance (Zhang and Li, 2012). UV-A and UV-B (wavelengths between 300 and 332 nm) exposure reduced germination compared to controls up to 50%. The germinated seedlings also showed reduced hypocotyl length and anomalous leaf structure and appearance. Exogenous phenylalanine reduced damage from exposure to UV radiation (Warpeha, 2012).

Agronomic Management to Mitigate Environmental Stress in Soybeans

Since the environmental stresses are unavoidable, finding the ways to alleviate the stress-induced damages is the important task for plant scientists. It may be achieved by proper nutrient management, irrigation management, and changes in cropping system. For example, drought stress in soybeans can be mitigated by the efficient use of water resources. This can be attained by minimizing water input; reducing water loss from irrigation systems and the field; and increasing crop water use efficiency (WUE). Therefore under field conditions, WUE can be increased by manipulating the irrigation scheduling, which ultimately avoids drought stress (Ku et al., 2008; Liu, 2009). A loss of more than 50% of irrigated water happens in these irrigation systems through evaporation, leakage, seepage, and percolation, especially when the water source is far away from the field (Barta et al., 2004). A well-managed pipe system can achieve 90–100% conveyance efficiency (Barta et al., 2004). Traditional mulching involves covering the field with straw or other harvest leftovers. The mulch can trap moisture and hence retain soil water. The degrading organic mulch also adds humus to the soil and improves the water-holding capacity of the soil.

Nutrient management often plays a vital role in mitigating environmental stress in soybeans under field conditions. Environmental stresses cause ionic imbalances in roots and shoots of soybean seedlings by decreasing or increasing mineral nutrient content. Bellaloui et al. (2013b) reported that the mineral nutrient K, P, Ca, Cu, and B contents decreased under drought stress condition. Under salt stress condition, K content and the ratio of K/Na decreased in roots and shoots of soybean seedlings (Essa, 2002; Ma et al., 2014). The plant nutrient N, K, Ca, Mg, and S decreased in salt-induced soybean seedlings (Essa, 2002; Bellaloui et al., 2013b). On the other hand, micronutrient Fe, Mn, and Cu contents increased, causing an ionic imbalance in salt-treated soybean seedlings (Tuncturk et al., 2008). Short-term Cd-exposure decreased N, P, Ca, Fe, Zn, Cu, and Mn contents and increased Mg and Cd contents in roots of soybean seedlings and ultimately disrupted ion homeostasis in both roots and shoots (Dražić et al., 2004).

As environmental stress caused ionic imbalance by increasing and decreasing nutrient and mineral uptake, plants also suffered from toxicity and deficiency problems of that nutrient along with environmental stress-induced damage in soybean seedlings. So, the management of nutrients in stress-induced soybean seedlings might be one important strategy for environmental stress tolerance. Numerous studies showed that the maintenance of nutrients improved stress-tolerance capacity in soybean seedlings under environmental stress conditions. External nutrients in stress-induced soybean seedlings also helped to enhance stress tolerance by maintaining nutrient balance. Manaf (2008) and Hashi et al. (2015) reported that K alleviated detrimental effects of salt stress in soybean by increasing K uptake and maintaining K/Na ratio. The exogenous application of Zn also improved salt stress tolerance in soybean seedlings by decreasing Na content and increasing K, Ca, P, and Fe contents (Weisany et al., 2014). Yang and Juang (2015) showed that external Ca alleviated Cd rhizotoxicity in soybeans by reducing Cd absorption and increasing Ca absorption. Mitigation of Al toxicity by Mg supplementation was also observed in soybean seedlings (Silva et al., 2001; Duressa et al., 2010). Bellaloui et al. (2013b) reported that the mineral nutrients (K, P, Ca, Cu, B) with other osmoregulators are beneficial for drought stress tolerance and remain higher in drought stress-tolerant soybean cultivars compared with check variety.

Conclusion and Outlook

Since ancient times, the soybean has been cultivated as a major crop due to its nutritional and economic importance. Every year the area for soybean production is increasing throughout the world. However, due to the climate change, this crop is now suffering from various environmental stresses. The adverse effects of environmental stress on soybean production has become a hot research topic. With the advancement of breeding programs and agronomic practices, the production of soybeans under drought can be improved by integrating all technologies and knowledge involved. Although many efforts have been placed on the enhancement of stress tolerance in soybeans, the development of a high yield stress-tolerant variety is still under investigation. The physiological and molecular basis of stress tolerance will be helpful for the breeder to work on these traits to develop a tolerant variety. Traditional breeding is a widely accepted strategy, which will combine desirable agronomic traits from soybean germplasms via repeated crossing and selection processes. However, advancements in genomics, genetics, and molecular biology will be needed to facilitate the identification of molecular markers and functional genes that are related to drought tolerance in soybeans.

References

Abbaspour, H., 2012. Effect of salt stress on lipid peroxidation, antioxidative enzymes, and proline accumulation in pistachio plants. Journal of Medicinal Plant Research 6, 526–529.

Abdo, F.A., Nassar, D.M.A., Gomaa, E.F., Nassar, R.M.A., 2012. Minimizing the harmful effects of cadmium on vegetative growth, leaf anatomy, yield and physiological characteristics of soybean plant [*Glycine max* (L.) Merrill] by foliar spray with active yeast extract or with garlic cloves extract. Research Journal of Agriculture and Biological Sciences 8, 24–35.

Abu-Muriefah, S.S., 2015. Effects of paclobutrazol on growth and physiological attributes of soybean (*Glycine max*) plants grown under water stress conditions. International Journal of Advanced Research in Biological Sciences 2, 81–93.

Adie, M.M., Krisnawati, A., 2015. Soybean yield stability in eight locations and its potential for seed oil source in Indonesia. Energy Procedia 65, 223–229.

Agarwal, S., Sairam, R., Srivastava, G., Meena, R., 2005. Changes in antioxidant enzymes activity and oxidative stress by abscisic acid and salicylic acid in wheat genotypes. Biologia Plantarum 49, 541–550.

Agboma, P.C., Sinclair, T.R., Jokinen, K., Peltonen-Sainio, P., Pehu, E., 1997. An evaluation of the effect of exogenous glycine betaine on the growth and yield of soybean: timing of application, watering regimes and cultivars. Field Crops Research 54, 51–64.

Amooaghaie, R., Moghym, S., 2011. Effect of polyamines on thermotolerance and membrane stability of soybean seedling. African Journal of Biotechnology 10, 9673–9679.

Anjum, S.A., Ehsanullah, L., Xue, L., Wang, L., Saleem, M.F., Huang, C., 2013. Exogenous benzoic acid (BZA) treatment can induce drought tolerance in soybean plants by improving gas-exchange and chlorophyll contents. Australian Journal of Crop Science 7, 555–560.

Anjum, S.A., Wang, L., Farooq, M., Khan, I., Xue, L., 2011. Methyl jasmonate-induced alteration in lipid peroxidation, antioxidative defence system and yield in soybean under drought. Journal of Agronomy and Crop Science 197, 296–301.

Anto, K.B., Jayaram, K.M., 2010. Effect of temperature treatment on seed water content and viability of green pea (*Pisum sativum* L.) and soybean (*Glycine max* L. Merr.) seeds. International Journal of Botany 6, 122–126.

Apel, K., Hirt, H., 2004. Reactive oxygen species: metabolism, oxidative stress, and signal transduction. Annual Review of Plant Biology 55, 373–399.

Bacanamwo, M., Purcell, L.C., 1999. Soybean root morphological and anatomical traits associated with acclimation to flooding. Crop Science 39, 143–149.

Baker, J.T., Allen, L.H., Boote, K.J., 1989. Response of soybean to air temperature and carbon dioxide concentration. Crop Science 29, 98–105.

Barnes, P.W., Flint, S.D., Caldwell, M.M., 1990. Morphological response of crop and weed species of different growth forms to UV-B radiation. American Journal of Botany 77, 1354–1360.

Baroniya, S.S., Kataria, S., Pandey, G.P., Guruprasad, K.N., 2011. Intraspecific variation in sensitivity to ambient ultraviolet-B radiation in growth and yield characteristics of eight soybean cultivars grown under field conditions. Brazilian Journal of Plant Physiology 23, 197–202.

Barta, R., Broner, I., Schneekloth, J., Waskom, R., 2004. Colorado high Plains Irrigation Practices Guide – Water Saving Options for Irrigators in Eastern Colorado. Colorado Water Resources Research Institute.

Bassman, J.H., 2004. Ecosystem consequences of enhanced solar ultraviolet radiation: secondary plant metabolites as mediators of multiple trophic interactions in terrestrial communities. Photochemistry and Photobiology 79, 382–398.

Bellaloui, N., Gillen, A., Mengistu, A., Kebede, H., Fisher, D., Smith, J.R., Reddy, K.N., 2013b. Responses of nitrogen metabolism and seed nutrition to drought stress in soybean genotypes differing in slow-wilting phenotype. Frontiers in Plant Science 4, 498. http://dx.doi.org/10.3389/fpls.2013.00498.

Bellaloui, N., Hu, Y., Mengistu, A., Kassem, M.A., Abel, C.A., 2013a. Effects of foliar boron application on seed composition, cell wall boron, and seed $\delta^{15}N$ and $\delta^{13}C$ isotopes in water stressed soybean plants. Frontiers in Plant Science 4, 270. http://dx.doi.org/10.3389/fpls.2013.00270.

Bîrsan, A., Rotaru, V., Jigau, G., Nagacevschi, T., Tofan, E., Sîtari, C., 2014. Influence of biological active substances on the proline content of soybean sorts with various resistances to drought. Soil Forming Factors and Processes from the Temperate Zone 13, 51–57.

Blokhina, O., Fagerstedt, K.V., 2010. Reactive oxygen species and nitric oxide in plantmitochondria: origin and redundant regulatory systems. Physiologia Plantarum 138, 447–462.

Blount, A.R., Wright, D.L., Sprenkel, R.K., Hewitt, T.D., Myer, R.O., 2013. Forage Soybeans for Grazing, Hay, and Silage. University of Florida, IFAS Extension. Publication #SS-AGR-180.

Boote, K.J., Allen, L.H., Prasad, P.V.V., Baker, J.T., Gesch, R.W., Snyder, A.M., Pan, D., Thomas, J.M.G., 2005. Elevated temperature and CO_2 impacts on pollination, reproductive growth, and yield of several globally important crops. Journal of Agricultural Meteorology 60, 469–474.

Boote, K.J., Jones, J.W., Hoogenboom, G., 1998. Simulation of crop growth: CROPGRO model. In: Peart, R.M., Curry, R.B. (Eds.), Agricultural Systems Modeling and Simulation. Marcel Dekker, New York, pp. 651–692.

Bramlage, W.J., Leopold, A.C., Parrish, D.T., 1978. Chilling stress to soybeans during imbibition. Plant Physiology 61, 525–529.

Brunes, A.P., Lemes, E.S., Dias, L.W., Gehling, V.M., Villela, F.A., 2014. Performance of soybean seeds treated with salicylic acid doses. Centro Científico Conhecer-goiânia 10, 1467–1474.

Calviño, P.A., Sadrasc, V.O., Andradeb, F.H., 2003. Development, growth and yield of late-sown soybean in the southern Pampas. European Journal of Agronomy 19, 265–275.

Catuchi, T.A., Vítolo, H.F., Bertolli, S.S., Souza, G.M., 2011. Tolerance to water deficiency between two soybean cultivars: transgenic versus conventional. Ciência Rural 31, 373–378.

Cheng, L., Gao, X., Li, S., Shi, M., Javeed, H., Jing, X., Yang, G., He, G., 2010. Proteomic analysis of soybean [Glycine max (L.) Meer.] seeds during imbibition at chilling temperature. Molecular Breeding 26, 1–17.

Chmielowska-Bąk, J., Lefèvre, I., Lutts, S., Kulik, A., Deckert, J., 2014. Effect of cobalt chloride on soybean seedlings subjected to cadmium stress. Acta Societatis Botanucorum Poloniae 83 (3), 201–207.

Chun-quan, L., Jiang-feng, S., Yuan, W., Da-jing, L., 2015. Effect of exogenous spermine on chilling injury and sucrose metabolism of post-harvest vegetable soybean. Scientia Agricultura Sinica 8, 1588–1596.

Dehghan, G., Rezazadeh, L., Habibi, G., 2011. Exogenous ascorbate improves antioxidant defense system and induces salinity tolerance in soybean seedlings. Acta Biologica Szegediensis 55 (2), 261–264.

Delgado, M.J., Ligero, F., Liuch, C., 1994. Effects of salt stress on growth and nitrogen fixation by pea, faba-bean, common bean and soybean plant. Soil Biology and Biochemistry 26, 371–376.

Desclaux, D., Huynh, T., Roumet, P., 2000. Identification of soybean plant charac-teristics that indicate the timing of drought stress. Crop Science 40, 716–722.

Djanaguiraman, M., Prasad, P.V.V., 2010. Ethylene production under high temperature stress causes premature leaf senescence in soybean. Functional Plant Biology 37, 1071–1084.

Djanaguiraman, M., Prasad, P.V.V., Al-Khatib, K., 2011b. Ethylene perception inhibitor 1-MCP decreases oxidative damage of leaves through enhanced antioxidant defense mechanisms in soybean plants grown under high temperature stress. Environmental and Experimental Botany 71, 215–223.

Djanaguiraman, M., Prasad, P.V.V., Boyle, D.L., Schapaugh, W.T., 2011a. High tempera-ture stress and soybean leaves: leaf anatomy and photosynthesis. Crop Science 51, 2125–2131.

Djanaguiraman, M., Prasad, P.V.V., Schapaugh, W.T., 2013. High day- or nighttime temperature alters leaf assimilation, reproductive success and phosphatidic acid of pollen grain in soybean [Glycine max (L.) Merr.]. Crop Science 53, 1594–1604.

Dolatabadian, A., Sanavy, S.A.M.M., Ghanati, F., 2011. Effect of salinity on growth, xylem structure and anatomical characteristics of soybean. Notulae Scientia Biologicae 3, 41–45.

Dornbos, D.L., Mullen, R.E., 1991. Influence of stress during soybean seed fill on seed weight, germination, and seedling growth rate. Journal of Plant Science 71, 373–383.

Dražić, G., Mihailović, N., Stojanović, Z., 2004. Cadmium toxicity: the effect on macro- and micro-nutrient contents in soybean seedlings. Biologia Plantarum 48, 605–607.

Duke, S.H., Schrader, L.E., Miller, M.G., 1977. Low temperature effects on soybean (Glycine max [L.] Merr. cv. Wells) mitochondrial respiration and several dehydrogenases during imbibition and germination. Plant Physiology 60, 716–722.

Duressa, D., Soliman, K.M., Chen, D., 2010. Mechanisms of magnesium amelioration of aluminum toxicity in soybean at the gene expression level. Genome 53, 787–797.

Edwards, T.I., 1934. Relations of germinating soybeans to temperature and length of incubation time. Plant Physiology 9, 1–30.

Egbichi, I., Keyster, M., Ludidi, N., 2014. Effect of exogenous application of nitric oxide on salt stress responses of soybean. South African Journal of Botany 90, 131–136.

Egbichi, I., Keyster, M., Jacobs, A., Klein, A., Ludidi, N., 2013. Modulation of antioxidant enzyme activities and metabolites ratios by nitric oxide in short-term salt stressed soybean root nodules. South African Journal of Botany 88, 326–333.

Egli, D.B., Bruening, W.P., 2006. Fruit development and reproductive survival in soybean: position and age effects. Field Crops Research 98, 195–202.

Egli, D.B., Wardlaw, I.F., 1980. Temperature response of seed growth characteristics of soybeans. Agronomy Journal 72, 560–564.

El-Sabagh, A., Sorour, A., Ueda, A., Saneoka, H., Barutcular, C., 2015. Evaluation of salinity stress effects on seed yield and quality of three soybean cultivars. Azarian Journal of Agriculture 2, 138–141.

Embrapa, 2011. Sistema de Produção 15. Exigências Climáticas. In: Tecnologia de pro-dução de soja – Região central do Brasil – 2012 e 2013. Embrapa Soja, Londrina. pp. 11–12.

Essa, T.A., 2002. Effect of salinity stress on growth and nutrient composition of three soybean (*Glycine max* L. Merrill) cultivars. Journal of Agronomy and Crop Science 188, 86–93.

Fanaei, H.R., Sadegh, H.N., Yousefi, T., Farmanbar, M., 2015. Influence of drought stress on some characteristics of plants. Biological Forum 7, 1732–1738.

FAOSTAT, 2014. http://*faostat*.fao.org/ last accessed November 2015.

Feng, H.Y., An, L.Z., Xu, S.J., Qiang, W.Y., Chen, T., Wang, X.L., 2001a. Effect of enhanced ultraviolet-B radiation on growth, development, pigments and yield of soybean (*Glycine max* (L.) Merr.). Acta Agronomica Sinica 27, 319–323.

Feng, H.Y., Chen, T., Xu, S.J., An, L.Z., Qiang, W.Y., Zhang, M.X., Wang, X.L., 2001b. Effect of enhanced UV-B radiation on growth, yield and stable carbon isotope composition in *Glycine max* cultivars. Acta Botanica Sinica 43, 709–713.

Ferris, R., Wheeler, T.R., Ellis, R.H., Hadley, P., 1999. Seed yield after environmental stress in soybean grown under elevated CO_2. Crop Science 39, 710–718.

Ganesh, K.S., Baskaran, L., Chidambaram, A.A., Sundaramoorthy, P., 2009. Influence of chromium stress on proline accumulation in soybean (*Glycine max* L. Merr. Genotypes Global Journal of Environmental Research 3, 106–108.

Gibson, L.R., Mullen, R.E., 1996. Influence of day and night temperature on soybean seed yield. Crop Science 36, 98–104.

Gill, S.S., Anjum, N.A., Gill, R., Yadav, S., Hasanuzzaman, M., Fujita, M., Mishra, P., Sabat, S.C., Tuteja, N., 2015. Superoxide dismutase–mentor of abiotic stress tolerance in crop plants. Environmental Science and Pollution Research 22, 10375–10394.

Gill, S.S., Tuteja, N., 2010. Reactive oxygen species and antioxidant machinery in abiotic stress tolerance in crop plants. Plant Physiology and Biochemistry 48, 909–930.

Gobetti, S.T.C., Neuman, M., Oliveira, M.R., Oliboni, R., 2011. Production and use of the ensilage of entire soy plant (*Glycine max*) for ruminants. Ambiência 7, 603–616.

Gruber, H., Heijde, M., Heller, W., Albert, A., Seidlitz, H.K., Ulm, R., 2010. Negative feedback regulation of UV-B-induced photomorphogenesis and stress acclimation in *Arabidopsis*. Proceedings of the National Academy of Sciences, USA 107, 20132–20137.

Gu-wen, Z., Sheng-chun, X., Qi-zan, H., Wei-hua, M., Ya-ming, G., 2014. Putrescine plays a positive role in salt-tolerance mechanisms by reducing oxidative damage in roots of vegetable soybean. Journal of Integrative Agriculture 13 (2), 349–357.

Hamayun, M., Khan, S.A., Khan, A.L., Shin, J.H., Ahmed, B., Shin, D.H., Lee, I.J., 2010. Exogenous gibberellic acid reprograms soybean to higher growth and salt stress tolerance. Journal of Agricultural and Food Chemistry 58, 7226–7232.

Hasanuzzaman, M., Fujita, M., 2012a. Heavy metals in the environment: current status, toxic effects on plants and possible phytoremediation. In: Anjum, N.A., Pereira, M.A., Ahmad, I., Duarte, A.C., Umar, S., Khan, N.A. (Eds.), Phytotechnologies: Remediation of Environmental Contaminants. CRC Press, Boca Raton, FL, pp. 7–73.

Hasanuzzaman, M., Fujita, M., 2012b. Selenium and plants' health: the physiological role of selenium. In: Aomori, C., Hokkaido, M. (Eds.), Selenium: Sources, Functions and Health Effects. Nova Science Publishers, New York, pp. 101–122.

Hasanuzzaman, M., Hossain, M.A., da Silva, J.A.T., Fujita, M., 2012. Plant responses and tolerance to abiotic oxidative stress: antioxidant defense is a key factor. In: Bandi, V., Shanker, A.K., Shanker, C., Mandapaka, M. (Eds.), Crop Stress and its Management: Perspectives and Strategies. Springer, Berlin, pp. 261–316.

Hasanuzzaman, M., Nahar, K., Gill, S.S., Fujita, M., 2014. Drought stress responses in plants, oxidative stress and antioxidant defense. In: Tuteja, N., Gill, S.S. (Eds.), Climate Change and Plant Abiotic Stress Tolerance, Biological Techniques, vol. 1. first ed. Wiley, Weinheim, pp. 209–250.

Hashem, H.A., 2014. Cadmium toxicity induces lipid peroxidation and alters cytokinin content and antioxidant enzyme activities in soybean. Botany 92, 1–7.

Hashi, U.S., Karim, A., Saikat, H.M., Islam, R., Islam, M.A., 2015. Effect of salinity and potassium levels on different morpho-physiological characters of two soybean (*Glycine max* L.) genotypes. Journal of Rice Research 3, 143. http://dx.doi.org/10.4172/2375-4338.1000143.

Hatfield, J., Boote, K., Fay, P., Hahn, L., Izaurralde, C., Kimball, B.A., Mader, T., Morgan, J., Ort, D., Polley, W., Thomson, A., Wolf, D., 2008. Agriculture. In: The Effects of Climate Change on Agriculture, Land Resources, Water Resources, and Biodiversity in the United States. U.S. Climate Change Science Program and the Subcommitee on Global Change Research, Washington, DC, USA. p. 362.

Heuzé, V., Tran, G., Hassoun, P., Lebas, F., 2015. Soybean Forage. Feedipedia, a programme by INRA, CIRAD, AFZ and FAO http://feedipedia.org/node/294 Last updated on May 11, 2015, 14:31.

Hosseini, M.K., Powell, A.A., Bingham, I.J., 2002. Comparison of the seed germination and early seedling growth of soybean in saline conditions. Seed Science Research 12, 165–172.

Howarth, C.J., 2005. Genetic improvements of tolerance to high temperature. In: Ashraf, M., Harris, P.J.C. (Eds.), Abiotic Stresses: Plant Resistance through Breeding and Molecular Approaches. Howarth Press, New York, pp. 277–300.

Hu, Y., Burucs, Z., Tucher, S.V., Schmidhalter, U., 2007. Short-term effects of drought and salinity on mineral nutrient distribution along growing leaves of maize seedlings. Environmental and Experimental Botany 60, 268–275.

Hymowitz, T., 2008. The History of soybean. In: Johnson, L.A., White, P.J., Gallowa, R. (Eds.), Soybean: Chemistry, Production, Processing, and Utilization. AOCS Press, Urbana. pp. 1–23.

Hymowitz, T., Newell, C.A., 1981. Taxonomy of the genus *Glycine*, domestication and uses of soybeans. Economic Botany 35, 272–288.

Imtiyaz, S., Agnihotri, R.K., Ahmad, S., Sharma, R., 2014a. Effect of cobalt and lead induced heavy metal stress on some physiological parameters in *Glycine max*. International Journal of Agriculture and Crop Sciences 7, 26–34.

Imtiyaz, S., Agnihotri, R.K., Ganie, S.A., Sharma, R., 2014b. Biochemical response of *Glycine max* (L.) Merr. to cobalt and lead stress. Journal of Stress Physiology and Biochemistry 10, 259–272.

Janas, K.M., Cvikrová, M., Pałągiewicz, A., Eder, J., 2000. Alterations in phenylpropanoid content in soybean roots during low temperature acclimation. Plant Physiology and Biochemistry 38, 587–593.

Jianguo, W., Xiaomin, L., Xiaoting, Z., Lin, C., Zhiguo, L., 2006. Effect of SA on the seedling growth and waterlogging resistance of *Glycine max* Merr. under water stress. Chinese Agricultural Science Bulletin 2006–01.

Kang, S., Radhakrishnam, R., Khan, A.L., Kim, M., Park, J., Kim, B., Shin, D., Lee, I., 2014. Gibberellin secreting rhizobacterium, *Pseudomonas putida* H-2-3 modulates the hormonal and stress physiology of soybean to improve the plant growth under saline and drought conditions. Plant Physiology and Biochemistry 84, 115–124.

Kasmakoglu, H., 2004. Fao Production Earbook-2003. vol. 57, Rome, Italy.

Keerio, M.I., Chang, S.Y., Mirjat, M.A., Lakho, M.H., Bhatti, I.P., 2001. The rate of nitrogen fixation in soybean root nodules after heat stress and recovery period. International Journal of Agriculture and Biology 3, 512–514.

Keyser, H.H., Li, F., 1992. Potential for increasing biological nitrogen fixation in soybean. Plant and Soil 141, 119–135.

Khan, R., Srivastata, R., Abdin, M.Z., Manzoor, N., Mahmooduzzafar, 2013. Effect of soil contamination with heavy metals on soybean seed oil quality. European Food Research and Technology 236 (4), 707–714.

Kim, E.H., Seguin, P., Lee, J.E., Yoon, C.G., Song, H.K., Ahn, J.K., Chung, I.M., 2011. Ele-vated ultraviolet-B radiation reduces concentrations of isoflavones and phenolic compounds in soybean seeds. Journal of Agronomy and Crop Science 197, 75–80.

Kitano, M., Saitoh, K., Kuroda, T., 2006. In: Effect of High Temperature on Flowering and Pod Set in Soybean, 95. Scientific Report of the Faculty of Agriculture, Okayama University, pp. 49–55.

Klein, A., Keyster, M., Ludidi, N., 2015. Response of soybean nodules to exogenously applied caffeic acid during NaCl-induced salinity. South African Journal of Botany 96, 13–18.

Koti, S., Reddy, K.R., Reddy, V.R., Kakani, V.G., Zhao, D., 2005. Interactive effects of carbon dioxide, temperature, and ultraviolet-B radiation on soybean (*Glycine max* L.) flower and pollen morphology, pollen production, germination, and tube lengths. Journal of Experimental Botany 56, 725–736.

Koti, S.K., Reddya, V.R., Kakani, G., Zhao, A.D., Gaoc, W., 2007. Effects of carbon dioxide, temperature and ultraviolet-B radiation and their interactions on soybean (*Glycine max* L.) growth and development. Environmental and Experimental Botany 60, 1–10.

Ku, Y.-S., Au-Yeung, W.-K., Yung, Y.-L., Li, M.-W., Wen, C.-Q., Liu, X., Lam, H.-M., 2008. Drought stress and tolerance in soybean. In: Board, J.E. (Ed.), A Comprehensive Survey of International Soybean Research – Genetics, Physiology, Agronomy and Nitrogen Relationships. Intech, Rijeka, pp. 209–237.

Kumar, A., Pandey, V., Shekh, A.M., Kumar, M., 2008a. Growth and yield response of soybean (*Glycine max* L.) in relation to temperature, photoperiod and sunshine duration at Anand, Gujarat, India. American-Eurasian Journal of Agronomy 1, 45–50.

Kumar, V., Rani, A., Chauhan, G.S., 2008b. Nutritional value of soybean. In: Johnson, L.A., White, P.J., Gallowa, R. (Eds.), Soybean: Chemistry, Production, Processing, and Utilization. AOCS Press, Urbana, pp. 375–403.

Lecube, M.L., Noriega, G.O., Santa Cruz, D.M., Tomaro, M.L., Batlle, A., Balestrasse, K.B., 2014. Indole acetic acid is responsible for protection against oxidative stress caused by drought in soybean plants: the role of hemeoxygenase induction. Redox Report 19, 242–250.

Li, P., Cheng, L., Gao, H., Jiang, C., Peng, T., 2009. Heterogeneous behavior of PSII in soybean (*Glycine max*) leaves with identical PSII photochemistry efficiency under different high temperature treatments. Journal of Plant Physiology 166, 1607–1615.

Li-jun, Z., Guang-zhou, S., Shuang, B., Ying, W., Zhen-hai, C., Fu-ti, X., 2008. Effect of exogenous H_2O_2 on germination rate and activities of main antioxidase in soybean (*Glycine max* L.) seedlings under low temperature. Soybean Science 2008–2101.

Lima, R., Diaz, R.F., Castro, A., Fievez, V., 2011. Digestibility, methane production and nitrogen balance in sheep fed ensiled or fresh mixtures of sorghum-soybean forage. Livestock Science 141 (1), 36–46.

Lindsey, L., Thomson, P., 2012. High temperature effects on corn and soybean. C.O.R.N Newsletter 2012 23–26.

Linkemer, G., Board, J.E., Musgrave, M.E., 1998. Waterlogging effects on growth and yield components in late-planted soybean. Crop Science 38, 1576–1584.

Liu, B., Liu, X.B., Li, Y.S., Herbert, S.J., 2013. Effects of enhanced UV-B radiation on seed growth characteristics and yield components in soybean. Field Crops Research 154, 158–163.

Liu, X., 2009. Drought. In: Lam, H.M., Chang, R., Shao, G., Liu, Z. (Eds.), Research on Tolerance to Stresses in Chinese Soybean. China Agricultural Press, Beijing.

Liu, X., Jian, J., Guanghua, W., Herbert, S.J., 2008. Soybean yield physiology and development of high- yielding practices in Northeast China. Field Crops Research 105, 157–171.

Luan, Z.Q., Cao, H.C., Yan, B.X., 2008. Individual and combined phytotoxic effects of cadmium, lead and arsenic on soybean in Phaeozem. Plant Soil and Environment 54, 403–411.

Ma, H., Song, L., Huang, Z., Yang, Y., Wang, Z., Tong, J., Gu, W., Ma, H., Xiao, L., 2014. Comparative proteomic analysis reveals molecular mechanism of seedling roots of different salt tolerant soybean genotypes in responses to salinity stress. EuPA Open Proteomics 4, 40–57.

Maksymiec, W., 2007. Signaling responses in plants to heavy metal stress. Acta Physiologiae Plantarum 29, 177–187.

Malan, H.L., Farrant, J.M., 1998. Effects of the metal pollutants cadmium and nickel on soybean seed development. Seed Science Research 8, 445–453.

Manaf, H.H., 2008. The role of potassium nitrate in aleviating deterimental effects of salt stress of soybean roots *in vitro*. Journal of Biological Chemistry and Environmental Sciences 3, 363–380.

Manafi, E., Modarressanavy, S.A.M., Aghaalikhani, M., Dolatabadian, A., 2015. Exogenous 5-aminolevulenic acid promotes antioxidative defence system, photosynthesis and growth in soybean against cold stress. Notulae Scientia Biologicae 7, 486–494.

Markhart, A.H., 1984. Amelioration of chilling-induced water stress by abscisic acid-induced changes in root hydraulic conductance. Plant Physiology 74, 81–83.

Masoumi, H., Masoumi, M., Darvish, F., Daneshian, J., Nourmohammadi, G., Habibi, D., 2010. Change in several antioxidant enzymes activity and seed yield by water deficit stress in soybean (*Glycine max* L.) cultivars. Notulae Botanicae Horti Agrobotanici Cluj-Napoca 38, 86–94.

McKenzie, R.L., Conner, B., Bodeker, G., 1999. Increased summertime UV radiationin four aquatic macrophytes. Chemosphere 74, 642–647.

Mishra, M., Kumar, U., Prakash, V., 2013. Influence of salicylic acid pre-treatment on water stress and its relationship with antioxidant status in *Glycine max*. International Journal of Pharmacology Biological Sciences 4, 81–97.

Mittler, R., 2002. Oxidative stress, antioxidants and stress tolerance. Trends in Plant Science 7, 405–410.

Mittona, F.M., Ferreira, J.L.R., Gonzalez, M., Miglioranza, K.S.B., Monserrat, J.M., 2015. Antioxidant responses in soybean and alfalfa plants grown in DDTs contaminated soils: useful variables for selecting plants for soil phytoremediation. Pesticide Biochemistry and Physiology. http://dx.doi.org/10.1016/j.pestbp.2015.12.005.

Moghym, S., Amooaghaie, R., Shareghi, B., 2010. Protective role of polyamines against heat shock in soybean seedlings. Iranian Journal of Plant Biology 2, 31–39.

Morse, W.J., Cartter, J.L., 1952. Soybeans for Feed, Food and Industrial Products. USDA. Farmer's Bulletin No. 2038.

Munns, R., 2002. Comparative physiology of salt and water stress. Plant Cell and Environment 25, 239–250.

Munns, R., Tester, M., 2008. Mechanisms of salinity tolerance. Annual Review of Plant Biology 59, 651–681.

Nahar, K., Hasanuzzaman, M., Alam, M.M., Fujita, M., 2015. Exogenous glutathione confers high temperature stress tolerance in mung bean (*Vigna radiata* L.) by modulating antioxidant defense and methylglyoxal detoxification system. Environmental and Experimental Botany 112, 44–54.

Ndunguru, B.J., Sununerfield, R.J., 1975. Comparative laboratory studies of cowpea (*Vigna unguiculata*) and soybean (*Glycine max*) under tropical temperature conditions. I. Germination and hypocotyl conditions. East African Agricultural and Forestry Journal 41, 58–64.

Neto, M., Saturnino, S.M., Bomfim, D.C., Custodio, C.C., 2004. Water stress induced by mannitol and sodium chloride in soybean cultivars. Brazilian Archives of Biology and Technology 47, 521–529.

Noctor, G., De Paepe, R., Foyer, C.H., 2007. Mitochondrial redox biology and homeostasis in plants. Trends in Plant Science 12, 125–134.

Noctor, G., Gomez, L., Vanacker, H., Foyer, C.H., 2002b. Interactions between biosynthesis, compartmentation and transport in the control of glutathione homeostasis and signalling. Journal of Experimental Botany 53, 1283–1304.

Noctor, G., Veljovic-Jovanovic, S., Driscoll, S., Novitskaya, L., Foyer, C.H., 2002a. Drought and oxidative load in the leaves of C_3 plants: a predominant role for photorespiration? Annals of Botany 89, 841–850.

Olsen, L.J., Harada, J.J., 1995. Peroxisomes and their assembly in higher plants. Annual Review of Plant Physiology and Plant Molecular Biology 46, 123–146.

Palanisamy, K., Lenin, M., Mycin, T.R., 2012. Effect of mercuric chloride on growth and cytotoxicity of soybean *Glycine max* (L.) Hepper K. International Journal Toxicology and Applied Pharmacology 2, 37–41.

Pan, D., 1996. Soybean Responses to Elevated Temperature and Doubled CO_2. Ph.D. dissertation. University of Florida, Gainesville. p. 227.

Peng, Q., Zhou, Q., 2010. Effects of enhanced UV-B radiation on the distribution of mineral elements in soybean (*Glycine max*) seedlings. Chemosphere 78, 859–863.

Phang, T., Shao, G., Lam, H., 2008. Salt tolerance in soybean. Journal of Integrative Plant Biology 50, 1196–1212.

Pimentel, D., Patzek, T.D., 2005. Ethanol production using corn, switchgrass, and wood; biodiesel production using soybean and sunflower. Natural Resources Research 14, 65–76.

Pradhan, A., Shrestha, D.S., van Gerpen, J., Duffield, J., 2008. The energy balance of soybean oil biodiesel production: a review of past studies. American Society of Agricultural and Biological Engineers 51, 185–194.

Prasad, P.V.V., Boote, K., Allen, L.H., Jean, M.G., 2002. Effects of elevated temperature and carbon dioxide on seed-set and yield of kidney bean. Global Change Biology 8, 710–721.

Prasad, P.V.V., Boote, K.J., Allen, L.H., Sheehy, J.E., Thomas, J.M.G., 2006. Species, ecotype and cultivar differences in spikelet fertility and harvest index of rice in response to high temperature stress. Field Crops Research 95, 398–411.

Prasad, P.V.V., Pisipati, S.R., Mutava, R.N., Tuinstra, M.R., 2008. Sensitivity of grain sorghum to high temperature stress during reproductive development. Crop Science 48, 1911–1917.

Puteh, A.B., ThuZar, M., Mondal, M.M.A., Abdullah, N.A.P.B., Halim, M.R.A., 2013. Soybean [*Glycine max* (L.) Merrill] seed yield response to high temperature stress during reproductive growth stages. Australian Journal of Crop Science 7, 1472–1479.

Rasolohery, C.A., Berger, M., Lygin, A.V., Lozovaya, V.V., Nelson, R.L., Dayde´, J., 2008. Effect of temperature and water availability during late maturation of the soybean seed on germ and cotyledon isoflavone content and composition. Journal of the Science of Food and Agriculture 88, 218–228.

Rezaei, M.A., Kaviani, B., Masouleh, A.K., 2012. The effect of exogenous glycine betaine on yield of soybean [*Glycine max* (L.) Merr.] in two contrasting cultivars Pershing and DPX under soil salinity stress. Plant Omics Journal 5, 87–93.

Rigueira, J.P., Pereira, O.G., Leão, M.I., Valadares Filho, S.C., Garcia, R., 2008. Intake and total and partial digestibility of nutrients, ruminal pH and ammonia concentration in beef cattle fed diets containing soybean silage. In: ADSA ASAS Joint Annual Meeting, July 7–11, 2008 Indianapolis, USA.

del Río, L.A., Sandalio, L.M., Corpas, F.J., Romero-Puertas, M.C., Palma, J.M., 2009. Peroxisomes as a cellular source of ROS signal molecules. In: Rio, L.A., Puppo, A. (Eds.), Reactive Oxygen Species in Plant Signaling. Signaling and Communication in Plants. Springer, New York, pp. 95–111.

Salem, M.A., Kakani, V.G., Koti, S., Reddy, K.R., 2007. Pollen-based screening of soybean genotypes for high temperature. Crop Science 47, 219–231.

Salem, M.S., Koti, S., Kakani, V.G., Reddy, K.R., 2005. Glycine betaine and its role in heat tolerance as studied by pollen germination techniques in soybean genotypes. In: The ASA-CSSA-SSSA International Annual Meetings, 6–10 November 2005, Salt Lake City, Utah, USA.

Salimi, S., 2015. Evaluation of soybean genotypes (*Glycine max* L) to drought tolerance at germination stage. Research Journal of Environmental Sciences 9, 349–354.

Sallam, A., Scott, H.D., 1987. Effects of prolonged flooding on soybeans during early vegetative growth. Soil Science 144, 61–68.

Sandalio, L.M., Rodríguez-Serrano, M., Romero-Puertas, M.C., del Río, L.A., 2013. Role of peroxisomes as a source of reactive oxygen species (ROS) signaling molecules. In: del Río, L.A. (Ed.), Peroxisomes and Their Key Role in Cellular Signaling and Metabolism, Subcellular Biochemistry69, pp. 231–255.

Sanghera, G.S., Wani, S.H., Hussain, W., Singh, N.B., 2011. Engineering cold stress tolerance in crop plants. Current Genomics 12, 30–43.

Santa-Cruz, D.M., Pacienza, N.A., Polizio, A.H., Balestrasse, K.B., Tomaro, M.L., Yannarelli, G.G., 2010. Nitric oxide synthase-like dependent NO production enhances heme oxygenase up-regulation in ultraviolet-B-irradiated soybean plants. Phytochemistry 71, 1700–1707.

Santos, R.F., Carlesso, R., 1988. Déficit hídrico e os processos morfológicos e fisioló-gicos das plantas. Revista Brasileira de Engenharia Agrícola e Ambiental 2 (3), 287–294.

Sapra, V.T., Anaek, A.O., 1991. Screening soya bean genotypes for drought and heat tolerance. Journal of Agronomy and Crop Science 167, 96–102.

Sarma, H., 2011. Metal hyperaccumulation in plants: a review focusing on phytoremediation technology. Journal of Environmental Science and Technology 4, 118–138.

Scott, H.D., De Angulo, J., Daniels, M.B., Wood, L.S., 1989. Flood duration effects on soybean growth and yield. Agronomy Journal 81, 631–636.

Scott, W.O., Aldrich, S.R., 1983. Modern Soybean Production. S & A Publication, Champaign, IL. p. 209.

Shamsi, I.H., Wei, K., Jilani, G., Zhang, G., 2007. Interactions of cadmium and aluminum toxicity in their effect on growth and physiological parameters in soybean. Journal of Zhejiang University Science B 8, 181–188.

Sharma, P., Jha, A.B., Dubey, R.S., Pessarakli, M., 2012. Reactive oxygen species, oxidative damage, and antioxidative defense mechanism in plants under stressful conditions. Journal of Botany. http://dx.doi.org/10.1155/2012/217037.

Sheeren, A., Ansari, R., 2001. Salt tolerance in soybean (*Glycine max* L.): effect on growth and water relations. Pakistan Journal of Biological Sciences 4, 1212–1214.

Sheirdil, R.A., Bashir, K., Hayat, R., Akhtar, M.S., 2012. Effect of cadmium on soybean (*Glycine max* L) growth and nitrogen fixation. African Journal of Biotechnology 11, 1886–1891.

Shen, X., Li, J., Duan, L., Li, Z., Eneji, A.E., 2009. Nutrient acquisition by soybean treated with and without silicon under ultraviolet-B radiation. Journal of Plant Nutrition 32, 1731–1743.

Shena, X., Zhoua, Y., Duana, L., Li, Z., Enejib, A.E., Li, J., 2010. Silicon effects on photosynthesis and antioxidant parameters of soybean seedlings under drought and ultraviolet-B radiation. Journal of Plant Physiology 167, 1248–1252.

Sheteawi, S.A., 2007. Improving growth and yield of salt-stressed soybean by exogenous application of jasmonic acid and ascobin. International Journal of Agriculture and Biology 9, 473–478.

Shinoda, O.A., 1971. Treatise on Tofu (Bean Curd). 42. Zhe Continent Magazine, pp. 172–178 (In Chinese, translation by E. Wolft).

Silva, I.R., Smyth, T.J., Israel, D.W., Raper, C.D., Rufty, T.W., 2001. Magnesium ameliorates alumunium rhizotoxicity in soybean by increasing citric acid production and exudation by roots. Plant and Cell Physiology 42, 546–554.

Silva, M.L.S., Vitti, G.C., Trevizam, A.R., 2014. Heavy metal toxicity in rice and soybean plants cultivated in contaminated soil. Revista Ceres 61, 248–254.

Simaei, M., Khavari-nezad, R.A., Bernard, F., 2012. Exogenous application of salicylic acid and nitric oxide on the ionic contents and enzymatic activities in NaCl-stressed soybean plants. American Journal of Plant Science 3, 1495–1503.

Sinclair, T.R., Weisz, P.R., 1985. Response to soil temperature of dinitrogen fixation (Acetylene reduction) rates by field grown soybean. Agronomy Journal 77, 685–688.

Sionit, N., Strain, B.R., Flint, E.P., 1987. Interaction of temperature and CO_2 enrichment on soybean: growth and dry matter partitioning. Canadian Journal of Plant Science 67, 59–67.

Snider, J.M., Oosterhuis, D.M., Skulman, B.W., Kawakami, E.M., 2009. Heat-induced limitations to reproductive success in *Gossypium hirsutum*. Physiologia Plantarum 137, 125–138.

Souza, G.M., Catuchi, T.A., Bertolli, S.C., Soratto, R.P., 2013. Soybean under water deficit: physiological and yield responses. In: Board, J.E. (Ed.), A Comprehensive Survey of International Soybean Research-genetics, Physiology, Agronomy and Nitrogen Relationships. Intech. pp. 273–298.

Sundarmoorthy, P., Sankarganesh, K., Selvaraj, M., Baskaran, L., Chidambaram, A.A., 2015. Chromium induced changes in soybean (*Glycine max* L.) metabolism. World Scientific News 10, 145–178.

Sung, F.J.M., 1993. Waterlogging effect on nodule nitrogenase and leaf nitrate reductase activities in soybean. Field Crops Research 35, 183–189.

Sunita, K., Guruprasad, K.N., 2012. Solar UV-B and UV-A/B exclusion effects on intraspecific variations in crop growth and yield of wheat varieties. Field Crops Research 125, 8–13.

Taiz, L., Zeiger, E., 2009. Fisiologia Vegetal, fourth ed. Artmed Editora, Porto Alegre.

Takashi, S., Murata, N., 2008. How do environmental stresses accelerate photoinhibition? Trends in Plant Science 13, 178–182.

Tambussi, E.A., Bartoli, C.G., Guiamet, J.J., Beltrano, J., Araus, J.L., 2004. Oxidative stress and photodamage at low temperatures in soybean (*Glycine max* L. Merr.) leaves. Plant Science 167, 19–26.

Tanguilig, V.C., Yambao, E.B., O'Toole, J.C., De datta, S.K., 1987. Water stress effects on leaf elongation, leaf water potential, transpiration, and nutrient uptake of rice, maize, and soybean. Plant and Soil 103, 155–168.

Thakur, A.K., Singh, K.J., 2014. Heavy metal Cd affecting nodulation and leghaemoglobin proteins in soybean and chickpea. Journal of Plant and Agriculture Research 1, 1–6.

Thomas, A.L., Sodek, L., 2006. Amino acid and ureide transport in the xylem of symbiotic soybean plants during short-term flooding of the root system in the presence of different sources of nitrogen. Brazilian Journal of Plant Physiology 18, 333–339.

Thomas, J.M.G., 2001. Impact of Elevated Temperature and Carbon Dioxide on Development and Composition of Soybean Seed. University of Florida, Gainesville, Florida, USA, pp. 185. Ph.D. dissertation.

Tobia, C., Villalobos, E., Rojas, A., Soto, H., Moore, K.J., 2008. Nutritional value of soybean (*Glycine max* L. Merr.) silage fermented with molasses and inoculated with *Lactobacillus brevis 3*. Livestock Research for Rural Development 20, 106.

Tuncturk, M., Tuncturk, R., Yasar, F., 2008. Changes in micronutrients, dry weight and plant growth of soybean (*Glycine max* L. Merrill) cultivars under salt stress. African Journal of Biotechnology 7, 1650–1654.

Tuteja, N., Gill, S.S., Tuteja, R., 2011. Plant responses to abiotic stresses: shedding light on salt, drought, cold and heavy metal stress. In: Tuteja, N., Gill, S.S., Tuteja, R. (Eds.), Omics and Plant Abiotic Stress Tolerance. Bentham Science Publishers, UAE, pp. 39–64.

Vantoai, T.T., Beuerlein, J.E., Schmitthenner, A.F., St Martin, S.K., 1994. Genetic variability for flooding tolerance in soybeans. Crop Science 34, 1112–1115.

Vašková, J., Vaško, L., Kron, I., 2012. Oxidative processes and antioxidative metaloenzymes. In: El-Missiry, M.A. (Ed.), Antioxidant Enzyme. InTech, Rijeka. http://dx.doi.org/10.5772/50995.

Villamil, M.B., Davis, V.M., Nafziger, E.D., 2012. Estimating factor contributions to soybean yield from farm field data. Agronomy Journal 104, 881–887.

Wang, L., Ma, H., Song, L., Shu, Y., Gu, W., 2012. Comparative proteomics analysis reveals the mechanism of pre-harvest seed deterioration of soybean under high temperature and humidity stress. Journal of Proteomics 75, 2109–2127.

Warpeha, K., 2012. Supplemental supply of phenylalanine to soybean seeds reduces damage by ultraviolet radiation in etiolated seedlings. In: 97th ESA Annual Meeting. PS 49–97.

Wei, S., Zhou, Q., 2008. Trace elements in agro-ecosystem. In: Prasad, M.N.V. (Ed.), Trace Elements as Contaminants and Nutrients: Consequences in Ecosystems and Human Health. Wiley, Hoboken, pp. 55–79.

Wei, W., Li, Q., Chu, Y., Reiter, R.J., Yu, X.M., Zhu, D.H., Zhang, W., Ma, B., Lin, Q., Zhang, J., Chen, S., 2014. Melatonin enhances plant growth and abiotic stress tolerance in soybean plants. Journal of Experimental Botany 66, 695–707.

Weisany, W., Sohrabi, Y., Heidari, G., Siosemardeh, A., 2014. Effects of zinc application on growth, absorption and distribution of mineral nutrients under salinity stress in soybean (*Glycine max* L.). Journal of Plant Nutrition 37, 2255–2269.

Widiati, R., Musa, Y., Ala, A., Bdr, F.M., 2014. Stomatal performance of soybean genotypesdue to drought stress and acidity. International Journal of Scientific and Technology Research 3, 270–275.

Yang, C., Juang, K., 2015. Alleviation effects of calcium and potassium on cadmium rhizotoxicity and absorption by soybean and wheat roots. Journal of Plant Nutrition and Soil Science 178, 748–754.

Yang, S., Xie, J., Li, Q., 2012. Oxidative response and antioxidative mechanism in germinating soybean seeds exposed to cadmium. International Journal of Environmental Research and Public Health 9, 2827–2838.

Yathavakilla, S.K.V., Caruso, J.A., 2007. A study of Se-Hg antagonism in *Glycine max* (soybean) roots by size exclusion and reversed phase HPLC-ICPMS. Analytical and Bioanalytical Chemistry 389, 715–723.

Yin, Y., Yang, R., Han, Y., Gu, Z., 2015. Comparative proteomic and physiological analyses reveal the protective effect of exogenous calcium on the germinating soybean response to salt stress. Journal of Proteomics 113, 110–126.

Youn, J.T., Van, K., Lee, J.E., Kim, W.H., Yun, H.T., Kwon, Y.U., Ryu, Y.H., Lee, S.H., 2008. Waterlogging effects on nitrogen accumulation and N_2 fixation of super nodulating soybean mutants. Journal of Crop Science and Biotechnology 11, 111–118.

Zabihi, A., Satei, A., Ghorbanli, M., 2014. Evaluation effect of putrescine treatment on growth factors of soybean (*Glycine max* L.) under drought stress induced by polyethylene glycol. Journal of Applied Environmental and Biological Sciences 4, 24–32.

Zhang, F., Lynch, D.H., Smith, D.L., 1995. Low root temperature and nodulation, nitrogen fixation, photosynthesis and growth by soybean [*Glycine max* (L.) Merr.]. Environmental and Experimental Botany 35, 279–285.

Zhang, J., 2011. China's success in increasing per capita food production. Journal of Experimental Botany 62, 3707–3711.

Zhang, L., Li, X., 2012. Exogenous treatment with salicylic acid attenuates ultraviolet-B radiation stress in soybean seedlings. Advances in Intelligent and Soft Computing 134, 889–894.

Zhou, Q., Dong, Y., Bian, Y.J., Han, L.L., Tian, Y.D., Xing, H., Jiang, H.D., 2012. Influence of different duration of waterlogging on the growth and C and N metabolism of soybean at seedling and flowering stages. Ying Yong Sheng Tai Xue Bao 23, 1577–1584.

Soybean (*Glycine max* [L.] Merr.) Production Under Organic and Traditional Farming

F. Zaefarian[1] and M. Rezvani[2]
[1]Faculty of Crop Sciences, Sari Agricultural Sciences and Natural Resources University, Sari, Iran; [2]Islamic Azad University (Qaemshahr Branch), Qaemshahr, Iran

Introduction

Global trade and the production of organic crop plants, including soybeans, are increasing (Willer and Kilcher, 2009). The consumption of soybeans has grown, mainly due to a need for healthy food. Organic soybeans are a basic food, providing natural protein in the European market; for example, soybean products such as tofu are greatly demanded by consumers of organic food (Elena and Iancu, 2014). The soybean organic management is comprised of three parts of nutrient, pest, and weed management. The soybean, as a Fabacea family crop, is able to develop a symbiotic association with rhizobium. This is a unique ability that can significantly decrease the use of nitrogenous chemical fertilizer. Also, the use of other kinds of biofertilizers in fortifying the soil and increasing crop productivity will eliminate the detrimental effects of the chemical fertilizers in organic soybean agroecosystems. Our successful management of pests and weeds highly depend on our knowledge about pest and weed biology and their interaction with the environment. Controlling pest is mainly dependent on the use of biological and cultural control. In such an approach the population of pests and diseases is diminished under economical threshold. Also, the biodiversity conservation of natural pest enemies is applied as a way to achieve a better management of organically controlling field pests. Organic weed management is an ecological base process consisting of a high range of practices. Therefore not only our knowledge about above ground cover is important, but we should also have information about soil microorganisms such as mycorrhizal fungi and their interactions with weeds and soybeans. As an important part of controlling weeds in organic farming, soil ecology knowledge is of significance.

This chapter covers the latest details related to the production of soybeans in organic and traditional systems, achieved by efficiently managing nutrients, pests, and weeds.

Soybean Nutrient Management

Organic farming is a method in which the ecological processes are utilized for providing plants with their essential nutrients. The goal of organically producing systems is the elimination of agrichemicals and the reduction of other external resources, which

Environmental Stresses in Soybean Production. http://dx.doi.org/10.1016/B978-0-12-801535-3.00005-X

results in the improvement of the environment and is profitable to farmers (Pimentel and Burgess, 2016).

Nonmicrobial Compounds

Different organic sources have indicated the potential to provide reasonable amounts of nutrients to plants (Diacono and Montemurro, 2010) and act as alternatives for chemical fertilizers. As a rule, and compared to the use of mineral fertilization, this practice is both more profitable and sustainable (Bhattacharyya et al., 2010).

Soybean crops have been tested with different organic wastes. Biosolid products as means of fertilizer are more feasible and are agronomically more efficient than mineral fertilizers (Lemainski and da Silva, 2006). In this sense, soybean yield can be increased by using amendments including lime, flue dust, and sewage sludge. In addition, the dynamics of nutrients in a soybean/sorghum intercropping system indicated that the use of 75% mineral nitrogen (N), phosphorus (P), and potassium (K) combined with a chicken manure/farmyard manure/phosphocompost compound is an appropriate method of providing a soybean with its essential nutrients (Corrêa et al., 2008). Also, rock phosphate could also be used as a potential alternative for phosphorous nutrition.

Microbial Compounds

Organic farming for sustainable agriculture production is dependent on the effects of microbial activities in the soil (Lee and Kim, 2011). Lee and Kim (2011) also stated that an organic farming system significantly affected the microbial composition of soybean fields compared with a conventional farming system. The soil microbial community is comprised of rhizobium, which are able to biologically fix nitrogen; mycorrhizal fungi; phosphate solubilizing microorganisms (PSM); plant growth-promoting rhizobacteria (PGPR); and macrofauna (such as earthworms as the producers of vermicompost), which can undertake a key role in the mineral nutrition of soybeans under organic and traditional farming.

Rhizobial Symbiosis

The fabaceae family has the ability to biologically fix atmospheric nitrogen via rhizobial symbionts. Under optimal environmental conditions, most of their nitrogen demand can be supplied by nitrogen fixation, and nitrogen resources in the soil may even be increased (Unkovich and Pate, 2000). The production of a high nodule number is essential for the production of high grain yield and protein (Gretzmacher et al., 1994). Different parameters, including soybean genotype, rhizobium strains, and the environment, can affect the efficiency of the symbiosis and consequently soybean yield (Montanez et al., 1995; Ayisi et al., 2000).

Biological nitrogen fixation is an important source of new nitrogen under organic farming systems. Due to the adverse effects of nitrogen chemical fertilization on the process of biological N fixation, the use of synthetic N sources are prohibited under organic standards. Accordingly, the farmers are partially dependent upon the fixation of atmospheric N_2 to provide N for crop growth and development. The nitrogen-fixing bacteria, collectively called rhizobia, are able to develop root nodules with fabaceae family plants by a symbiotic association resulting in the fixation of N. The soybean is the main grain legume grown in North America and is an important component of organic grain rotations due to its ability to acquire its own essential N via N fixation. Further, including soybeans in the crop rotation enables the farmers to meet national standards for organic certification. For the standard production of soybeans, crop rotation must be used by the farmers to acquire organic certification. Hence the organic producers will often use nonlegume crops such as corn with soybeans or legume cover crops in their rotation (Grossman et al., 2011).

Although biological nitrogen fixation is of importance in organic cropping systems, almost nothing is known about how the use of techniques in organic production may affect the rhizobial persistence in the soil and their ability to form efficient symbioses with legumes in organic systems. The diversity of the soil rhizobial community has been shown to be affected by different agricultural practices such as tillage intensity (Batista et al., 2007; Kaschuk et al., 2006), cropping systems (Depret et al., 2004), and N fertilization (Caballero-Mellado and Martinez-Romero, 1999). According to these results the combination of agronomical practices used under organic agriculture may affect the populations of soil rhizobia present in a given field (Grossman et al., 2011).

Rhizobial strains used for legume root inoculation are from either a commercial inoculant applied to seeds at planting, or from the present indigenous populations of rhizobia in field soil. Most organic farmers use the commercial inoculants, specific to soybeans or other legumes, to inoculate seeds at planting. However, continuous use of rhizobium inoculums in many agricultural fields for legume production increases the populations of rhizobia in the soil, resulting from the past use of commercial inoculants (Batista et al., 2007), or by the presence of indigenous rhizobia (Grange and Hungria, 2004; Ferreira and Hungria, 2002). In some areas such indigenous rhizobia can be useful as they enhance legume production without using commercial inoculums. In contrast, a high number of indigenous rhizobia can be problematic, due to their competition with the commercial strain, and as a result the commercial strains would have a higher efficiency (Obaton et al., 2002).

Mycorrhizal Fungi

Mycorrhiza is the structure resulting from the symbiosis between fungi and plant roots enhancing the uptake of water and nutrients by the host plant (Ortas et al., 2001). Arbuscular mycorrhizas (AM) are the most usual type of mycorrhizal. The fungi are able to develop a symbiotic association with most terrestrial plants. Using DNA sequences the fungi have been reclassified into a separate fungal phylum, the Glomeromycota (Schüßler et al., 2001). Their major role is in the uptake of nutrients by the host plant, affecting the plant growth and yield production (Redecker et al., 2000; Heckman et al., 2001); their symbiotic association with the host plants have been affecting the plant interactions in ecosystems to the present day (Smith and Read, 2008).

Most horticultural and crop plants, including soybeans, develop a symbiosis with arbuscular mycorriza fungi (AMF). This root fungal association enhances soybean growth and development through the uptake of less mobile nutrients (Ortas et al., 2001), including phosphorus (P) and micronutrients like zinc (Zn) and copper (Cu). The symbiosis can also increase plant water uptake. AMF can also benefit the growth of their host plants by affecting plant morphology and physiology under different conditions, including stress. These include the production of growth-regulating substances, increased photosynthesis, improved osmotic adjustment under drought and salinity stresses, and alleviating the adverse effects of pests and soil-borne diseases on plant growth (Al-Karaki, 2006). These benefits are mainly due to the improved phosphorous nutrition (Plenchette et al., 2005).

Phosphate Solubilizing Microorganisms

Phosphate solubilizing microorganisms are among the ubiquitous microbes in the soil, affecting the uptake of phosphorous by plants in a more environmental friendly and sustainable manner. These microorganisms are able to change the insoluble phosphate into soluble form by the processes of acidification, chelation, and exchange reactions (Rodriguez and Fraga, 1999). These microorganisms can be considered as an important source of nonchemical plant phosphorous (biological P) in organic farming systems.

Results by Singh et al. (2010) indicated that PSM isolates can significantly affect soybean growth by increasing their field germination, root length, plant height, nodule dry weight, number of branches, nodules per plant, and finally economic yield compared with a control. The higher efficiency of bacterial isolates was attributed to their greater phosphate solubilization ability (Singh et al., 2010).

Plant Growth-Promoting Rhizobacteria

Plant growth-promoting rhizobacteria are the rhizosphere bacteria, which can ameliorate plant growth. These microorganisms are able to enhance the recycling of plant nutrients and decrease the use of chemical fertilization (Cakmakci et al., 2007). Hence these bacteria can be used as promising biofertilizers in organic farming and sustainable agriculture ecosystems. The PGPR comprises the following bacterial species: *Pseudomonas*, *Azospirillum*, *Azotobacter*, *Klebsiella*, *Enterobacter*, *Alcaligenes*, *Arthrobacter*, *Burkholderia*, *Bacillus*, and *Serratia*, which enhance plant growth and yield production (Okon and Labandera-Gonzalez, 1994; Glick, 1995). Zhang et al. (1996) reported the positive effects of PGPR on the number of soybean root nodules and the amount of biological nitrogen fixation when coinoculated with *Bradyrhizobium Japonicum*. The positive effects of coinoculation with PGPR and *B. japonicum* on soybean growth and yield production has also been demonstrated by Stefan et al. (2010).

Earthworms Producing Vermicompost

Organic materials, as ecofriendly compounds, are beneficial agricultural biofertilizers, which can increase crop yield and conserve soil productivity. Vermicompost is the product of microbial composting organic matters through earthworm activity, resulting in the production of organic fertilizer. Such types of fertilization contains higher levels of organic matter, organic carbon, total and available N, P, K, and micronutrients, with a greater rate of microbial and enzymatic activities (Parthasarathi et al., 2007). The effects of vermicompost on the improvement of the growth, productivity, and chemical characteristics of soybean straw (*Glycine max* L. Merril.) has been studied (Manna et al., 1996).

Vermicompost can affect plant growth and yield production by stimulating seed germination (Arancon et al., 2008; Lazcano et al., 2010) and vegetative growth (Edwards et al., 2004), as well as by altering seedling morphology such as increased leaf area and root branching (Lazcano et al., 2009). Vermicompost has also been shown to promote plant flowering, the number and the biomass of the flowers (Arancon et al., 2008), and the production of fruit yield (Singh et al., 2008).

Controlling Pests and Disease

The production of organic elements is largely dependent on the use of nonchemical strategies for controlling pests and diseases. Controlling pests and diseases is an ecological topic that relies on our knowledge about the ecology of pests and diseases. Linker et al. (2009) stated that the efficient controlling of pests under organic farming depends on learning about the biological, ecological, and behavioral details of both pests and beneficial insects. Hence in organic farming systems, farmers use preventive, cultural, biological, and integrated methods for controlling pests and diseases.

Prevention

The conservation of crop plants from infestation to pests and diseases is the major strategy of agronomical practices in ecological-based agroecosystems. The strategy of prevention is the use of disease-free seeds and plants, as well as field sanitation through removing and destroying the overwintering or breeding residences of the pests and preventing a new pest from establishing on the farm (eg, not allowing off-farm soil from farm equipment to spread pest or plant pathogens to your land) (www.attra.ncat.org).

Cultural Practices

Cultural controls are manipulations of the agroecosystem that make the cropping system less friendly to the establishment and proliferation of pest populations. Maintaining and increasing biological diversity of the farm system is a major strategy of cultural control (Altieri, 1994). Genetic diversity of crops and species diversity of the associated plant and animal community could be effective approaches in cultural practices (Zhu et al., 2000).

Crop rotations can change the environment of fields by altering both crops and the related agronomical practices. For example, in a corn/soybean rotation the pest complex, including soil and above ground organisms for the two crop plants, is quite different, and both crops generally provide a reasonable output. Also, rotations create a selection force that will ultimately alter pest genetics (Leslie and Cuperus, 1993).

Other cultural options, which could be used in organic and traditional farming systems, are multiple cropping, interplanting, intercropping, strip cropping, planting of resistant cultivars, spacing of crops, planting dates, optimizing crop growth conditions, trap cropping, and mulches.

Biological Control

Some insects that are natural enemies of pests are produced and sold as pest treatments. Controlling pests using an organic manner can be an important way to reduce damaging insects and fungi. Obtaining the best possible outcome depends on understanding how to develop an efficient method of biological control.

Effective biological control often requires a good understanding of the biology of the pest and its natural enemies, as well as the ability to recognize various life stages of relevant insects in the field. Field scouting is usually necessary to monitor natural enemy activity, evaluate the impact on pest populations, and anticipate the need for additional control strategies. Three different approaches to biological control are recognized including conservation, augmentation, and importation (Michaud et al., 2008).

Beneficial predators and parasitoids play an important role in decreasing the number of early season insect pests and therefore should be protected to act efficiently. Predators and parasites can often prevent pests from getting to treatable levels. *Binodoxys communis* is an approved parasitoid for release in the United States for controlling the soybean aphid under organic and traditional farming systems.

The use of mycorrhzal fungi is a suitable method for biocontrolling pests and diseases in soybean organic fields. Research work has indicated that AMF can reduce or prohibit the infection of crop roots by pathogens. Arbuscular mycorrhizal fungal induced resistance to the root pathogen *Phytophthora fragariae* in strawberries. In nonmycorrhizal plants, increased branching does enhance infection, and plant branching reduces infection by (for example) *Phytophthora* as a result of changing the pattern of exudates production (Norman et al., 1996). Cooper and Grandison (1987) demonstrated the suppressive effect of mycorrhizae on root-knot nematode (*Meloidogyne incognita*) reproduction and development in the tamarillo (*Cyphomandra betacea*) plant. Arbuscular mycorrhizal fungal prevented the infection of soybean plants by *Pseudomonas syringae* (Shalaby and Hanna, 1998) by decreasing the population density of the pathogen in the soybean rhizosphere (Xavier and Boyetchko, 2004). Soybean plants grown in the soil infested with root-infecting fungi indicated lower shoot and root weight and plant height compared to plants in soil inoculated with AM fungi (Zambolin and Schenck, 1983).

Controlling Weeds

Regardless of the crop production practice, the acceptable level of crop yield is achieved if the weeds are controlled at or below economic threshold. A number of factors determine the suitability of the weed control method, including crop prices, labor, fuel, machinery expenses, and the farmer's willingness to accept production risks (Diekmann and Batte, 2014).

In soybeans (*Glycine max* [L.] Merr.), weed infestation is an important issue, acting as a persistent and complex constraint in many parts of world. It is because weed influences soybean growth and development through competition for nutrients, water, and light. The soybean is mainly affected by annual, seed-propagated weeds grown at soybean planting time or during early stages of canopy establishment. Due to a high weed density, soybean crop yield and the harvesting efficiency decreases, and the increased moisture and damaged or diseased seeds subsequently result in significant economic reduction. Thus weed control is a suitable method for the efficient production of soybeans, and as a result, different weed practices have been developed for such a purpose (Buhler and Hartzler, 2004). In organic farming systems, soybean production has to exclusively rely on nonchemical weed control methods, which require sophisticated strategies of weed controlling (Vollmann et al., 2010).

When managing weeds in organic systems, producers use many of the similar techniques used in conventional systems, but with more emphasis on nonchemical control strategies. The major weed control strategies used under organic systems are cultural and mechanical, focusing on crop rotation, crop competition, prevention, and cultivation (Curran, 2004). To plan an effective weed-controlling strategy in organic systems, historical pest problems, properties of soil, crop rotation, machinery, markets, weather, and time and labor must be considered (Curran, 2004).

Prevention

Prevention is a method of controlling new weeds, including the prevention of the further spread of weed seeds or perennial plant parts. The important stage in controlling weeds is avoiding the addition of weed seed to the soil. The other important point in developing a preventive strategy is realizing the biology of weeds. The strengths and weaknesses of weed species make them vulnerable or resistant at different life stages. Accordingly, the important parameters determining the efficient controlling of weeds include the proper investigation and knowledge of the weed life cycle and reproduction and spread. For example, if a creeping perennial such as quackgrass or hedge bindweed is disked or field cultivated in the spring, this may not be a suitable method for controlling the weed, as the problem gets worse by spreading different weed parts including the underground rhizomes or other vegetative tissues (Curran, 2004).

Sanitation

The process of removing or controlling weeds in fields or near fields before the flowering stage and seed production is sanitation. Weed seeds can survive for a long time, depending on the species and the location of seeds in the soil. If necessary, the weeds can be removed from the field by hand before seed production. Weeds can also enter the fields by the use of different sources, including compost, manure, straw, hay, animal feed, contaminated crop seed, or other materials. Whenever a method is used in the field, it must be indicated that it does not contribute to the presence of weeds in the field; if so, the suitable method must be used to avoid the weed presence in the field (Curran, 2004).

It is possible to avoid the presence of many new weeds in the field and to prevent the existing weeds from producing large amounts of seeds. For example, among the useful methods for controlling weeds are (1) mowing weeds around the edges of fields, (2) using clean seeds, (3) preventing the weeds from going to the seed stage production, and (4) thoroughly composting manure before using it in the field. It is even possible to selectively control the new weeds by hand to effectively avoid their future infestations. However, it must be mentioned that selecting a suitable field for the production of organic soybeans is of great significance. For example, if the weeds of a field were not controlled in previous years, such a field is not a good choice for the production of organic soybeans. Clean seed not only means crop seed, which does not contain weed seeds, but it should also not be contaminated with viral or fungal disease pathogens. Crop seeds should also have a high germination rate, resulting in a fast and suitable plant growth, which will make weed control measures more efficient. If the choice of chemical seed treatments is not possible, it is essential to use healthy, vigorous seed, which can result in a high rate of germination. Inoculating the seeds may also be a good method to enhance the growth of seedlings, resulting in a more efficient competiveness with the weeds. For example, seed inoculation with the appropriate strain of rhizobium bacteria results in the homogenous process of N fixation, increasing the growth of soybean plants and their competitiveness with the field weeds (Martens, 2000).

Cultural Weed Control

With time and with respect to the cultural specifications, farmers have learned some techniques to efficiently control the weeds in their field. However, to most people, weed control is the use of tillage implements and rotary hoes. It must be mentioned that before any machinery is considered, controlling weeds using cultural methods is essential (Martens, 2000). Each tactic that increases crop competitiveness versus weeds is considered a cultural practice (Curran, 2004).

Planting Date

Planting date is an important parameter affecting the type and the number of weeds present in the field. Among the usual practices used by the organic farmers to avoid weeds is delaying the planting of spring-seeded crops. Accordingly, such a planting delay may sacrifice some yield potential, but higher soil temperatures will help the crop emerge more quickly and weeds that emerge earlier in the season can be killed before planting the crop to reduce the potential weed seed bank (Curran, 2004).

Stale Seedbed

The technique of stale seedbed, which is sometimes used for vegetable production systems, can also be used for the production of agronomic crops. In this technique, a seedbed is tilled several weeks before planting. When the seeds grow they are controlled, even if they are not large, using shallow cultivation or other nonselective methods. Weeds may grow one or more times before planting, depending on the time length before planting, and the weeds are controlled between the preparation of the seedbed and planting. The weed spectrum and planting time determine the success of a stale seedbed. The success of weed control is more likely if crop planting is delayed. However, the late growth of weeds cannot be favorable to the growth of crop plants.

Crop Competition

The other efficient method for controlling field weeds is crop competition suppressing weed growth. Accordingly, tactics, which result in the quick establishment and domination of crops, decrease weed growth and competition. Such strategies, which result in the vigorous growth of crop and canopy closure, are (1) the use of high quality and vigorous seeds, (2) adapted and resistant genotypes, (3) the proper placement of the crop seed, (4) optimal soil fertility, and (5) appropriate plant populations. The adverse effect of weeds is much less on a vigorous growing crop (Curran, 2004).

Planting Pattern, Row Spacing, and Crop Density

Wide rows are used for many crop plants, resulting in the following: (1) less seed expenses, (2) better controlling of stubble, (3) less work on the soil, and (4) more easily controlling weeds between rows. However, the disadvantages related to the use of wide row cropping are (1) lower rate of crop competition with weeds, (2) decreased yield in some situations, (3) higher rate of evaporation from the soil surface, and (4) less water efficiency. If the crop density is increased and the row spacing is decreased the competitive ability of crops with weeds increases, and weeds are controlled more easily (Lemerle et al., 2001; Mohler, 2001). Generally, wider row spacing will enhance the ability of crop plants for light, soil moisture, soil nutrients, and soil-borne pathogens (Blackshaw et al., 2007; Dunphy and Van Duyn, 2014).

Crop Rotation

The cultural practices, which can efficiently control weeds in organic farming, are (1) crop rotation, and (2) changing planting dates. The use of plant rotation on the suppression of weeds avoids the growth of specific weeds in subsequent years. This generally means the cocropping of summer annual crops with fall-seeded species or even perennials, which require different weed-controlling strategies (Curran, 2004). Accordingly, the use of diverse crop rotations can efficiently improve the strategy of weed controlling (Blackshaw et al., 2007). Using crop plants with different life cycles can disrupt the specific associations of weed crops, resulting in controlling weeds (Karlen et al., 1994; Derksen et al., 2002).

Well-planned crop rotation can efficiently disrupt the cycle of weeds and insects. If the continuous organic monoculture of crop species, including soybeans, is used, the related ecological and environmental selection results in the growth of weed populations, which are efficiently adapted to the present conditions. Accordingly, each year under such environmental conditions, the adapted species of weeds, which had not been controlled by the related measures, results in the production of thousands or even millions of seeds. The environmental changes in a proper crop rotation, each year, suppress the growth of weed seeds remaining in the soil from the favorable conditions of the previous year. Farmers strongly believe that the easiest way, which can efficiently control the growth of weeds in an organic farm, is to grow soybeans on a field that has been under hay or grass for a number of years. Although this may be the easiest method of controlling weeds, it could greatly decrease the acre number of soybean fields a farmer can plant each year. It also would indicate that hay is an important component of the farm rotation. The issue of weeds has been greatly controlled on such fields; however, under the conditions of using large amounts of animal manure in the field in the past, the high growth of different species of adapted weeds will decrease plant growth and yield production. Because the legumes including soybeans, alfalfa, and clover are able to fix nitrogen, their subsequent growth in the field using a heavy rate of manure may actually waste nitrogen, which can be used more efficiently by a nonleguminous crop plant with high N demand. Different factors affect planning a suitable and efficient crop rotation, which can be used in the field for a long time. Some crop plants are able to produce substances, which can suppress or inhibit weed growth by a phenomenon called allelopathy. For example, sowing oats with a new planting of alfalfa can result in allelopathy from the oats, preventing the alfalfa growth from being suppressed by weeds in the first year. Some other crop plants with allelopathic activities include sunflowers, oats, and wheat. Moreover, the risk of plant disease is diminished through suitable cropping, and due to the establishment of strong and vigorous crop stands, the growth of weeds is greatly controlled. Finally, a suitable cropping enhances the biological activity of soil, which significantly decreases the growth and the life cycle of weed seeds. However, much more must be learned from agricultural research conducted a long time ago, which has been dismissed by new agriculture (Martens, 2000).

Genotype Selection

Although various agronomic and crop production techniques have been developed for efficient weed control, the selection for the increased competitiveness of crop plants versus weeds has gained attention in various crop plants, providing an alternative method to the available weed control practices (Pester et al., 1999). There is a wide range of variability among crop species in their competitiveness with weeds. Such types of differences result from the morphology and physiology of different crop species, which can also strongly interact with the environment (Blackshaw et al., 2007).

The important crop parameters determining its competitiveness with weeds are (1) quick rate of seed germination and seedling growth, (2) early and vigorous growth of crop plants, (3) high leaf area index, (4) developing a dense canopy reducing the rate of light interception to the ground, (5) a high rate of tillering and branching, and (6) longer plant height (Pester et al., 1999). There is a high rate of variation among soybean genotypes in suppressing weed growth. For example, Rezvani et al. (2013) showed soybean variety Hill seemed to be more competitive than the other varieties on the basis of competition indices and weed biomass production. This intergenotypic variation in competitiveness ability versus weeds was confirmed by the weed biomass suppression ability of Hill. The development of more competitive soybeans could be a viable approach for reducing herbicide dependency and improving the profitability of soybean fields (Rezvani et al., 2013).

In soybeans, Jannink et al. (2000) suggested a selection index indicating the suppressiveness ability of weeds in which early plant height was an important trait highly correlated with weed suppression. Variation in early plant height was also indicated in wild soybeans (Chen and Nelson, 2006). Vollmann et al. (2010) declared the lower rate of yield loss in early maturity genotypes resulted by weed pressure than late maturity genotypes. Because there was no clear difference in weed suppression between the two genotypes, the difference in yield loss was attributed to the variation of weed tolerance in the two genotypes. For soybean cultivars grown under organic farming conditions, however, weed suppression is a more preferable choice over weed tolerance. It is because weed tolerance probably requires enhanced genetic variation in characters such as early ground cover development, leaf area, or others. It is also because weed tolerance results in a stronger weed development, and as a result the rate of weed seed in a given field significantly increases for the following years (Vollmann et al., 2010).

Variety selection in soybeans is largely dependent on the market choices. In an ideal situation, a precise selection of vigorous or crop varieties with a dense canopy can shade the soil surface and hence prevent the high biomass production of weeds and their subsequent growth above the crop. If soybean genotypes with less vigor are selected, the farmers may prefer to plant them at a high population rather than using wider rows, though this method of planting may not be suitable for using many types of mechanical weed control options. However, it is important to use a balanced rate of plant population with respect to the genotype properties. It is because a higher plant population of a competitive variety decreases the rate of crop yield due to the process of lodging (Martens, 2000).

Early Planting

The important factor determining the competition efficiency of crop plants with weeds is the early crop establishment. The crop seeds must be healthy to result in an optimum rate of seed germination and emergence (Dunphy and Van Duyn, 2014). Osmopriming or osmoconditioning has been examined to increase crop seed germinability, water uptake, emergence, and competitive ability with weeds (Blackshaw et al., 2007). The potential of osmoconditioning to enhance crop competitiveness with weeds has also been demonstrated with soybeans (Nunes et al., 2003).

Soil Fertility and Conditions

The presence of weeds can be a clear indicator of the chemical components, which are out of balance in the soil. For example, magnesium is the element that is commonly used for the regulation of weed population size and vigor. Many prevalent weed species in soybean fields, such as foxtail and summer annual grasses, throughout the

United States grow in hard compacted soils, which are also deficient in calcium and sufficient in magnesium. Accordingly, the use of calcium amendments can improve the strategy of weed controlling. The calcium:magnesium rates of 8:1 to 10:1 are optimal for weed control and crop plant growth. When the soil calcium:magnesium balance deviates from these optimal rates, due to the excessive use of calcium or magnesium, the higher growth of weeds results, though the concentrations of such elements determine the variation of weed species (Martens, 2000).

Crop competition and weed control can be increased by improving soil drainage, aeration, and the regular addition of organic material. The combination of a high rate of organic matter, with a proper calcium:magnesium balance and good aeration stimulates the activity of soil microbes. These organisms feed on weed seeds and decompose them, increase the availability of vital nutrients for crop growth, enhance the stability of soil aggregate, and decrease soil compaction (Martens, 2000).

There is also some kind of species-specific interactions between weed communities and composition of soil microbial community. Indeed, previous studies have demonstrated the significant effects of weed species (eg, *Centaurea maculosa*) on the abundance and community composition of soil microbes (Batten et al., 2006; Lutgen and Rillig, 2004). These changes in the composition of the microbial community are often considered as a new competitive process and resistance mechanism used by certain weedy and invasive species (Callaway and Ridenour, 2004; Marler et al., 1999). However, many of these observations have been conducted under the unmanaged ecosystems, on the invasive weeds, and there is not much research on the effects of agricultural weeds on the composition of the soil microbial community. Similar to the crops, which are influenced by specific species of soil microbes or functional groups, weeds are also subject to such effects. For example, AMF (*Glomus intraradices*) has been shown to decrease the growth and competitiveness of weeds compared with crop species (Veiga et al., 2011). Weeds, which often survive under stressful environmental conditions, can be unfavorably competitive with crop plants under the conditions of fertile and biologically "healthy" soil (Tollenaar et al., 1994).

Flooding

The effects of flooding on suppressing weeds is by decreasing the rate of oxygen under the anaerobic conditions of soil, which adversely effects the growth of different weed seeds, seedlings, and perennial grasses. McWhorter (1972) found that in the fields, subjected to a 2–4 weeks of flooding before planting in the Southern United States, the density of johnsongrass (*Sorghum halepense*) decreased, while the yield of soybeans remained unchanged. If the adapted crops, appropriate soils, and an adequate source of water are available, flooding can be used as an integrated component of weed-controlling strategies.

Mulches and Cover Crops

Because the sunlight, intercepted by the soil, enhances the growth of weeds, the use of mulch can effectively manage weeds under some organic production systems. The mulch acts as a physical barrier blocking the soil surface from all the sunlight and hence prevents the growth of the weeds growing beneath the mulch. Although the use of plastic mulches is suitable under some organic fields, such a method is not practical for crops with less value planted in large areas. Mulches of organic material, including cover crop residue, straw, and newspaper used on the soil surface, can also effectively prevent the weeds from the sunlight and are used most of the time in the production of organic row crops. Although cover crops have traditionally been used to decrease soil erosion and improve the quality of soil, more recently cover crops are being used as an effective method in controlling weeds under crop organic production (Wortman et al., 2013).

In agricultural fields, the structure of the soil microbial community is influenced by tillage and cover cropping, though some complex interactions control such alterations of the soil microbial community (Buckley and Schmidt, 2001; Carrera et al., 2007). For example, a suitable controlling method (eg, cover cropping) can substantially alter the community of subsequent weeds, increases the labile carbon of soil, as well as the soil moisture, and hence significantly affects the structure of the soil microbial community (Buyer et al., 2010).

Cover cropping can manage the weeds according to the following details (Curran, 2004). (1) Cover crops are an effective method in crop rotation and rapid change of weed seeds in the soil. (2) Cover crops are able to manage weeds by competing with weeds for different resources including light, moisture, nutrients, and space. This is especially helpful for controlling the growth of winter annual weeds or specific cool-season perennials. (3) Cover crops and their residues can also be used as mulches, suppressing the germination and growth of seeds by blocking light and decreasing soil temperatures. The larger the rates of the cover crop biomass and dry matter production, the more effectively the weeds are controlled. (4) The allelopathic compounds in cover crops produced by the living or decaying plant tissue can interfere with weed growth. However, the qualities of such biochemicals are determined by the quantity and quality of cover crops, as well as by the environment during the growing season. However, although the cover crops have these potential benefits, some disadvantages related to the physical and chemical effects of cover crops can result in the inefficient controlling of weeds in the field. The use of mechanical and cultural measures may be a suitable complement to the use of cover crops for controlling weeds (Curran, 2004).

Uchino et al. (2012) indicated that if main crops are intercropped with sufficiently fertilized cover crops under organic production farming, the weeds could be efficiently suppressed without decreasing yield. This weed suppression potential was mainly related to the role of intercropping cover crops. It was previously mentioned that the competition for different resources including light, soil moisture, and nutrients by the cover (Teasdale and Mohler, 2000; Barberi, 2002), combined with their ability of producing allelopathic compounds, can suppress the growth of weeds (Reberg-Horton et al., 2005). However, it has also been indicated that the competing ability of cover crops for light (Hooks and Johnson, 2001), nutrients (Feil et al., 1997), and water (Box et al., 1980) can also suppress the growth and the yield of main crops. Accordingly, a technique has been used to avoid or reduce such adverse effects of cover crops on the growth and yield of main crops, which is the seeding of cover crops into the established vegetation of main crops (Scott et al., 1987; Abdin et al., 2000). Interseeding cover crops is advantageous because the laborious intertillages and hand weeding, essential in organic farming, would not be used (Uchino et al., 2012). Among the important managing techniques of cover crops is the alteration of the sowing date. Uchino et al. (2009) stated weed number was not significantly affected by different treatments during the growing season (Fig. 5.1A), while the cover crops significantly decreased weed dry weight compared with the other treatments during the growing season (Fig. 5.1B), indicating how effectively the cover crops can suppress weed growth. However, as previously mentioned, the sowing date of cover crops is an important

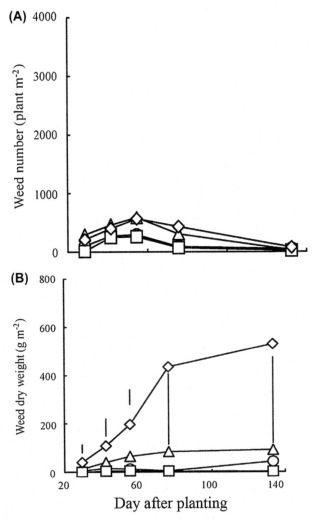

Figure 5.1 Weed number (A) and weed dry weight (B) in the experiment for soybeans. (^) Sole-main crop (MC); (*) Pre-MC; (~) Syn-MC; (&) Post-MC. *Vertical bars* indicate least significant difference values ($P=0.05$) for the comparison between treatments.

parameter, resulting in the high yield of main crops. Consequently, cover crop planting resulted in the highest yield of soybeans, although soybean yield decreased by 29, 18, and 7% in the treatments without cover crops using similar planting dates. The rate of seeds in the soil was also decreased by sowing cover crops. The authors accordingly indicated that sowing cover crops can effectively control weeds if the cover crops are planted after the main crops in organic farming production. Also, Uchino et al. (2015) found that if the cover crops are planted inside the rows rather than between the rows, they will control the weeds of soybeans more effectively.

Intercropping

Intercropping is the growth of two or more crops, simultaneously (Vandermeer, 1989). The major benefits of intercropping are (1) increasing the rate of crop production, with the advantage of simultaneously decreasing the risk of total crop reduction, and (2) controlling weeds (Liebman and Dyck, 1993). It must be mentioned that sometimes the second crop in intercropping farming is just planted for controlling weeds. Intercrop systems have been indicated to use resources differently and more efficiently, and more resources are used compared with monocropping; as a result the rate of resources available for weed utilization decreases (Liebman, 1988; Zimdahl, 1993). The allelopathic interactions are also of significance (Liebman and Dyck, 1993).

Intercropping increases the spatial diversification of crops, resulting in competitive interactions with weeds. Studies have reported that intercropping controls weeds by shading more significantly compared with monocropping (Liebman and Dyck, 1993; Itulya and Aguyoh, 1998). Intercropping also enhances the competitive ability of crops for nutrients and water related to monoculture systems (Hauggaard-Nielsen et al., 2001). Carruthers et al. (2000) stated intercropping with soybeans was also more efficient at reducing weed populations than monocropping. Also, Zaefarian et al. (2007) demonstrated soybean–maize intercropping diminished redroot pigweed and jimsonweed biomass.

Mechanical Weed Control

Although the precise use of cultural weed control can manage the weeds in the field, amazingly, a large population of weed seedlings can grow in a soybean field that has been recently planted. If the weeds are not controlled mechanically, the yield of soybeans decreases significantly (Martens, 2000).

It is essential that the weeds be controlled mechanically under organic systems. Mechanical control is especially essential for efficiently controlling weeds in organic row crops such as corn or soybeans. The use of preplant tillage, such as plowing, disking, and field cultivation, is among the most important practices in mechanical weed control. These types of tillage practices can help decrease the rate and spread of specific perennial weeds and can also control the germinated weed seedlings and place the weed seeds below the germination zone (Curran, 2004).

The preparation of conventionally tilled seedbeds is a practice usually done by soybean producers before planting their crops in the spring. Cultivation should follow immediately, a few days after planting. The implements such as a rotary hoe, chain-link harrow, or tine weeder are used in the field for controlling small weed seedlings located just beneath the soil surface or barely emerged. Accordingly, the small weeds get dried as they are displaced on the soil surface and subjected to the adverse effects of the wind and sunlight (Curran, 2004).

The use of implements, including rotary hoes, tine weeders, or similar implements, is among the most efficient methods for controlling weeds under the production of crop rows. The implement such as a rotary hoe is operated at 10 to 12 miles per hour, displacing the small seedlings and controlling their growth. It is essential that such an implement be used each 5–7 days as long as the weeds are sufficiently controlled and the crop is too big. The use of a rotary hoe for soybeans in the "hook" stage (during the exposure of the stem and before the opening of the cotyledons above the ground) is not suggested. It is also favorable to use rotary hoes or similar implements in the afternoon, with less turgor pressure and more flexibility of soybeans and corn. A maximum of three rotary hoeings may be conducted in a period of 2–3 weeks following planting (Curran, 2004).

The use of cultivation can be done for crop rows planted at 30 inches or a higher distance. At the growth stages of soybeans with two to three trifoliate leaves and corn beyond the two-leaf stage (V2) and 8–10 inches tall, the use of a row cultivator for controlling small weed seedlings is suggested. If the cultivation is done at a shallow level (1 to 2 inches deep), the crop roots will not be harmed. The use of cultivation must continue at 7–10-day intervals until the corn plants are too tall and the soybean canopy is at a suitable density. Depending on the species of weeds, their severity, and the rate of rainfall, one to three cultivations must be conducted for organic corn and soybeans. The best time for cultivation is during the hottest time of the day under bright sunlight, which results in the quick desiccation of weeds and their control under these conditions. However, the presence of rainfall or wet cloddy soils during or after cultivation can result in the regrowth and survival of weeds. The controlling of weeds by hand results in the maximum crop yield and prevents the production of weed seeds, which can affect the future controlling of weeds (Curran, 2004).

The use of mowing can also be important for controlling weeds in forage crops or noncrop areas. Using mowing repeatedly reduces the competitive ability of weeds, decreases the reserves of carbohydrate in the roots, and prevents the production of seeds. The mowing of some weeds, when they are young can make a suitable food for livestock. Mowing can effectively control or suppress annual and biennial weeds. Mowing can also control and suppress the growth of perennials, resulting in the restriction of their spread. Most weeds will not be controlled by a single mowing; however, mowing three or four times per year over several years can significantly decrease and sometimes control certain weeds efficiently, including Canada thistle. Also, mowing along fences and the side of the field can prevent the growth and spread of new weed seeds. If mowing is done on a regular basis, the establishing, spreading, and competing of weeds with desirable forage crops are prevented (Curran, 2004).

There are weaknesses and strengths for different weed species, which is due to their specific stages of life cycle or resulting from growth patterns. However, sometimes the use of mechanical control may intensify the growth of weeds, for example, if disking is used for a field infested with quackgrass, the growth of certain species of weeds may increase under specific conditions. The controlling of grassy weeds is done most of the time with different measures compared with broad-leafed weeds. The important stage in controlling weeds using the proper measures is that their species must be correctly identified. The other important point for controlling weeds is the area that will be under the cultivation of soybeans. If the area under soybeans is higher than a reasonable area, which can be controlled by farmers in a timely manner using the available machinery, it would not be possible to suppress the weeds, and they grow and infest the field quickly. It is accordingly possible to harvest more soybean yield from a smaller area of well managed and controlled soybeans than from a much larger planting that has not been controlled and managed properly (Martens, 2000).

Tillage

Tillage systems can affect the viability of weed seeds and their subsequent distribution in the soil (Carter and Ivany, 2006). Lutman et al. (2002) also stated that tillage operations can significantly influence the distribution and survival of weed seeds in the soil. Tillage has been explained by Cardina et al. (2002) as a filter or constraint affecting the weed species and the distribution and rate of their seeds in the soil. The use of a field pretillage a week or 10 days before planting will result in the germination of weeds and their subsequent suppression during the preparation of field for planting. In fields heavily infested with weeds, soil tillage during late spring will decrease the weed pressure by displacing large populations of germinating seeds in the soil. If tillage is done on a hot and warm day under sunlight conditions, especially for weeds, which are troublesome, with long underground rhizomes, such as quackgrass, they can be displaced to the surface using a spring tooth harrow to desiccate. This old technique can effectively control and suppress the infested field of quackgrass, providing that it is conducted several times. It is important to note that weather conditions importantly dictate the effectiveness of tillage practices in controlling weeds. If tillage is done under wet weather, the rerooting of weeds may result in the regrowth of the weeds, and hence they are not controlled and suppressed. The rate of weed seed germination decreases under cold wet conditions. Tillage practices will probably not damage weed seeds that are germinating but have not yet developed (Martens, 2000).

Blind Cultivation

After planting, and before the crop emerges, a number of tools can be used for the blind cultivation of weeds. Under blind cultivation, a surface tillage is used in the entire field using the proper implement, with a little attention to the place of the rows. However, the larger seeded soybean seeds germinate and are not damaged, because they are deeper than the level of the cultivation. The aim of blind cultivation is to acquire the biggest possible crop/weed size difference from the beginning of planting (Martens, 2000).

At the time of spraying the fields by conventional farmers, using a 45′ Kovar coil tine harrow, Klaas Martens is "blind cultivating." This implement is able to cultivate the fields faster than a sprayer and is really economical for using in the field. One evening in 1998, using a Kovar harrow, Martens was able to operate on 47 acres in less than 2 h. A week later, after the crop has emerged, a second blind cultivation was completed with a Lely weeder or rotary hoe. John Myer relied mainly on the rotary hoe for his blind tillage operations, the first time about 3–4 days after planting and the second time 5 days later. The important point in the efficient operation of a rotary hoe is speed. For the effective use of a rotary hoe, a high horsepower tractor with the speed of at least 8–12 mph is essential. Surprisingly, such an operation does little damage to the seedlings of young crop plants, but efficiently controls the weeds. The operation of a Lely weeder is done by shaking the loose soil and controlling small weeds but not harming the soybeans with larger and deeper roots. Blind cultivation is the easiest and the most efficient method for controlling the weeds that grow inside the rows and directly compete with the crop (Martens, 2000).

Weed species are different in their vulnerability to cultivation. The controlling of broad-leafed weeds with their growing point above ground are easily done because their tops are damaged during the cultivation; however, grasses, which have growing points below the soil surface must be uprooted and desiccated. During the early stage of germination, in the "white hair" stage, weeds are most sensitive to desiccation. However, it is not possible to suppress the growth of perennial weed species, which have deep roots and large reserves, using blind cultivation, and hence other methods must be used (Martens, 2000).

Between Row Cultivation

The optimum time for between row cultivation is when the plants are in the third trifoliate stage or later and at the time of clear visibility of the soybean rows. However, it may be essential to use an earlier cultivation, if there is not a good crop/weed size difference, but cultivation is not done quickly to prevent the crop plants from being displaced under the soil. It has been indicated by the farmers that if all the other suitable methods of controlling weeds are used, the cultivation of soybeans at least once during the season will control most weeds for the rest of the season. However, a second cultivation is essential in New York, for controlling the growth of weeds that were stimulated by the first cultivation and for a higher rate of soil aeration (Martens, 2000).

Thermal Weed Control

Using flame weeding is by far the most widely used method of thermal weed controlling. Flaming heats just rapidly rupture cells in plant tissues. It is a method currently and widely used for controlling weeds by organic farmers in Western Europe (Ascard et al., 2007). It has been indicated that if flaming before crop growth is combined with postemergence mechanical inter- and intrarow weeding, it can effectively control the weeds (Melander and Rasmussen, 2001).

Organic Farming Herbicides

The use of chemical weed control is generally prohibited in organic crop production systems. According to the USDA National Organic Program (NOP), it is possible to use certain nonsynthetic soap-based herbicides for the maintenance of a farmstead (right-of-ways, building perimeters, roadways, ditches) and in ornamental crops. In addition, several products with natural or nonsynthetic ingredients are allowed or regulated by the Organic Materials Review Institute (OMRI). Regulated substances are listed, as they have restriction on the USDA National List or in the NOP rule.

The listing of OMRI does not entail product approval by any federal or state government agency. However, the user is responsible to determine the compliance of a particular product. For example, although corn gluten meal is used as a preemergence herbicide in some production systems, because of the volume of product essential and the related expenses, corn gluten meal cannot be practically used under agronomic crop production systems. In addition, the requirement for the use of corn gluten for weed control must be explained in the Organic System Plan and the related sources must not be from genetically modified sources. More details are available about corn gluten on the corn gluten meal research web page at Iowa State University (www. gluten.iastate.edu) (Curran, 2004).

The nonsynthetic postemergence herbicides, as nonselective contact-type herbicides, have plant-based ingredients, including garlic, eugenol (clove oil), and citric acid. They effectively suppress the vegetation they come in contact with. It is essential that the requirement for the use of herbicides with plant or animal sources be explained in the Organic System Plan, and permissions must be obtained from the organic certifying agencies to use such compounds. Although acetic acid or vinegar is used as an ingredient in a number of products, it is not currently approved to be used as an herbicide under organic crop production systems. Some other products and ingredients are currently under review. Nonsynthetic adjuvants (such as surfactants and wetting products) can be used unless explicitly prohibited. All synthetic adjuvant, including the most adjuvant products on the market, are prohibited. However, a number of adjuvants with a plant source are available, including the derivatives of yucca (Natural Wet), pine resin (Nu-Film P), or other plant-based compounds. Some products contain acidifying products and other ingredients used to enhance the uptake of pesticide or nutrients.

Mycorrhizal Control of Weeds

Main agricultural crop families, including Poaceae, Asteraceae, and Fabaceae can establish symbiosis with AM fungi. Many troublesome agricultural weeds belong to families that appear to be predominantly nonhosting (Amaranthaceae, Brassicaceae, Caryophyllaceae, Chenopodiaceae, Cyperaceae, and Polygonaceae; Tester et al., 1987; Brundrett, 1991; Francis and Read, 1994).

Agricultural weeds from the commonly hosting AMF families (eg, Poaceae, Asteraceae) have been shown in some cases to be nonmycorrhizal (Harley and Harley, 1987; Feldmann and Boyle, 1999). A variety of weeds such as *Ambrosia artemisiifolia* L., *Avena fatua* L., *Abutilon theophrasti*, and *Setaria lutescens* appear to be host species (Koide and Lu, 1992; Koide et al., 1994). Heppel et al. (1998) showed AMF is able to improve the growth, seed production, and seed quality of agricultural host weeds.

However, this effect could be variable during the growing season and can diminish at higher plant densities (Shumway and Koide, 1994).

Mycorrhizal fungi are likely to influence the composition of weed communities and the relative abundance of species in such communities. Moreover, AMF may alter the agroecological functioning of weed communities to make the net effect of weeds more beneficial (Jordan et al., 2000). The authors indicated that mycorrhizal fungi can affect the weed flora of an agroecosystem through altering the mycotrophic and non-mycotrophic abundance. Jordan and Huerd (2008) suggested that the negative effects of AMF on weeds can be direct and indirect.

The negative and direct effects of AMF on weeds include their effects on shifting the host plant communities by increasing host plant nutrient uptake and growth while suppressing nonmycorrhizal species (Jordan and Huerd, 2008). The induction of plant resistance through production of toxic compounds by AMF is another aspect related to the negative and direct effects of AMF (Francis and Read, 1994). AM fungi suppress weed growth by exuding allelopathic compounds and inducing resistance responses (Francis and Read, 1994).

Also, there are several indirect mechanisms by which AMF can suppress weeds (Fig. 5.2):

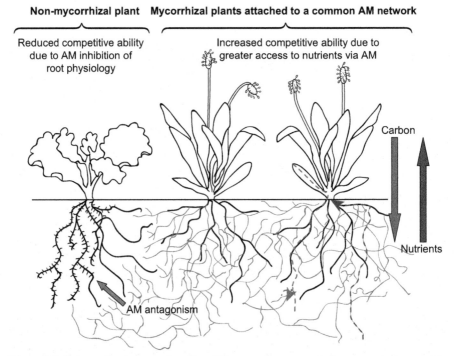

Non-mycorrhizal plant   Mycorrhizal plants attached to a common AM network

Reduced competitive ability
due to AM inhibition of
root physiology

Increased competitive ability due to
greater access to nutrients via AM

Carbon

Nutrients

AM antagonism

Figure 5.2 Schematic representation of an AM network composed of co-occurring mycorrhizal and nonmycorrhizal plant species, showing nutrient for carbon exchange in mycorrhizal species and AM suppression of nonmycorrhizal species (Smith and Read, 2008).

1. The access of nonmycorrhizal weeds to the nutrients already taken up by AMF and transported to crops or mycorrhizal weeds is limited.
2. AMF preferentially allocates nutrients to the crop.
3. Although weeds invest carbon in mycorrhizal networks, they receive no mutual benefit (Rinaudo et al., 2010). Reduced weed nutrient concentrations may be due to the potential allopathic effects of AM fungi on decreasing the numbers of root hairs and hence the surface area for nutrient uptake of nonmycorrhizal plant species.
4. The exudation of compounds by AM fungi may alter the morphological (and potentially physiological) properties of root structure in incompatible interactions (Abourghiba, 2005).

Conclusions

Organic farming has four main properties of health, ecology, fairness, and care (IFOAM, 2009). In this chapter, the main and most important approaches related to organic farming were analyzed. All the managing practices discussed here are of significance if used singly or integrated with other components in organic and traditional soybean production. The main aim of organic farming is the production of healthy food for humans while conserving the environment.

References

Abdin, O.A., Zhou, X.M., Cloutier, D., Coulman, D.C., Faris, M.A., Smith, D.L., 2000. Cover crops and interrow tillage for weed control in short season maize (*Zea mays*). European Journal of Agronomy 12, 93–102.

Abourghiba, T.Y., 2005. Comparative Analysis of the Impacts of AMF on 'Host' and 'Non-host' Plants (Ph.D. thesis). University of Sheffield, UK.

Al-Karaki, G.N., 2006. Nursery inoculation of tomato with arbuscular mycorrhizal fungi and subsequent performance under irrigation with saline water. Scientia Horticulture 109, 1–7.

Altieri, M.A., 1994. Biodiversity and Pest Management in Agroecosystems. The Haworth Press, Binghamton, NY, p. 185.

Arancon, N.Q., Edwards, C.A., Babenko, A., Cannon, J., Galvis, P., Metzger, J.D., 2008. Influences of vermicomposts, produced by earthworms and microorganisms from cattle manure, food waste and paper waste, on the germination, growth and flowering of petunias in the greenhouse. Applied Soil Ecology 39, 91–99.

Ascard, J., Hatcher, P.E., Melander, B., Upadhyaya, M.K., 2007. Thermal weed control. In: Upadhyaya, M.K., Blackshaw, R.E. (Eds.), Non-chemical Weed Management. Principles, Concepts and Technology. CABI North American. Cambridge University Press, Cambridge, UK.

Ayisi, K.K., Nkgapele, R.J., Dakora, F.D., 2000. Nodule formation and function in six varieties of cowpea (*Vigna unguiculata* L. Walp.) grown in a nitrogen-rich field soil in South Africa. Symbiosis 28, 17–31.

Barberi, P., 2002. Weed management in organic agriculture: are we addressing the right issues? Weed Research 42, 177–193.

Batista, J.S.S., Hungria, M., Barcellos, F.G., Ferreira, M.C., Mendes, I.C., 2007. Variability in *Bradyrhizobium japonicum* and *B. elkanii* seven years after introduction of both the exotic microsymbiont and the soybean host in a cerrados soil. Microbial Ecology 53, 270–284.

Batten, K.M., Scow, K.M., Davies, K.F., Harrison, S.P., 2006. Two invasive plants alter soil microbial community composition in serpentine grasslands. Biological Invasions 8, 217–230.

Bhattacharyya, R.P.S.C., Chandra, S., Kundu, S., Supradip, S., Mina, B.L., Srivastva, A.K., Gupta, H., 2010. Fertilization effects on yield sustainability and soil properties under irrigated wheat-soybean rotation of an indian Himalayan upper valley. Nutrient Cycling in Agroecosystems. 86, 255–268. http://dx.doi.org/10.1007/s10705-009-9290-7.

Blackshaw, R.E., Anderson, R.L., Lemerle, D., 2007. Cultural weed management. In: Upadhyaya, M.K., Blackshaw, R.E. (Eds.), Non-chemical Weed Management. Principles, Concepts and Technology. CABI North American. Cambridge University Press, Cambridge, UK.

Box Jr., J.E., Wilkinson, S.R., Dawson, R.N., Kozachyn, J., 1980. Soil water effects on no-till corn production in strip and completely killed mulches. Agronomy Journal 72, 797–802.

Brundrett, M.C., 1991. Mycorrhizas in natural ecosystems. In: MacFayden, A., Begon, M., Fitter, A.H. (Eds.), Advances in Ecological Research. Academic Press, London, UK, pp. 171–133.

Buckley, D.H., Schmidt, T.M., 2001. The structure of microbial communities in soil and the lasting impact of cultivation. Microbial Ecology 42, 11–21.

Buhler, D.D., Hartzler, R.G., 2004. Weed biology and management. In: Boerma, H.R., Specht, J.E. (Eds.), Soybeans: Improvement, Production, and Uses, Series Agronomy, vol. 16. third ed. American Society of Agronomy, Madison, WI, pp. 883–918.

Buyer, J.S., Teasdale, J.R., Roberts, D.P., Zasada, I.A., Maul, J.E., 2010. Factors affecting soil microbial community structure in tomato cropping systems. Soil Biology & Biochemistry 42, 831–841.

Caballero-Mellado, J., Martinez-Romero, E., 1999. Soil fertilization limits the genetic diversity of *Rhizobium* in bean nodules. Symbiosis 26, 111–121.

Cakmakci, R., Dönmez, M.F., Erdoğan, Ü., 2007. The effect of plant growth promoting rhizobacteria on barley seedling growth, nutrient uptake, some soil properties, and bacterial counts. Turkish Journal of Agriculture and Forestry 31, 189–199.

Callaway, R.M., Ridenour, W.M., 2004. Novel weapons: invasive success and the evolution of increased competitive ability. Frontiers in Ecology and the Environment 2, 436–443.

Cardina, J., Herms, C.P., Doohan, D.J., 2002. Crop rotation and tillage system effects on weed seedbanks. Weed Science 50, 448–460.

Carrera, L.M., Buyer, J.S., Vinyard, B., Abdul-Baki, A.A., Sikora, L.J., Teasdale, J.R., 2007. Effects of cover crops, compost, and manure amendments on soil microbial community structure in tomato production systems. Applied Soil Ecology 37, 247–255.

Carruthers, K., Fe, Q., Cloutier, D., Smith, D.L., 2000. Intercropping corn with soybean, lupin and forages: weed control by intercrops combined with interrow cultivation. European Journal of Agronomy 8, 225–238.

Carter, M.R., Ivany, J.A., 2006. Weed seed bank composition under three long-term tillage regimes on a fine sandy loam in Atlantic Canada. Soil & Tillage Research 90, 29–38.

Chen, Y., Nelson, R.L., 2006. Variation in early plant height in wild soybean. Crop Science 46, 865–869.

Cooper, K.M., Grandison, G.S., 1987. Demonstrated that mycorrhizae suppressed root-knot nematodes (*Meloidogyne incognita*) reproduction and development in tamarillo (*Cyphomandra betacea*). Plant and Diseases 71, 1101–1106.

Corrêa, J.C., Büll, L.T., Crusciol, C.A.C., Tecchio, M.A., 2008. Aplicação superficial de escória, lama cal, lodos de esgoto e calcário na cultura da soja. Pesquisa Agropecuária Brasileira. 43, 1209–1219. http://dx.doi.org/10.1590/S0100-204X2008000900016.

Curran, W., 2004. Weed Management in Organic Cropping Systems. The Pennsylvania State University.

Depret, G., Houot, S., Allard, M.R., Breuil, M.C., Nouaim, R., Laguerre, G., 2004. Longterm effects of crop management on *Rhizobium leguminosarum* biovar *viciae* populations. FEMS Microbiology Ecology 51, 87–97.

Derksen, D.A., Anderson, R.L., Blackshaw, R.E., Maxwell, B., 2002. Weed dynamics and management strategies for cropping systems in the northern Great Plains. Agronomy Journal 94, 174–185.

Diacono, M., Montemurro, F., 2010. Long-term effects of organic amendments on soil fertility. A review. Agronomy for Sustainable Development. 30, 401–422. http://dx.doi.org/10.1051/agro/2009040.

Diekmann, F., Batte, M.T., 2014. Economics of technology for precision Weed control in conventional and organic systems. In: Young, S.L., Pierce, F.J. (Eds.), Automation: The Future of Weed Control in Cropping Systems. Springer Science+Business Media, Dordrecht.

Dunphy, J., Van Duyn, J., 2014. Crop production management – organic soybeans. In: Hamilton, M. (Ed.), North Carolina Organic Grain Production Guide. North Carolina State University.

Edwards, C.A., Domínguez, J., Arancon, N.Q., 2004. The influence of vermicomposts on plant growth and pest incidence. In: Shakir, S.H., Mikhaïl, W.Z.A. (Eds.), Soil Zoology for Sustainable Development in the 21st Century, pp. 397–420 Cairo.

Elena, B., Iancu, P., 2014. Study regarding influence of the organic fertilization on the morphological and productivity traits to the soybean (*Glycine max*). Analele Universității Din Craiova, Seria Agricultură – Montanologie – Cadastru (Annals of the University of Craiova – Agriculture, Montanology, Cadastre Series) XLIV, 28–32.

Feil, B., Garibay, S.V., Ammon, H.U., Stamp, P., 1997. Maize production in a grass mulch system-seasonal patterns of indicators of the nitrogen status of maize. European Journal of Agronomy 7, 171.

Feldmann, F., Boyle, C., 1999. Weed-mediated stability of arbuscular-mycorrhizal effectiveness in maize monocultures. Journal of Applied Botany 73, 1–5.

Ferreira, M.C., Hungria, M., 2002. Recovery of soybean inoculant strains from uncropped soils in Brazil. Field Crops Research 79, 139–152.

Francis, R., Read, D.J., 1994. The contributions of mycorrhizal fungi to the determination of plant community structure. Plant and Soil 159, 11–25.

Glick, B.R., 1995. The enhancement of plant growth by free-living bacteria. Canadian Journal of Microbiology 41, 109–117.

Grange, L., Hungria, M., 2004. Genetic diversity of indigenous common bean (*Phaseolus vulgaris*) rhizobia in two Brazilian ecosystems. Soil Biology & Biochemistry 36, 1389–1398.

Gretzmacher, R., Schahbazian, N., Pourdavai, N., 1994. Einfluss von symbiontischem, organischem und anorganischem Stickstoff auf Ertrag und Qualität von Sojabohnen. Die Bodenkultur 45 (3), 253–267.

Grossman, J.M., Schipanskib, M.E., Sooksanguana, T., Seehavera, S., Drinkwater, L.E., 2011. Diversity of rhizobia in soybean [*Glycine max* (Vinton)] nodules varies under organic and conventional management. Applied Soil Ecology 50, 14–20.

Harley, J.L., Harley, E.L., 1987. A checklist of mycorrhiza in the British flora. New Phytologist 105 (Suppl.), 1–102.

Hauggaard-Nielsen, H., Ambus, P., Jensen, E.S., 2001. Interspecific competition, N use and interference with weeds in pea–barley intercropping. Field Crops Research 70, 101–109.

Heckman, D.S., Geiser, D.M., Eidell, B.R., Stauffer, R.L., Kardos, N.L., Hedges, S.B., 2001. Molecular evidence for the early colonization of land by fungi and plants. Science 293, 1129–1133.

Heppel, K.B., Shumway, D.L., Koide, R.T., 1998. The effect of mycorrhizal infection of *Abutilon theophrasti* on competitiveness of offspring. Functional Ecology 12, 171–175.

Hooks, C.R.R., Johnson, M.W., 2001. Broccoli growth parameters and level of head infestations in simple and mixed plantings: impact of increased flora diversification. Annals of Applied Biology 138, 269–280.

IFOAM, 2009. Principles of Organic Agriculture. International Federation of Organic Agriculture Movements (IFOAM), p. 4. Online at: http://www.ifoam.org/organicfacts/principles/pdfs/IFOAM-FS-Principles-for Website.pdf.

Itulya, F.M., Aguyoh, J.N., 1998. The effects of intercropping kale with beans on yield and suppression of redroot pigweed under high altitude conditions in Kenya. Experimental Agriculture 34, 171–176.

Jannink, J.-L., Orf, J.H., Jordan, N.R., Shaw, R.G., 2000. Index selection for weed suppressive ability in soybean. Crop Science 40, 1087–1094.

Jordan, N.R., Zhang, J., Huerd, S., 2000. Arbuscular-mycorrhizal fungi: potential roles in weed management. Weed Research 40, 397–410.

Jordan, N., Huerd, S., 2008. Effects of soil fungi on weed communities in a corn-soybean rotation. Renewable Agriculture and Food Systems 23, 108–117.

Karlen, D.L., Varvel, G.E., Bullock, D.G., Cruse, R.M., 1994. Crop rotations for the 21st century. Advances in Agronomy 53, 1–45.

Kaschuk, G., Hungria, M., Andrade, D.S., Campo, R.J., 2006. Genetic diversity of rhizobia associated with common bean (*Phaseolus vulgaris* L.) grown under no-tillage and conventional systems in Southern Brazil. Applied Soil Ecology 32, 210–220.

Koide, R.T., Lu, X.H., 1992. Mycorrhizal infection of wild oats: maternal effects on offspring growth and reproduction. Oecologia 90, 218–226.

Koide, R.T., Shumway, D.L., Mabon, S.A., 1994. Mycorrhizal fungi and reproduction of field populations of *Abutilon theophrasti* (Malvaceae). New Phytologist 126, 123–130.

Lazcano, C., Arnold, J., Tato, A., Zaller, J.G., Domínguez, J., 2009. Compost and vermicompost as nursery pot components: effects on tomato plant growth and morphology. Spanish Journal of Agricultural Research 7, 944–951.

Lazcano, C., Sampedro, L., Zas, R., Domínguez, J., 2010. Vermicompost enhances germination of the maritime pine (*Pinus pinaster* Ait.). New Forests 39, 387–400.

Lee, Y.H., Kim, H., 2011. Response of soil microbial communities to different farming systems for upland soybean cultivation. Journal of the Korean Society for Applied Biological Chemistry 54 (3), 423–433.

Lemainski, J., da Silva, J.E., 2006. Avaliação agronômica e econômica da aplicação de biossólido na soja. Pesquisa Agropecuária Brasileira. 41, 1477–1484. http://dx.doi.org/10.1590/S0100-204X2006001000004.

Lemerle, D., Gill, G.S., Murphy, C.E., Walker, S.R., Cousens, R.D., Mokhtari, S., Peltzer, S.J., Coleman, R., Luckett, D.J., 2001. Genetic improvement and agronomy for enhanced wheat competitiveness with weeds. Australian Journal of Agricultural Research 52, 527–548.

Leslie, A.R., Cuperus, G., 1993. Successful Implementation of Integrated Pest Management for Agricultural Crops. CRC Press, Boca Raton, FL, p. 193.

Liebman, M., 1988. Weed suppression in intercropping: a review. In: Altieri, M.A., Liebman, M. (Eds.), Weed Management in Agroecosystems: Ecological Approaches. CRC Press, Boca Raton, FL, pp. 197–210.

Liebman, M., Dyck, E., 1993. Crop rotation and intercropping strategies for weed management. Ecological Applications 3, 92–122.

Lutgen, E.R., Rillig, M.C., 2004. Influence of spotted knapweed (*Centaurea maculosa*) management treatments on arbuscular mycorrhizae and soil aggregation. Weed Science 52, 172–177.

Linker, H.M., Orr, D.B., Barbercheck, M.E., 2009. Insect Management on Organic Farms. North Carolina Cooperative Extension Service, NC University, NC, USA.

Lutman, P.J.W., Cussans, G.W., Wright, K.J., Wilson, B.J., McN Wright, G., Lawson, H.M., 2002. The persistence of seeds of 16 weed species over six years in two arable fields. Weed Research 42, 231–241.

Manna, M.C., Singh, M., Kundu, S., Tripathi, A.K., Takkar, P.N., 1996. Growth and reproduction of the vermicomposting earthworm *Perionyx excavatus* as influenced by food materials. Journal Biology and Fertility of Soils 24, 129–132.

Marler, M.J., Zabinski, C.A., Callaway, R.M., 1999. Mycorrhizae indirectly enhance competitive effects of an invasive forb on a native bunchgrass. Ecology 80, 1180–1186.

Martens, M.H.R., 2000. Organic Soybean Weed Control. Acres USA.

McWhorter, C.G., 1972. Flooding for johnsongrass control. Weed Science 20, 238–241.

Melander, B., Rasmussen, G., 2001. Effects of cultural methods and physical weed control on intrarow weed numbers, manual weeding and marketable yield in direct-sown leek and bulb onion. Weed Research 41, 491–508.

Michaud, J.P., Sloderbeck, P.E., Nechols, J.R., 2008. Biological Control of Insect Pests on Field Crops in Kansas. Kansas State University, USA.

Mohler, C.L., 2001. Enhancing the competitive ability of crops. In: Liebman, M., Mohler, C.L., Staver, C.P. (Eds.), Ecological Management of Agricultural Weeds. Cambridge University Press, Cambridge, UK, pp. 269–321.

Montanez, A., Danso, S.K.A., Hardarson, G., 1995. The effect of temperature on nodulation and nitrogen fixation by five *Bradyrhizobium japonicum* strains. Applied Soil Ecology 2, 165–174.

Norman, J.R., Atkinson, D., Hooker, J.E., 1996. Arbuscular mycorrhizal fungal induced alteration to root architecture in strawberry and induced resistance to the root pathogen *Phytophthora fragariae*. Plant and Soil 185, 191–198.

Nunes, U.R., Silva, A.A., Reis, M.S., Sediyama, C.S., Sediyama, T., 2003. Soybean seed osmoconditioning effect on the crop competitive ability against weeds. Planta Daninha 21, 27–35.

Obaton, M., Bouniols, A., Piva, G., Vadez, V., 2002. Are *Bradyrhizobium japonicum* stable during a long stay in soil? Plant and Soil 245, 315–326.

Okon, Y., Labandera-Gonzalez, C.A., 1994. Agronomic applications of *Azospirillum*. In: Ryder, M.H., Stephens, P.M., BOWen, G.D. (Eds.), Improving Plant Productivity with Rhizosphere Bacteria. Commonwealth Scientific and Industrial Research Organization, Adelaide, Australia, pp. 274–278.

Ortas, I., Kaya, Z., Cakmak, I., 2001. Influence of VA-Mycorrhiza inoculation on growth of maize and green pepper plants in phosphorus and zinc deficient soils. In: Horst, W.J., Schenk, M.K., Bürkert, A., Claassen, N., Flessa, H., Frommer, W.B., Goldbach, H.E., Olfs, H.W., Römheld, V., Sattelmacher, B., Schmidhalter, U., Schubert, S., von Wirén, N., Wittenmayer, L. (Eds.), Plant Nutrition-Food Security and Sustainability of Agro-Ecosystems, Kluwer Academic Publishers, Dordrecht, Netherlands, pp. 632–633.

Parthasarathi, K., Ranganathan, L.S., Anandi, V., Zeyer, J., 2007. Diversity of microflora in the gut and casts of tropical composting earthworms reared on different substrates. Journal of Environmental Biology 28, 87–97.

Pester, T.A., Burnside, O.C., Orf, J.H., 1999. Increasing crop competitiveness to weeds through crop breeding. Journal of Crop Production 2, 59–76.

Pimentel, D., Burgess, M., 2016. An Environmental, Energetic and Economic Comparison of Organic and Conventional Farming Systems. Springer Science+Business Media, Dordrecht, Netherland.

Plenchette, C., Clermont-Dauphin, C., Meynard, J.M., Fortin, J.A., 2005. Managing arbuscular mycorrhizal fungi in cropping systems. Canadian Journal of Plant Science 85 (1), 31–40.

Reberg-Horton, S.C., Burton, J.D., Danehower, D.A., Ma, G.Y., Monks, D.W., Murphy, J.P., Ranells, N.N., Williamson, J.D., Creamer, N.G., 2005. Changes over time in the allelochemical content of ten cultivars of rye (*Secale cereale* L.). Journal of Chemical Ecology 31, 179–193.

Redecker, D., Kodner, R., Graham, L.E., 2000. Glomalean fungi from the Ordovician. Science 289, 1920–1921.

Rezvani, M., Zaefarian, F., Jovieni, M., 2013. Weed suppression ability of six soybean [*Glycine max* (L.) Merr.] varieties under natural weed development conditions. Acta Agronomica Hungarica 61 (1), 43–53.

Rinaudo, V., Bàrberi, P., Giovannetti, M., van der Heijden, M.G.A., 2010. Mycorrhizal fungi suppress aggressive agricultural weeds. Plant and Soil 333, 7–20.

Rodriguez, H., Fraga, R., 1999. Phosphate solubilizing bacteria and their role in plant growth promotion. Biotechnology Advances 17, 319–339.

Schüßler, A., Schwarzott, D., Walker, C., 2001. A new fungal phylum, the *Glomeromycota*: phylogeny and evolution. Mycological Research 105, 1413–1421.

Scott, T.W., Pleasant, J.M., Burt, R.F., Otis, D.J., 1987. Contributions of ground cover, dry-matter, and nitrogen from intercrops and cover crops in a corn polyculture system. Agronomy Journal 79, 792–798.

Shalaby, A.M., Hanna, M.M., 1998. Preliminary studies on interactions between VA mycorrhizal fungus *Glomus mosseae*, *Bradyrhizobium japonicum* and *Pseudomonas syringae* in soybean plants. Acta Biochimica Polonica 47, 385–391.

Shumway, D.L., Koide, R.T., 1994. Within-season variability in mycorrhizal benefit to reproduction in *Abutilon theophrasti*. Plant Cell and Environment 17, 821–827.

Singh, A.V., Shah, S., Prasad, B., 2010. Effect of phosphate solubilizing bacteria on plant growth promotion and nodulation in soybean (*Glycine max* (L.) Merr.). Journal of Hill Agriculture 1 (1), 35–39.

Singh, R., Sharma, R.R., Kumar, S., Gupta, R.K., Patil, R.T., 2008. Vermicompost substitution influences growth, physiological disorders, fruit yield and quality of strawberry (*Fragaria x ananassa* Duch.). Bioresource Technology 99, 8507–8511.

Smith, S.E., Read, D.J., 2008. Mycorrhizal Symbiosis. Academic Press, London.

Stefan, M., Dunca, S., Olteanu, Z., Oprica, L., Ungureanu, E., Hritcu, L., Marius Mihasan, M., Cojocaru, D., 2010. Soybean [*Glycine max* (L.) Merr.] inoculation with *Bacillus pumilus* RS3 promotes plant growth and increases seed protein yield: relevance for environmentally-friendly agricultural applications. Carpathian Journal of Earth and Environmental Sciences 5 (1), 131–138.

Teasdale, J.R., Mohler, C.L., 2000. The quantitative relationship between weed emergence and the physical properties of mulches. Weed Science 48, 385–392.

Tester, M., Smith, S.E., Smith, F.A., 1987. The phenomenon of "non-mycorrhizal" plants. Canadian Journal of Botany 65, 419–431.

Tollenaar, M., Nissanka, S.P., Aguilera, A., Weise, S.F., Swanton, C.J., 1994. Effect of weed interference and soil nitrogen on four maize hybrids. Agronomy Journal 86, 596–601.

Uchino, H., Iwama, K., Jitsuyama, Y., Ichiyama, K., Sugiura, E., Yudate, T., Nakamura, S., Gopal, J., 2012. Effect of interseeding cover crops and fertilization on weed suppression under an organic and rotational cropping system. 1. Stability of weed suppression over years and main crops of potato, maize and soybean. Field Crops Research 127, 9–16.

Uchino, H., Iwama, K., Jitsuyama, Y., Yudate, T., Nakamura, S., Gopal, J., 2015. Interseeding a cover crop as a weed management tool is more compatible with soybean than with maize in organic farming systems. Plant Production Science 18 (2), 187–196.

Uchino, H., Iwama, K., Jitsuyama, Y., Yudate, T., Nakamura, S., 2009. Yield losses of soybean and maize by competition with interseeded cover crops and weeds in organic-based cropping systems. Field Crops Research 113, 342–351.

Unkovich, M.J., Pate, J.S., 2000. An appraisal of recent field measurements of symbiotic N_2 fixation by annual legumes. Field Crops Research 65, 211–228.

Vandermeer, J., 1989. The Ecology of Intercropping. Cambridge University Press, Cambridge, UK.

Veiga, R.S.L., Jansa, J., Frossard, E., Van der Heijden, M.G.A., 2011. Can arbuscular mycorrhizal fungi reduce the growth of agricultural weeds? PLoS One 6, e27825.

Vollmann, J., Wagentristl, H., Hartl, W., 2010. The effects of simulated weed pressure on early maturity soybeans. European Journal of Agronomy 32, 243–248.

Willer, H., Kilcher, L., 2009. The World of Organic Agriculture. Statistics and Emerging Trends 2009. IFOAM FiBL Report. IFOAM, Bonn, FiBL, Frick and ITC, Geneva, p. 304.

Wortman, S.E., Drijber, R.A., Francis, C.A., Lindquist, J.L., 2013. Arable weeds, cover crops, and tillage drive soil microbial community composition in organic cropping systems. Applied Soil Ecology. 72, 232–241. www.attra.ncat.org.

Xavier, L.J.C., Boyetchko, S.M., 2004. Arbuscular Mycorrhizal Fungi in Plant Disease Control. Marcel Dekker, Inc.

Zaefarian, F., Aghaalikhani, M., Rahimian Mashhadi, H., Zand, E., Rezvani, M., 2007. Yield and yield components response of corn/soybean intercrop to simultaneous competition of redroot pigweed and jimson weed. Iranian Journal of Weed Science 3 (1–2), 39–58.

Zambolin, L., Schenck, N.C., 1983. Reduction of the effects of pathogenic root-infecting fungi on soybean by the mycorrhizal fungus, *Glomus mosseae*. Phytopathology 73, 1402–1405.

Zhang, F., Dashti, N., Hynes, R.K., Smith, D.L., 1996. Plant growth promoting rhizobacteria and soybean [*Glycine max* (L.) Merr.] nodulation and nitrogen fixation at suboptimal root zone temperatures. Annals of Botany 77, 453–459.

Zhu, Y.H., et al., August 17, 2000. Genetic diversity and disease control in rice. Nature 718–722.

Zimdahl, R.L., 1993. Fundamentals of Weed Science. Academic Press, New York.

Soybeans and Plant Hormones

M. Miransari
AbtinBerkeh Scientific Ltd. Company, Isfahan, Iran

Introduction

The soybean [*Glycine max* (Merr.) L.] is among the most important legumes, feeding a large number of people in the world. The soybean is used for the production of food, oil, protein, and biodiesel, and hence its production in different parts of the world is of great importance. The soybean is able to develop a symbiotic association with its symbiotic rhizobium, *Bradyrhizobium japonicum*, to acquire most of its essential nitrogen (N) for growth and yield production (Miransari, 2016).

Environmental stresses adversely affect plant growth and yield production. Large parts of agricultural fields are under stress or are subjected to some kind of stress, decreasing their potential yield production. Plants utilize different morphological and physiological mechanisms to alleviate stress. A set of plant factors including plant hormones, auxin, abscisic acid (ABA), ethylene, gibberellins, cytokinins, salicylic acid (SA), strigolactones, brassinosteroids (BRs), and nitrous (nitric) oxide are essential to make the plant tolerate the stress (Farooq et al., 2010; Hasanuzzaman et al., 2013).

Plant hormones were first defined by the German botanist Julius von Sachs (1832–97), who indicated that plants are able to produce substances, which are transported and hence perceived by plant tissues, resulting in the formation of different plant parts. Plant hormones are biochemicals with a wide range of functions in the plant. Plant hormones are effective in really small concentrations and make the communication between different plant cells, tissues, and parts likely (Kucera et al., 2005).

Soybean growth and its symbiotic association with rhizobium are adversely affected by environmental stresses. The soybean and rhizobium are not tolerant under stress, and hence the morphological and physiological evaluation of plants and bacteria under stress make the production of tolerant soybean plants more likely. Plant hormones are able to regulate plant morphology and physiology so that the plant can survive the stress. The important point about plant hormones is their interactions and cross talk affecting their activity and hence their effects on plant growth (Caba et al., 2000).

Seed germination is the most important morphological and physiological alteration, which results in the production of a new seedling and hence a complete plant. Such a process is regulated by different plant hormones and their interactions are reviewed by different researchers (Kucera et al., 2005; Miransari and Smith, 2014). The expression of genes in seeds, which results in the release of dormancy and germination of seeds, is regulated by environmental and hormonal signaling. ABA is a positive regulator of seed dormancy and maintenance, and results in the induction of dormancy; however, it is a negative regulator of seed germination. Gibberellins are able to result in the release of seed dormancy and induce the germination of seeds,

Environmental Stresses in Soybean Production. http://dx.doi.org/10.1016/B978-0-12-801535-3.00006-1

counteracting the activity of ABA. Similarly, ethylene and BRs are able to induce the germination of seeds (Davies, 2010).

It has been indicated that soybean nodulation is affected by plant hormones (Fig. 6.1); for example, plant hormones are able to affect the process of morphogenesis in soybean nodules (Hirsch, 1992; Hirsch and Fang, 1994; Long, 1996). The process of nodulation is controlled by two different mechanisms, including the one related to the process of inoculation and the one related to the process of autoregulation, which is the regulation of nodule production by plant hormones (Delves et al., 1986). Accordingly, for the proper performance of nodulation, the timing and concentration of plant hormones and other signaling molecules must be regulated.

Thimann (1936) was among the first who indicated the role of plant hormones, specifically auxin, in the process of nodulation. The role of soil microbes, including rhizobium in the production of hormones, was also expressed by researchers, although

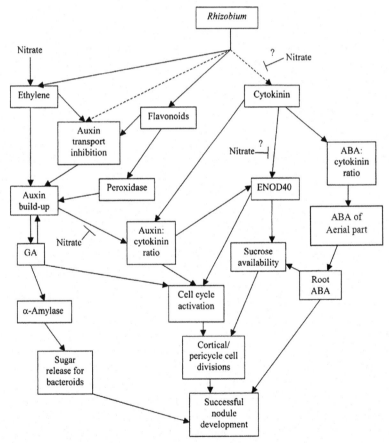

Figure 6.1 How the initiation of cell division and the development of nodules is regulated by the interaction of hormones and other signaling molecules (Ferguson and Mathesius, 2003). With kind permission from Springer, License number 3771320255851.

this did not indicate the specificity of the process of N fixation by the two symbionts. For example, Hamayun et al. (2010a) investigated the level of plant hormones in soybeans as affected by drought stress. The drought stress was induced using polyethylene glycol (8% and 16%) in a 14-day period at the pre- and postflowering stage. The maximum reduction of soybean growth resulted in polyethylene glycol at 16% during the preflowering stage. The rate of plant gibberellins (GA1 and GA4) decreased by stress; however, the rates of other plant hormones including jasmonate, ABA, and salicylate increased in plants under stress.

Different plant hormones including auxin, ABA, ethylene, cytokinin, strigolactones, and gibberellins affect phosphorous sensing and the related pathway (Chiou et al., 2006; Chiou and Lin, 2011; Zhang et al., 2014). Under P-deficient conditions, the sensitivity, production, and transport of hormones in plants change (Chiou and Lin, 2011).

Hamayun et al. (2010b) also investigated the effects of different levels of salinity on the growth and hormone levels of soybean genotype Hwangkeumkong. Salinity stress (70 and 140mM) significantly affected soybean growth attributes including biomass, plant length, weight of 100 seeds, number of pods, yield, and chlorophyll content. Salinity stress decreased the rate of gibberellins and SA and significantly increased the rate of ABA and jasmonate. The authors accordingly indicated that the adverse effects of salinity on plant growth are by affecting the level of plant hormones.

Under biotic stresses, such as the presence of pathogenic microbes, plants induce a systemic tolerance (ISR) to alleviate the stress (Schenk et al., 2012). The levels of plant hormones and hence ISR in *Arabidopsis thaliana* as affected by *Pseudomonas fluorescens* was investigated by Pieterse et al. (2000). Because the level of jasmonate and ethylene remained constant in *A. thaliana*, the authors accordingly indicated that the induction of ISR by *P. fluorescens* is not a function of jasmonate and ethylene changes in *A. thaliana*. However, the rate of 1-aminocyclopropane-1-carboxylate production in plants increased, indicating that the bacteria is able to enhance the level of ethylene in plants in the presence of pathogenic microbes.

The process of nodulation and the presence of nitrate results in the alteration of plant hormones in legumes (Caba et al., 2000; Kinkema and Gresshoff, 2008). It has been indicated that the mutation of *GmNARK* in soybean plants affects plant hormones (Gresshoff, 2003; Kinkema and Gresshoff, 2008), and it may also affect the other plant biochemicals including signals and metabolites used by the microbes in plant aerial parts. Accordingly, it may be possible to determine the presence and combination of microbes in plant aerial parts with respect to the presence and concentrations of biochemicals in plant aerial parts.

It has been indicated that the degree of nodulation in legumes is controlled by different biochemicals including plant hormones (Ferguson and Mathesius, 2003). For example, ABA and ethylene suppress the number of nodules (Oldroyd et al., 2001; Ferguson and Mathesius, 2003). Such effects of plant hormones have been attributed to their alteration of early root response. For example, ABA suppresses the production of lateral roots. The rate of ABA/cytokinin also affects the process of autoregulation (Caba et al., 2000). It has been accordingly indicated that the longevity of nodules is controlled in a similar manner to the mechanisms, which control the production and the number of nodules by plant hormones and other signaling metabolites (Oldroyd et al., 2001; Ferguson and Mathesius, 2003).

The main function of rhizobium for the enhancement of plant growth is by the process of N fixation in symbiotic association with their host plant. However, some other functions are also done by such plant growth-promoting rhizobacteria including the production of different biochemicals such as falvonoid, lipochitooligosaccharides (LCOs), organic products such as phenol, and plant hormones including auxin, gibberellins, cytokinins, etc. (Table 6.1). (Avis et al., 2008; Mabood et al., 2014). Such activities result in controlling plant pathogens, increasing the solubility of organic and inorganic phosphate, and producing iron-chelating siderophore (Dakora, 2003; Mishra et al., 2006).

The effects of waterlogging on the vegetative growth of soybean plants was investigated by Alam et al. (2010) using the proteomic analyses of plant roots. The soybean seedlings were subjected to the stress of waterlogging for 3 and 7 days. Stress increased the in vivo level of lipid peroxidation and the H_2O_2 content of the roots. A total of 24 proteins were recognized in soybean roots, among which 14 were upregulated, 5 were downregulated, and 5 were newly induced the soybean roots under stress. The recognized proteins include the well-known proteins, which are produced under anaerobic conditions, and new proteins, which have not been recognized previously under waterlogging conditions.

The newly recognized proteins can regulate different plant activities including signaling processes, redox homeostasis, processing of RNA, and cellular metabolism. The increased production of proteins, including glycolysis and fermentation enzymes, under waterlogging conditions suggests that due to the absence of oxygen, soybean plants acquire the essential ATP for their activities by the process of fermentation. The proteomic analyses indicated the presence of new proteins under waterlogging stress including auxin-amidohydrolase, apyrase, and coproporphyrinogen oxidase, which can make the plant alleviate the stress (Alam et al., 2010).

Table 6.1 The Positive (+) or Negative (−) Role of Plant Hormones in the Process of Nodulation

Hormone	Nodulation	References
Auxin	+	Deinum et al. (2012)
Ethylene	−	Penmetsa and Cook (1997) and Oldroyd et al. (2001)
Gibberellins	+ At optimum range	Ferguson et al. (2005)
Abscisic acid	+/−	Biswas et al. (2009)
Brassinosteroid	+ At optimum range	Ferguson et al. (2005)
Strigolactone	+	Tirichine et al. (2007)
Cytokinin	+	Foo and Davies (2011)
Jasmonic acid	+/−	Kinkema and Gresshoff (2008)
Salicylic acid	−	Sato et al. (2002)

Zhang et al. (2013) investigated the effects of shade and drought stress on the level of plant hormones in soybeans under greenhouse conditions. The pots were subjected to the shade and normal light treatments. In the first year of the experiment the plants were subjected to the control and a moderate level of irrigation, and in the second year and at the stage of branching the plants were subjected to the drought stress. The stress of shade enhanced the height of seedling and the length of first internode and decreased the stem diameter; the concentration of auxin and gibberellins increased, and the concentration of ABA and zeatin decreased. The authors accordingly indicated that the yield reduction under shade and drought stress was due to the reduction of main stem yield.

In the following some of the most important findings related to the effects of plant hormones on the growth and activities of soybeans and its symbiotic rhizobium are presented, reviewed, and analyzed.

Auxins and Stress

Auxin is a plant hormone with multiple roles in plant growth and development, including the division and differentiation of cells and production of vascular bundle, also taking place during the production of nodulation. The highest rate of auxin is produced in the aerial part and is allocated to the roots by an active process of transport. The hormone is imported to the cell by an auxin import protein (AUX1) and exported from the cell by an export protein (PIN1). The other important role of auxin is regulating the expression of aquaporins in the sites of root lateral production. The hormone is also able to suppress the transport of water at the cellular and the root level. Accordingly, auxin is able to facilitate the direction of water to the plant lateral roots (Péret et al., 2012; Li et al., 2014).

The effects of nitrate and inoculation on the rate of different plant hormones including auxin, ABA, and cytokinins in the roots of different soybean genotypes including Bragg and its super nodulating mutant *nts382* was investigated by Caba et al. (2000) for the first time. Significant differences in the rate of hormones in the roots of the two soybean genotypes were found 48 h following inoculation with *B. japonicum*. A higher rate of cytokinins (30–196%) was found in the roots of the soybean mutant. However, the rate of ABA was higher at a twofold time in Bragg than the mutant genotype, and the rate of auxin was not different in the soybean genotypes. The rate of auxin, ABA, and cytokinins increased in the roots of Bragg fed with 1 mM NO_3^-, 48 h following inoculation with *B. japonicum*, in relation to the uninoculated treatment.

However, in the mutant, the rate of cytokinins increased. The rate of root auxin was significantly decreased by the high concentration of NO_3^- (8 mM). As a result, there were no changes in the rate of auxin in the two genotypes and by inoculation in Bragg genotype. However, there were little differences in the concentration of ABA and cytokinins by 8 mM NO_3^-. The rate of auxin/cytokinins and ABA/cytokinins was higher in Bragg than the mutant genotype and was affected by inoculation (just in Bragg) and nitrate (in both genotypes). The results are consistent with respect to the process of soybean supernodulation and autoregulation. However, more investigation is essential with respect to the adverse effects of nitrate on the process of soybean nodulation.

The allocation of auxin in plants is regulated by the presence of flavonoids (Peer and Murphy, 2007). Flavonoids are able to regulate the transport of auxin in plant (Buer and Muday, 2004; Buer et al., 2010). Among the alleviating effects of auxin in plants under stress is the production of new roots. For example, the formation of adventitious roots at the base of the plant aerial part is an important and quick plant response to stress under waterlogging conditions. Under such conditions the higher rate of auxin may be produced and translocated to the roots; the increased production of auxin-amidohydrolase under such conditions can result in the higher release of auxin, and hence the plant can be more tolerant under stresses such as waterlogging.

Abscisic Acid and Stress

ABA is among the most important plant hormones regulating different plant activities under different conditions including biotic and abiotic stresses. Such activities include seed growth and development, delayed seed germination and dormancy, stomatal activities, protein and lipid synthesis, and leaf senescence (Tuteja, 2007). For example, under drought and salinity stress conditions, ABA regulates the activity of the aquaporins, which affect the leaf hydraulic conductance (Li et al., 2014). The hormone is able to reduce the leaf hydraulic conductivity in *Arabidopsis* by the downregulation of aquaporins in the bundle sheath cells. However, auxin decreases the hydraulic activity of root cortical cells. SA is also able to decrease the activity of aquaporins by a mechanism related to reactive oxygen species. Gibberellins and BRs are also able to regulate the expression of aquaporins by unknown mechanisms (Li et al., 2009; Peret et al., 2012).

The important function of ABA is under stresses such as salinity and drought. Although not under all stress conditions, the ABA level increases in plants, and it has been indicated that ABA is able to regulate the activities and expression of different stress genes under drought and salinity stress. The ABA mutants are able to grow under nonstress conditions; however, under stress conditions, their growth significantly decreases (Hasanuzzaman et al., 2013).

It has been indicated that ABA suppresses the process of nodulation. Use of ABA decreased the rate of nodulation in wild soybeans and its supernodulating mutant line NOD1-3 by decreasing the number and weight of nodules and the level of flavonoid. It was also indicated that ABA suppresses the initiation, functioning, and development of nodules in wild soybeans and NOD1-3. The increased concentration of ABA in different parts of mycorrhizal soybeans and in the nodules has been indicated. Although ABA can suppress the process of nodulation, it is able to allocate the organic products including photosynthates to the nodules to be used by the bacteria and hence for the process of biological N fixation. The bacterial production of ABA can make the plant allocate a higher rate of photosynthates to the nodules (Ferguson and Mathesius, 2003).

With respect to the above-mentioned details, ABA activity in the process of biological N fixation can be evaluated from two different sides: (1) ABA can suppress the process of nodulation in soybean and (2) deficient rates of ABA in soybean may decrease the efficiency of N fixation by adversely affecting the activity of bacteria. The interactions of these two processes may explain why in some cases ABA does not affect the process of nodulation and hence biological N fixation. However, it has been shown that increasing levels of ABA decrease the activity of enzymes essential for the process of biological N fixation (Fig. 6.2).

Exogenous application of ABA decreases the rate of biological N fixation; such a reduction may also decrease the content of nodule hemoglobin and limits the rate of oxygen essential for bacterial activity; as a result, this reduces the level of biological N fixation. The increased rate of ABA in the older nodules related to the younger nodules indicates that besides its other roles in the process of biological N fixation, ABA may also have a role in nodule senescence (Hancock et al., 2011).

Ethylene and Stress

Ethylene as a gaseous plant hormone is able to affect different plant growth and developmental activities, including the regulation of induced nodulation activities in legumes (Gresshoff et al., 2009). Among the most important activities of ethylene during the process of nodulation is the autoregulation of nodulation. Such a process interacts with the signaling of ethylene and cytokinin.

The stages of nodule development, including the activities of plant hormones, are (1) rhizobium inoculation of root hairs, with the production of ethylene, reactive oxygen species, and ENOD40 in the precycle cells just a few hours following inoculation; (2) in the cortex cells, which will eventually produce nodules, the expression of GH3

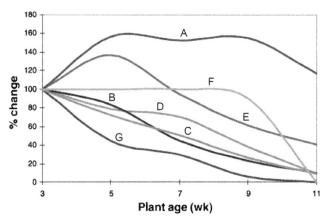

Figure 6.2 Time development of legume root nodules at different growth stages for pea nodules. A: nitrogenase activity, B: total ascorbate content, C: total glutathione content, D: total protein content, E: catalase activities, F: relative transcript of leghemoglobin level (isoform 120-2), G: glutathione reductase activities (Puppo et al., 2005).
With kind permission from Wiley, License number 3771330005326.

and ENOD40 and the concentration of flavonoid increases; (3) the division of cortical cells at the early stages, resulting in the increased expression of GH3, AUX1, and ENOD40 and the enhanced concentration of cytokinin and flavonoids; and (4) at the stage of nodule differentiation, the level of auxin, ABA, gibberellins, and nitrous oxide increases. The expression of GH3, AUX1, and ENOD40 is done in the peripheral (probably vascular) tissue. The level of ABA and cytokinin increases in the nodule cells (Ferguson et al., 2010).

It has been indicated that the rate of auxin/cytokinin in the roots initiates the division of cortical cells and the production of nodules. The rate of auxin:cytokinin decreased in the *nts*386 genotype compared with the wild type, indicating that such a balance is important for the regulation of nodule numbers. Accordingly, the symbiosis with rhizobium can affect the auxin level and hence auxin:cytokinin in plant roots. It has been indicated that auxin has two different functions in the process of nodulation: (1) during the early stages of nodulation, the suppression of auxin allocation decreases the rate of auxin:cytokinin and hence causes the start of the cellular division; and (2) at the later stages of nodulation the process of cellular division is suppressed by the increased levels of super optimal auxin (Spaepen and Vanderleyden, 2011) (Fig. 6.3).

Interestingly and similar to the effect of ethylene, Nod factors or oligosaccharides affect the affinity of transporters to the binding sites or Nod factors, which induces the production of transporter inhibitors. Similarly, other products such as ethylene, flavonoid, and cytokinin are also able to suppress the transfer of auxin and regulate the activities of different peroxidases and IAA oxidase, which is able to catalyze auxin (Murphy et al., 2000).

Ethylene and nitrate decreases the number of nodules (Fig. 6.4). Under stress the production of ethylene significantly increases. It is possible that under stress the presence of a mechanism prevents the use of assimilates by nodules (Gresshoff et al., 2009). It is also likely that because the main nutrient, which is fixed by the process of symbiosis between the plant and the bacteria, is N, there is a mechanism, which prevents the plants from forming a symbiotic association with rhizobium under sufficient levels of N. If the plant's ability to perceive ethylene and nitrate is mutated, the plant will be able to alleviate the adverse effects of such factors on the number of nodules, and hence the number of nodules increases. The genes, which are affected by mutation, are the ones that determine ethylene response and sensitivity including *LjEIN2/MtEIN2*.

Jasmonate and Stress

Methyl jasmonate is a plant signal regulating plant morphogenesis and responses to abiotic and biotic stresses. Accordingly, the role of methyl jasmonate in the development of genetically modified (GM) soybeans was investigated by Xue and Zhang (2007). The growth patterns of the leaf and root in the GM soybean and wild-type soybean were significantly different. The leaf of the GM soybean was a little longer in length but significantly wider in width compared with the wild-type soybean. The growth of main roots was also suppressed in GM soybeans related to wild types;

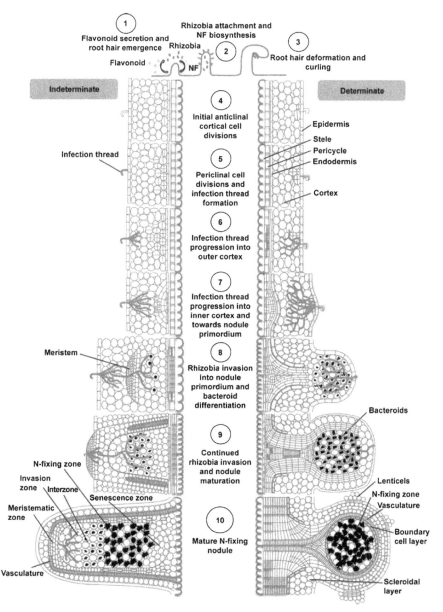

Figure 6.3 The developmental stages of soybeans (determinate: right) and peas (indeterminate: left) nodules (Ferguson et al., 2010). *NF*, nod factor.
With kind permission from Wiley, License number 3771351402285.

however, the growth of the lateral roots was stimulated in GM soybeans compared with the wild types. There was a 2–2.5 times higher rate of methyl jasmonate in the GM soybean than the wild type, indicating the role of methyl jasmonate in the process of morphogenesis in soybeans (Xue and Zhang, 2007).

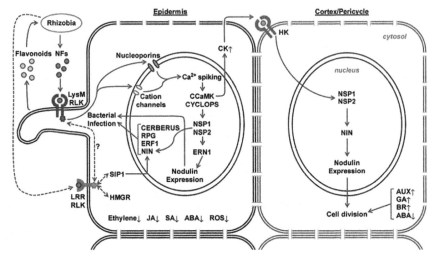

Figure 6.4 Molecular processes related to the early stages of nodulation (Ferguson et al., 2010). *NF*, nod factor.
With kind permission from Wiley, License number 3771360438745.

The effects of jasmonate on the alleviation of salinity stress in soybean plants were evaluated using a hydroponic medium (Yoon et al., 2009). Soybean seedlings, which had been treated with methyl jasmonate at 20 and 30 μM, were subjected to salinity stress using 60 mM NaCl for 14 days. Salinity stress significantly decreased plant growth, the rate of gibberellins, and the rate of photosynthesis and transpiration. However, the stress resulted in a significant increase in the rate of ABA and proline content. The results of the experiment indicated that methyl jasmonate was able to alleviate the stress of salinity on plant growth by positively affecting plant morphology and physiology including plant growth, plant hormones, plant photosynthesis rate, chlorophyll and proline content, and rate of transpiration.

The number of nodules in soybeans is regulated by a signaling mechanism from the aerial part to the roots. It has been indicated that the signaling in the roots inoculated with rhizobia suppress the process of nodulation in the uninoculated roots (Delves et al., 1986). However, the mutation of the nitrate-tolerant symbiosis (NTS) gene in soybeans, which activates the GmNARK kinase, results in the excessive development of nodules, a process called hypernodulation. Seo et al. (2007) examined the mutation of NTS on the gene activation of the soybean leaf. Interestingly, the expression of jasmonate-responsive genes including *vspA* and *vspB* were upregulated. The results also indicated that the level of jasmonate increased two or three times in the mutant compared with the wild type under natural conditions. The expression and activation of jasmonate persisted in the soybean leaf although the roots were not inoculated with rhizobia. The authors accordingly indicated that mutation of the NTS gene in soybeans increases the level of jasmonate and is not related to the process of root hypernodulation.

Jasmonate is also able to suppress plant response to rhizobia by affecting the process of calcium signaling, which induces the production of LCO, and results in the suppression of plant genes, which are essential for the earlier stages of nodulation (Sun et al., 2006). Although such results may indicate that methyl jasmonate may be essential for the process of nodule autoregulation in nodules including soybeans, as previously indicated, methyl jasmonate is not essential for the process of nodule auto-regulation in soybeans (Seo et al., 2007).

The effects of jasmonates on the process of *nod* gene induction in *B. japonicum*, with a genetically modified plasmid, were investigated (Mabood and Smith, 2005). The jasmonates were able to strongly induce the expression of the nod genes. The jasmonates did not suppress the activity of *B. japonicum* although genistein suppressed their activities. The authors also investigated the effects of *B. japonicum* induced by jasmonates on the process of soybean nodulation and biological N fixation under optimal (25°C) and suboptimal (17°C) root zone temperature.

The induction of *B. japonicum* was conducted using a liquid yeast extract mannitol medium used for the growth of bacteria by genistein and jasmonates. Soybean seedlings grown under the temperature of 25 and 17°C and the constant air temperature of 25°C were inoculated with the *B. japonicum* induced with different treatments. Their results indicated that genistein and jasmonates were able to enhance the process of nodulation, N fixation, and soybean growth under suboptimal root zone temperature (Mabood and Smith, 2005).

Genistein at a higher concentration suppressed soybean growth under natural conditions although it stimulated plant growth at 17°C. However, preincubated *B. japonicum* increased soybean nodulation and N fixation under both optimal and suboptimal root zone temperatures. The authors accordingly indicated that jasmonates, as a new class of signals in the process of symbiosis between *B. japonicum* and soybeans, are able to enhance soybean nodulation, N fixation, and growth under optimal and suboptimal root zone temperatures if used for the preincubation of *B. japonicum* (Mabood and Smith, 2005).

Gibberellins and Stress

Gibberellins are plant hormones with a wide range of activities including seed germination and cell elongation (Miransari and Smith, 2009; Hayashi et al., 2014). The hormones are also able to regulate the production of legume nodules indicated recently. The hormone is essential at two different stages of nodulation including the early stage of root colonization and the late stage of nodule production and maturity (Hayashi et al., 2014).

It has been indicated that the activity of gibberellin oxidase is induced in *B. japonicum* under the symbiotic conditions, and hence the bacteria are able to produce gibberellins under such conditions (Méndez et al., 2014). Gibberellin has also an important role in nodule senescence. The regulation of nodulation is by a loop of signaling from the roots to the aerial part and from the aerial part to the roots (Ferguson and Mathesius, 2014). Plant hormones including auxin, gibberellin, ethylene, BR, strigolactone, cytokinin, ABA, jasmonic acid, and SA can positively or negatively affect the process of nodulation in legumes.

The seed germination of wheat plants was enhanced by gibberellins under salinity stress; although wheat growth and yield was decreased by salinity stress, gibberellins were able to alleviate the stress (Hasanuzzaman et al., 2013). The foliar use of gibberellins (100 mg/kg) increased sugarcane nutrient uptake by improving plant morphology and physiology under salt stress. It has been indicated that the exogenous use of gibberellins can alleviate the adverse effects of salinity on the growth and yield of soybeans by regulating the level of other hormones in soybeans. Accordingly, it was indicated that in the treated soybean plants the level of bioactive gibberellins and jasmonate increased and the level of ABA and SA decreased. Although the exact details related to the alleviating effects of gibberellins on plant growth under salinity stress have not yet been elucidated, hormonal homeostasis may be the most important reason for enhancing plant growth under salinity stress (Hasanuzzaman et al., 2013).

Cytokinins and Stress

The senescence of soybean flowers during plant development is a prevalent process. The number of pods in a plant is affected by the presence of assimilates and cytokinin. Yashima et al. (2005) investigated the effects of source/sink rate using cytokinin at nodes on the number of pods. With increasing the rate of source/sink, the number of pods per node of plants increased curvilinearly. By contrast, using cytokinins increased their number by increasing the number of flowers or by increasing the percentage of pod-set depending on the year. The authors accordingly indicated that the plant hormone cytokinin is able to increase the number of pods in soybean plants with a high rate of assimilate (Yashima et al., 2005).

It has also been indicated that with respect to the effects of cytokinin on the rate of source/sink photosynthates, source leaf senescence is delayed by cytokinin, compared with plants that had not been treated with cytokinin. The combination of cytokinin and nutrients result in the total prevention of leaf senescence. It is accordingly indicated that if the growth and activity of the root is maintained during the reproductive stage, the plant leaf can grow and be active for a longer time. The main effect of cytokinin on delaying the process of leaf senescence is by retarding the process of protein degradation rather than increasing the process of protein synthesis (Liu et al., 2008).

Experiments indicated that the level of auxin, ABA, and cytokinin was high in young bean seedlings and decreased quickly with plant growth. The development of the plant leaf is a strong sink for plant photosynthates and is similar to the effect of drought stress. ABA decreases leaf development and increases the translocation of photosynthates. The permeability of cellular membrane to solutes and water is improved by ABA, and under high levels of ABA the unloading of solutes from the phloem increases; however, ABA may suppress the cell development. The most probable reason for the transfer of ABA from the sink to the source results from a decrease in auxin activity, which as a result decreases the extension of the cell wall and the sink strength of the leaf (Liu et al., 2008). The level of gibberellins and cytokinin increased due to shading following flowering; however, the level of auxin was not affected.

With increasing the seed dry weight of soybeans, the rate of cytokinin increased. A high rate of ABA resulted in a high rate of soybean grain yield with a high rate of cell division. Gibberellins and auxin are the highest 15 days following the R5 stage, when the soybean rate of dry matter is the highest and the activity of sink and the unloading of phloem are the highest. During the process of seed development, there is a significant difference in the rate of ABA and auxin between different seed parts. However, it has not been indicated that the developing seeds are dependent on the presence of such hormones. The role of plant hormones is important during the period of environmental stress, as they act as signals causing soybean seeds to tolerate the stress. Accordingly, with respect to the effects of plant hormones on the growth of soybeans under different conditions, including stress, it is pertinent to use bioregulators, affecting the production of hormones by soybean plants (Liu et al., 2008).

Under stresses such as drought the role of phosphorous is important and can make the plant tolerate the stress. It is because phosphorous increases root growth, and as a result the roots will be able to grow in the soil and absorb water and nutrients. Jin et al. (2005) indicated that under the lower rate of rainfall, the rate of phosphorous must increase, so that plants can tolerate the stress by increasing its root growth, phosphorous uptake, and yield production.

The effects of drought stress at the R4 stage are more significant than the other stages of plant growth. The role of plant hormones is also significant under such conditions; for example, increased production of ABA enhances root growth although the growth of the aerial part decreases. Accordingly, root biomass and hence the rate of the root/aerial part increases (Lynch and Brown, 2008; Valliyodan and Nguyen, 2008).

A cytokinin receptor is essential for the development of nodules, indicating that cytokinins are essential for the organogenesis of nodules. Accordingly, it has been suggested that cytokinins act as the communicator of epidermal perception of nod factor to the inner parts of the roots. ABA has also been indicated a mobile signal during the process of nodulation. ABA is typically a suppressor of nodule development and can act in the epidermis and cortex (Ding and Oldroyd, 2009). The other plant hormones, including auxin, gibberellins, and BRs, can act as the positive regulator of nodule development and reactive oxygen species, and ethylene and jasmonate can act as the negative regulator of nodule development. However, more details related to the role of these hormones in the process of nodule development must be illustrated (Ferguson and Mathesius, 2003; Ferguson et al., 2005; Sun et al., 2006; Mathesius, 2008; Kinkema and Gresshoff, 2008), as the role of some of these hormones are in the division of cells, differentiation, and maintenance (Ferguson et al., 2010).

Different cell types must synchronize their activities so that the process of organogenesis results in the formation of nodules. When the nod factors are perceived in the epidermis, there will be fast responses in the inner root. The rearrangement of cellular structure and the expression of ENOD40 have been detected in the root cells 16–24 h following the perception of the Nod factor by plant cells. However, it is likely that such fast responses to the Nod factor is controlled by molecular signaling (Indrasumunar et al., 2011).

During the process of cell division in the roots, the functioning of a cytokinin receptor is essential. The receptor has a domain of histidine kinase and is activated by *MtCRE1/LjLHK1*. However, it has been indicated that if the activity of such histidine kinase is downregulated, the number of nodules significantly decreases, which is due to malfunctioning of the plant in the formation of nodules. Under such conditions, although the bacteria is able to colonize plant roots, due to the loss of directionality, the nodules grow to the sides rather than toward the root cortex (Ferguson et al., 2010).

If the activity of *Mtdmi3* is mutated, a nonnodulating phenotype is produced, because the activity of CCaMK in the epidermis is downregulated. However, if the CCaMK is activated, the controlled cell divisions in the cortex results in the spontaneous nodulation. The mutation also results in the activation of *ENOD11* similar to the time that the plant is colonized by rhizobium. Accordingly, CCaMK is essential for controlling the processes in the epidermis and the cortex in the presence of the essential molecular pathway (Libault et al., 2010).

The other important gene controlling the process of nodulation is NIN. The higher rate of root hair curling and suppressing the process of infection by rhizobia in the epidermis are the results of NIN mutation. In the cortex of the NIN mutant plants the process of cell division is mutated, and hence the nodules are not produced. The activity of NIN is also essential for the spontaneous nodulation of CCaMK and cytokinin receptor mutants. It has also been indicated that NIN is activated following the activity of CCaMK and the cytokinin receptor (Ferguson et al., 2010).

NIN is not essential for the *ENOD11* expression activated by the Nod factor. Hence, it may negatively regulate the Nod factor signaling, which is able to modify the activity of *ENOD11* in the root epidermis. NIN is also induced by using the Nod factor or cytokinin, indicating the positive regulation of cortical cell divisions. However, more details must be elucidated on the role of NIN in the process of cell divisions. That the presence of the cytokinin receptor is essential for the process of nodulation indicating cytokinin is an essential plant hormone for the process of nodulation and nodule organogenesis (Ferguson and Mathesius, 2003; Ferguson et al., 2005; Sun et al., 2006; Mathesius, 2008; Kinkema and Gresshoff, 2008).

Roots are the major site of cytokinin production, which can suppress the process of senescence resulting from flooding in *Arabidopsis*. Under waterlogging conditions, the reduced rate of oxygen is the main cause of decreased plant growth; however, the activation of some other process such as the increased concentration of the stress hormone, ethylene, may affect soybean morphology and physiology including root growth (Alam et al., 2010).

Salicylic Acid and Stress

SA is a phenolic plant hormone regulating different plant activities, including fruit ripening, seed germination, flowering, leaf senescence, and induced plant resistance under biotic and abiotic stress conditions. The hormone is able to enhance plant tolerance under different types of stresses, including salinity and heavy metals, by the regulation of redox homeostasis (Jibran et al., 2013). Among the important effects of SA is the regulation of leaf senescence. For example, the mutants of *Arabidopsis* plants, which are deficient in the production and signaling of SA, indicate the delay of leaf senescence. It must also be mentioned that the rate of SA increases at the time of decreased leaf content, which is an important controlling process during the latter part of leaf senescence.

Interestingly, it has been indicated that from the total of 18 genes in plant-producing SA and the related signaling pathway 12 are the genes essential for the process of leaf senescence, and they were accordingly upregulated during such a period. Besseau et al. (2012) have indicated that SA not only controls the progression of leaf senescence, it can also act as the negative and positive regulator of leaf senescence. The other important process regulated by SA in the process of leaf senescence is autophgy (ATG). ATG is an intercellular process for the degradation of vacuole in the cellular cytoplasm and is induced by SA. The interaction of ATG and SA results in the regulation of pathogen response, activity, and development (Table 6.2). In the mutants of ATG, including *atg5* and *atg2*, which are without ATG, the rate of SA and reactive oxygen species increases and the plant senesce early. SA regulates the onset and progression of leaf senescence by controlling the presence of positive and negative regulators.

Khan et al. (2003) investigated the effects of SA on the photosynthetic rates and growth of two different crop plants including soybeans (C3) and corn (C4) in a greenhouse experiment. The use of SA in both plants increased the photosynthetic rate, stomatal conductance, and the rate of transpiration. However, SA did not have effects on the rate of chlorophyll content. Although leaf area index and plant biomass were also affected by SA, plant height and root growth remain unaffected.

The exogenous use of SA can induce systemic induce resistance and activate the related genes. With respect to the properties of SA, Lian et al. (2000) investigated the effect of exogenous SA including (5, 1, 0.5, 0.1, and 0 mM) on the growth and nodulation of soybean seedlings by watering the seedlings' roots with SA or soaking the seedlings with SA. The use of 5 mM SA and not the other SA concentrations had adverse effects on the growth of seedlings. The soaking of seedlings also did not have negative effects on the growth and development of seedlings. The use of 5 mM SA suppressed the process of biological N fixation and hence the growth of seedlings grown in sterile soil.

Table 6.2 Species of Rhizobium-Controlling Plant Pathogens (Avis et al., 2008)

Rhizobium	Actions		
	IR	Sid	AC
R. leguminosarum bv. trifolii	+	+	+
R. leguminosarum bv. phaseoli	+	+	
R. leguminosarum bv. viceae	+	+	
R. etli	+		
R. meliloti	+	+	

IR, induced resistance; *Sid*, siderophore production; *AC*, production of antimicrobial compounds.

The hormone, along with ABA, methyl jasmonate, and polyamine, is able to regulate the process of nodule autoregulation in legumes. Different signals are essential for the process of autoregulation. Using the nonsymbiotic mutants is a suitable method for the determination of signal molecules, which are activated during the process of nodule autoregulation (Sato et al., 2002; Tanabata and Ohyama, 2014).

Strigolactones and Stress

Although the role of strigolcatones in the process of plant mycorrhization have been indicated (Bonfante and Genre, 2010), their important functions in the process of nodulation have also been shown. The rates of soil N and P are as important regulators of the hormone production activities in plants. However, N deficiency in certain species may result in the production of the strigolcatones hormone (Foo et al., 2013). Nitrate presence in soil regulates the number of nodules in legumes. Such a process is affected by the activity of nodulation genes in plants. Interestingly, in nonlegumes with a dependency on the process of mycorrhization, the increased production of strigolcatones may increase the process of mycorrhization. Because under nutrient-deficient conditions, there is a correlation between the high levels of strigolcatones and the high number of nodules, it is interesting to find how the presence of strigolcatones may affect the process of nodulation in legumes under such conditions.

It was indicated by Foo and Reid (2013) that strigolcatones are not essential for pea plants to suppress the number of nodules under a high nitrate level. Although under a high rate of nitrate, the rate of nodulation decreases in legumes, and decreased levels of P results in the reduction of nodulation in legumes (Tsvetkova and Georgiev, 2007). It has been indicated that strigolcatones are plant hormones and can also act as a rhizosphere signal; however, they are also able to regulate nutrient acquisition by plants. The results of Foo et al. (2013) showed that strigolcatones are essential for mycorrhization and nodulation; however, they do not affect plant response to the level of soil nutrients.

The effects of strigolactones include the germination of parasitic weed seeds, the suppression of plant stems, and the regulation of plant growth and development. The hormone can also affect root growth; although it suppresses the growth of lateral roots, it enhances the elongation of root hairs (Kapulnik et al., 2011). Auxin has an important role in the production of lateral roots, and auxin distribution determines the positioning, initiation, and elongation of root hairs. The effects of strigolactones on the formation of lateral roots are by altering the auxin efflux in the roots. Accordingly, by using mutants, it has been indicated that auxin signaling downregulate strigolactone signaling during the process of lateral root hair production (Kapulnik et al., 2011).

However, cytokinin can adversely affect the production of lateral roots, which is due to its inference with the transfer of auxin to the lateral roots indicating that cytokinin and strigolactones can act similarly adversely, affecting the production of lateral roots. Like ethylene, which can regulate the production and transfer of auxin and positively affect the production of root hairs, strigolactones can similarly affect the growth of root hairs (Kapulnik et al., 2011).

It has also been indicated that cytokinin and strigolactones can suppress the formation of adventitious nonroot tissues (Li et al., 2009). In the strigolactones mutants of pea and *Arabidopsis*, the level of cytokinin decreased in the xylem (Beveridge et al., 1996, 1997; Foo et al., 2001), which is the reason for the higher production of adventitious roots. More research must indicate how the network of auxin, strigolactones, and cytokinin may act in the formation of adventitious roots.

Under stress conditions, including the phosphorous deficiency, the levels of strigolactones in the plant increase and hence make the plant grow under such conditions (Umehara et al., 2008; Umehara, 2011). Plant roots produce strigolactones and are able to enhance the hyphal branching in mycorrhizal fungi (Yoneyama et al., 2007). Accordingly, the increased production of strigolactones enhance the rate of mycorrhizal symbiosis and hence the plant will be able to acquire more phosphorous under P-deficient conditions.

The plant by itself also indicates some physiological and morphological changes in P-deficient conditions. For example, the increased production of lateral roots in the soil surface make the plant absorb a higher rate of phosphorus from the soil and hence tolerate the stress. The roles of root hair length and density, affecting root surface area, are also of importance under P-deficient conditions. The other mechanisms used by plants under P-deficient conditions are (1) the suppression of the aerial part branching, (2) leaf senescence, and (3) the redirection of plant essential photosynthate to the roots. In most cases such mechanisms are affected by the presence of strigolactones (Brewer et al., 2013).

Plant mutants in the absence of strigolactones are not able to tolerate the stresses such as P deficiency, and hence plant growth significantly decreases. This is the important reason indicating the importance of strigolactones in plants and the conserved signaling pathway, even in plants such as *Arabidopsis*, which are not able to develop a symbiotic association with mycorrhizal fungi. With respect to plant species, nitrogen deficiency can also affect the behavior of strigolactones in plants by enhancing its production by plant roots. Under N-deficient conditions the level of strigolactones is changed in the plant, which is due to the altered rate of P in the plant aerial part (Yoneyama et al., 2012).

The interactions between strigolactones, auxin, and cytokinins may determine plant response to different nutrients; accordingly, understanding such a cross talk is of significance for managing plants under nutrient-deficient conditions. For example, under nutrient-deficient conditions the level of cytokinin in the plant decreases and using cytokinin may adversely affect root response under P-deficient conditions (Salama and Wareing, 1979; Martin et al., 2000). Under P-deficient conditions the level of auxin increases in plants, but not at a high rate; however, under the optimal levels of P, ethylene is able to modulate the root hair response to strigolactones (Kapulnik et al., 2011). Accordingly, strigolactones may be important modulators among different plant hormones, affecting plant root response under different conditions including nutrient-deficient conditions.

Brassinosteroids and Stress

BRs have a wide range of activities in plants regulating plant growth and development. Plants, which are deficient in the hormone or lack the related signaling pathways, indicate symptoms of abnormal growth and hence yield reduction. Using the new molecular techniques, the components of the BR-signaling pathways have been recognized. The perception and transduction of BR is conducted in cytoplasm, and the related gene is expressed in the center of the cell. The related cellular, physiological, and molecular mechanisms affecting plant growth processes and yield production by BR are also being illustrated. Such growth processes include the development and production of roots and pollen, differentiation of vasculature, elongation of stems, and biosynthesis of cellulose, indicating the diverse role of BR in plant growth and development (Zhang et al., 2009; Yang et al., 2011).

The important functions of BR in plants include regulating the vegetative and reproductive growth, germination of seed, senescence, and response to biotic and abiotic stresses. Research has indicated that BR has important functions under different types of stresses including salinity, suboptimal root zone temperature, drought, and pathogens. There is also a cross talk between BR and the other plant hormones including ABA, ethylene, and jasmonates. For example, BR is able to alleviate the stress of suboptimal root zone temperature by increasing plant growth and maintaining the leaf chlorophyll content, and under the high rate of temperature, the increased production of proteins can alleviate the stress (Zhang et al., 2006; Yang et al., 2011).

BR was able to alleviate the stress of salinity on the rice seed germination by the production of proteins. It has also been indicated that BR is able to alleviate the stress of drought on plant growth by enhancing the osmotic permeability. The alleviating effects of BR on the growth of plants in the presence of different pathogens have been elucidated. However, in the presence of ABA, the hormone may not affect the process of rice seed germination.

Using BR, the effects of salinity on the growth of maize seedlings were alleviated by enhancing the activity of antioxidant enzymes including superoxide dismutase, guaiacol peroxidase, catalase, glutathione reductase, and ascorbate peroxidase, which resulted in the decreased rate of lipid peroxidation. In wheat (*Triticum aestivum* L.), using BR increased plant biomass, crop yield, and photosynthesis pigment, but it did not increase the uptake of nutrients by plants (Eleiwa and Ibrahim, 2011).

Although salinity decreased the rate of total sugar, proteins, and carbohydrates of wheat grains, the exogenous use of BR gradually increased all such components. The uptake of macro- and micronutrients was also increased by BR in wheat straw and grains under salinity stress (Eleiwa and Ibrahim, 2011). The water use efficiency and the rate of photosynthesis were also enhanced by BR under salinity by increasing stomatal conductance.

Ferguson et al. (2005) investigated the effects of BR on the process of nodulation in pea plants using the mutant phenotypes. Such mutants exhibited a reduction in the process of nodule organogenesis. The authors accordingly indicated that there is a mechanism in the plant aerial part by BR controlling the process of nodulation, and the root levels of auxin and gibberellins are also a part of such mechanism. They also showed that there was a high correlation between the number of nodules and the number of lateral roots.

Upreti and Murti (2004) investigated the effects of BR (1 and 5 μM) before drought stress on the root nodulation of beans and the endogenous content of ABA, cytokinin, and the N fixation enzyme. BR increased root nodulation in nonstressed plants, as well as the rate of cytokinin and the enzyme content, and also ameliorated their decrease in the nodulated roots. BR did not affect the root content of ABA under nonstress and stress conditions. However, the weight of the pod increased by BR 5 μM under nonstress and stress conditions. Among the different types of BR, epibrassinolide was the more effective one. The use of BR affected the process of nodulation in peanuts, peas, and soybeans. The hyper nodulating mutant of the soybean foliar use of BR increased stem height and suppressed nodulation depending on the rate of BR (Nagata and Suzuki, 2014).

Nitrous Oxide and Stress

Nitrous oxide has a wide range of favorable and unfavorable activities in plants, depending on the concentration and location of nitrous oxide in plant cells. Under abiotic stresses the production of nitrous oxide in plants decreases. The molecule is able to alleviate the adverse effects of reactive oxygen species on plant growth, interact with the other plant hormones, and regulate the activities of different stress responsive genes (Zhang et al., 2006; Qiao and Fan, 2008; Akaike et al., 2011; Esim et al., 2014).

The activities of nitrous oxide in plants include the enhancement of seed germination, decreasing the dormancy of seed (Beligni and Lamattina, 2000), regulation of plant growth and senescence, plant flowering, and stomatal activities affecting ABA signaling (Qiao and Fan, 2008). The molecule is also able to regulate plant growth and yield production under different types of stresses including drought, salt, heat, heavy metals, pathogens, etc. (Kopyra et al., 2006; Zhao et al., 2007a,b; Khan et al., 2012). Biotic and abiotic stresses alter the production of nitrous oxide, and exogenous use of nitrous oxide can enhance plant tolerance under stress (Delledonne et al., 1998; Uchida et al., 2002; Zhao et al., 2001, 2007a,b; Karpets et al., 2015). When treating with 2,2'-(hydroxynitrosohydrazono)bis-ethanimine (nitric oxide donor) the activity of ascorbate peroxidase increased in the soybean root nodules (Yang et al., 2015).

The effects of drought stress on rice (*Oryza sativa* L.) growth and yield were investigated by Farooq et al. (2010). Among the most important effects of drought on plant growth is by decreasing the uptake of water and water use efficiency and adversely affecting the cellular membrane functioning. Accordingly, the exogenous use of polyamines and different plant hormones including nitrous oxide (100 mmol/L nitroprusside sodium as nitrous oxide donor), glycinebetaine (150 mg/L), BRs (0.01 μm 24-epibrassinolide), and SA (100 mg/L) were foliarly sprayed at the five-leaf stage (5 weeks following emergence) were evaluated on the growth of rice plants. The plants were subjected to the drought stress at the four-leaf stage (4 weeks following emergence), while the soil moisture was maintained at 50% of the field capacity.

Farooq et al. (2010) also used control treatments at the well water and drought conditions. Drought stress significantly decreased water uptake and plant growth. Under drought stress the level of H_2O_2 and malondialdehyde as well as the relative permeability of the membrane increased; however, the foliar application of chemicals enhanced plant growth by increasing the rate of carbon assimilation and the synthesis of metabolites, maintaining the plant water potential and enhancing the antioxidant level in plants. In the treated plants the rate of H_2O_2 and malondialdehyde decreased.

During different stages of nodule development, nitric oxide is produced. In the mutants, the absence of nitric oxide adversely affects the process of nodulation

(Scheler et al., 2013). Nitric oxide also has a role in the functioning of nodules. Under oxygen deficiency conditions, $NO_2{}^-$ can act as a terminal electron acceptor instead of oxygen and produce nitric oxide, which is converted to nitrate by an oxidation mechanism of cytosolic hemoglobin-dependent. Nitric oxide is also essential for providing nodules with ATP during the process of N fixation. It has also been recently indicated that senescence during plant development and during stress is controlled by nitric oxide. Interestingly, it has been indicated that a bacterial response to detoxification prevented the early senescence of nodules. Nitric oxide is able to regulate the expression of genes essential for the process of nodulation. During such activities there is a cross talk between nitric oxide and reactive oxygen species (Cam et al., 2012; Scheler et al., 2013).

Conclusion and Future Perspectives

The effects of different plant hormones including auxin, ABA, ethylene, gibberellins, cytokinins, SA, strigolactones, BRs, and nitrous (nitric) oxide on the growth and development of soybeans (*Glycine max* L.) and its symbiotic association with the N fixing bacteria, *B. japonicum*, under environmental stresses have been presented, reviewed, and analyzed (Fig. 6.5). The findings, including the new ones, illustrate the important effects of plant hormones on the growth of soybeans and its symbiosis with *B. japonicum* including the development of symbiosis, the bacterial activities in the nodules, and eventually the nodule senescence. Such findings can be useful for

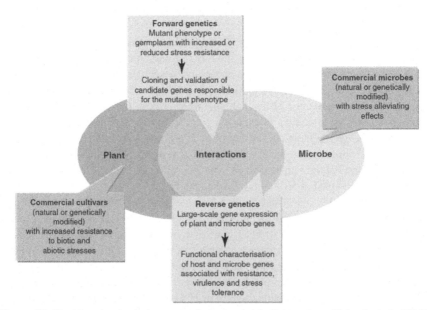

Figure 6.5 The biotechnological aspects of plant–microbe interactions (Schenk et al., 2012). With kind permission from Elsevier, License number 3771370895374.

the production of tolerant soybean genotypes and *B. japonicum* species under stress conditions. Future research may focus on the modification of soybean hormonal activities during environmental stresses, resulting in the enhanced growth and activity of soybeans and *B. japonicum* under different conditions including stress.

References

Akaike, T., van der Vliet, A., Eaton, P., 2011. Frontiers in nitric oxide and redox signaling. Nitric Oxide 25, 57–58.

Alam, I., Lee, D.G., Kim, K.H., Park, C.H., Sharmin, S.A., Lee, H., et al., 2010. Proteome analysis of soybean roots under waterlogging stress at an early vegetative stage. Journal of Biosciences 35, 49–62.

Avis, T.J., Gravel, V., Antoun, H., Tweddell, R.J., 2008. Multifaceted beneficial effects of rhizosphere microorganisms on plant health and productivity. Soil Biology and Biochemistry 40, 1733–1740.

Beligni, M.V., Lamattina, L., 2000. Nitric oxide stimulates seed germination, de-etiolation, and inhibits hypocotyl elongation, three light-inducible responses in plants. Planta 210, 215–221.

Besseau, S., Li, J., Palva, E.T., 2012. WRKY54 and WRKY70 co-operate as negative regulators of leaf senescence in *Arabidopsis thaliana*. Journal of Experimental Botany 63, 2667–2679.

Beveridge, C.A., Ross, J.J., Murfet, I.C., 1996. Branching in pea: action of genes *Rms3* and *Rms4*. Plant Physiology 110, 859–865.

Beveridge, C.A., Symons, G.M., Murfet, I.C., Ross, J.J., Rameau, C., 1997. The *rms1* mutant of pea has elevated indole-3-acetic acid levels and reduced root-sap zeatin riboside content but increased branching controlled by graft-transmissible signal(s). Plant Physiology 115, 1251–1258.

Biswas, B., Chan, P.K., Gresshoff, P.M., 2009. A novel ABA insensitive mutant of *Lotus japonicus* with a wilty phenotype displays unaltered nodulation regulation. Molecular Plant 2, 487–499.

Bonfante, P., Genre, A., 2010. Mechanisms underlying beneficial plant–fungus interactions in mycorrhizal symbiosis. Nature Communications 1, 48.

Brewer, P.B., Koltai, H., Beveridge, C.A., 2013. Diverse roles of strigolactones in plant development. Molecular Plant 6, 18–28.

Buer, C., Muday, G., 2004. The *transparent testa4* mutation prevents flavonoid synthesis and alters auxin transport and the response of *Arabidopsis* roots to gravity and light. Plant Cell 16, 1191–1205.

Buer, C., Imin, N., Djordjevic, M., 2010. Flavonoids: new roles for old molecules. Journal of Integrative Plant Biology 52, 98–111.

Caba, J.M., Centeno, M.L., Fernández, B., Gresshoff, P.M., Ligero, F., 2000. Inoculation and nitrate alter phytohormone levels in soybean roots: differences between a supernodulating mutant and the wild type. Planta 211, 98–104.

Cam, Y., Pierre, O., Boncompagni, E., Herouart, D., Meilhoc, E., Bruand, C., 2012. Nitric oxide (NO): a key player in the senescence of *Medicago truncatula* root nodules. New Phytologist 196, 548–560.

Chiou, T.J., Aung, K., Lin, S.I., Wu, C.C., Chiang, S.F., Su, C.L., 2006. Regulation of phosphate homeostasis by microRNA in *Arabidopsis*. Plant Cell 18, 412–421.

Chiou, T.J., Lin, S.I., 2011. Signaling network in sensing phosphate availability in plants. Annual Reviews of Plant Biology 62, 185–206.

Dakora, F.D., 2003. Defining new roles for plant and rhizobial molecules in sole and mixed plant cultures involving symbiotic legumes. New Phytologist 158, 39–49.

Davies, P.J., 2010. The plant hormones: their nature, occurrence, and functions. In: Plant Hormones. Springer Netherlands, pp. 1–15.

Deinum, E.E., Geurts, R., Bisseling, T., Mulder, B.M., 2012. Modeling a cortical auxin maximum for nodulation: different signatures of potential strategies. Frontiers in Plant Science 3, 96.

Delledonne, M., Xia, Y., Dixon, R.A., Lamb, C., 1998. Nitric oxide functions as a signal in plant disease resistance. Nature 394, 585–588.

Delves, A., Mathews, A., Day, D., Carter, A., Carroll, B., Gresshoff, P., 1986. Regulation of soybean-rhizobium nodule symbiosis by shoot and root factors. Plant Physiology 82, 588–590.

Ding, Y.L., Oldroyd, G.E.D., 2009. Positioning the nodule, the hormone dictum. Plant Signaling and Behavior 4, 89–93.

Eleiwa, M.E., Ibrahim, S.A., 2011. Influence of brassinosteroids on wheat plant (*Triticum aestivum* L.) production under salinity stress conditions. II. Chemical constituent and nutritional status. Australian Journal of Basic Applied Science 5, 49–57.

Esim, N., Atici, O., Mutlu, S., 2014. Effects of exogenous nitric oxide in wheat seedlings under chilling stress. Toxicology and Industrial Health 30, 268–274.

Farooq, M., Wahid, A., Lee, D.J., Cheema, S.A., Aziz, T., 2010. Drought stress: comparative time course action of the foliar applied glycinebetaine, salicylic acid, nitrous oxide, brassinosteroids and spermine in improving drought resistance of rice. Journal of Agronomy and Crop Science 196, 336–345.

Ferguson, B.J., Mathesius, U., 2003. Signaling interactions during nodule development. Journal of Plant Growth Regulation 22, 47–72.

Ferguson, B.J., Ross, J.J., Reid, J.B., 2005. Nodulation phenotypes of gibberellin and brassinosteroid mutants of pea. Plant Physiology 138, 2396–2405.

Ferguson, B.J., Indrasumunar, A., Hayashi, S., Lin, M.H., Lin, Y.H., Reid, D.E., Gresshoff, P.M., 2010. Molecular analysis of legume nodule development and autoregulation. Journal of Integrative Plant Biology 52, 61–76.

Ferguson, B., Mathesius, U., 2014. Phytohormone regulation of legume-rhizobia interactions. Journal of Chemical Ecology 40, 770–790.

Foo, E., Turnbull, C.G.N., Beveridge, C.A., 2001. Long-distance signaling and the control of branching in the *rms1* mutant of pea. Plant Physiology 126, 203–209.

Foo, E., Davies, N.W., 2011. Strigolactones promote nodulation in pea. Planta 234, 1073–1081.

Foo, E., Reid, J.B., 2013. Strigolactones: new physiological roles for an ancient signal. Journal of Plant Growth Regulation 32, 429–442.

Foo, E., Yoneyama, K., Hugill, C.J., Quittenden, L.J., Reid, J.B., 2013. Strigolactones and the regulation of pea symbioses in response to nitrate and phosphate deficiency. Molecular Plant 6, 76–87.

Gresshoff, P.M., 2003. Post-genomic insights into nodulation. Genome Biology 4, 201.

Gresshoff, P.M., Lohar, D., Chan, P.K., Biswas, B., Jiang, Q., Reid, D., Ferguson, B., Stacey, G., 2009. Genetic analysis of ethylene regulation of legume nodulation. Plant Signaling & Behavior 4, 818–823.

Hamayun, M., Khan, S.A., Shinwari, Z.K., Khan, A.L., Ahmad, N., Lee, I.J., 2010a. Effect of polyethylene glycol induced drought stress on physio-hormonal attributes of soybean. Pakistan Journal of Botany 42, 977–986.

Hamayun, M., Khan, S.A., Khan, A.L., Shinwari, Z.K., Iqbal, I., Sohn, E.Y., Khan, M., Lee, I.J., 2010b. Effect of salt stress on growth attributes and endogenous growth hormones of soybean cultivar Hwangkeumkong. Pakistan Journal of Botany 42, 3103–3112.

Hancock, J.T., Neill, S.J., Wilson, I.D., 2011. Nitric oxide and ABA in the control of plant function. Plant Science 181, 555–559.

Hasanuzzaman, M., Nahar, K., Fujita, M., 2013. Plant response to salt stress and role of exogenous protectants to mitigate salt-induced damages. In: Ecophysiology and Responses of Plants under Salt Stress. Springer New York, pp. 25–87.

Hayashi, S., Gresshoff, P., Ferguson, B., 2014. Mechanistic action of gibberellins in legume nodulation. Journal of Integrative Plant Biology 56, 971–978.

Hirsch, A.M., 1992. Developmental biology of legume nodulation. New Phytologist 122, 211–237.

Hirsch, A.M., Fang, Y., 1994. Plant hormones and nodulation: what's the connection. Plant Molecular Biology 26, 5–9.

Indrasumunar, A., Searle, I., Lin, M.H., Kereszt, A., Men, A., Carroll, B.J., Gresshoff, P.M., 2011. Nodulation factor receptor kinase 1α controls nodule organ number in soybean (Glycine max L. Merr). The Plant Journal 65, 39–50.

Jibran, R., Hunter, D.A., Dijkwel, P.P., 2013. Hormonal regulation of leaf senescence through integration of developmental and stress signals. Plant Molecular Biology 82, 547–561.

Jin, J., Wang, G.H., Liu, X.B., Pan, X.W., Herbert, S.J., 2005. Phosphorus regulates root traits and phosphorus uptake to improve soybean adaptability to water deficit at initial flowering and full pod stage in a pot experiment. Soil Science and Plant Nutrition 51, 953–960.

Kapulnik, Y., Resnick, N., Mayzlish-Gati, E., Kaplan, Y., Wininger, S., Hershenhorn, J., Koltai, H., 2011. Strigolactones interact with ethylene and auxin in regulating root-hair elongation in Arabidopsis. Journal of Experimental Botany 62, 2915–2924.

Karpets, Y.V., Kolupaev, Y.E., Vayner, A.A., 2015. Functional interaction between nitric oxide and hydrogen peroxide during formation of wheat seedling induced heat resistance. Russian Journal of Plant Physiology 62, 65–70.

Khan, W., Prithiviraj, B., Smith, D.L., 2003. Photosynthetic responses of corn and soybean to foliar application of salicylates. Journal of Plant Physiology 160, 485–492.

Khan, M.N., Siddiqui, M.H., Mohammad, F., Naeem, M., 2012. Interactive role of nitric oxide and calcium chloride in enhancing tolerance to salt stress. Nitric Oxide 27, 210–218.

Kinkema, M., Gresshoff, P.M., 2008. Investigation of downstream signals of the soybean autoregulation of nodulation receptor kinase GmNARK. Molecular Plant Microbe Interactions 21, 1337–1348.

Kopyra, M., Wilk, S., Gwózdzacute, E.A., 2006. Effects of exogenous nitric oxide on the antioxidant capacity of cadmium-treated soybean cell suspension. Acta Physiologiae Plantarum 28, 526–536.

Kucera, B., Cohn, M.A., Leubner-Metzger, G., 2005. Plant hormone interactions during seed dormancy release and germination. Seed Science Research 15, 281–307.

Li, G., Santoni, V., Maurel, C., 2014. Plant aquaporins: roles in plant physiology. Biochimica et Biophysica Acta 1840, 1574–1582.

Li, S.W., Xue, L.G., Xu, S.J., Feng, H.Y., An, L.Z., 2009. Mediators, genes and signaling in adventitious rooting. Botanical Reviews 75, 230–247.

Lian, B., Zhou, X., Miransari, M., Smith, D.L., 2000. Effects of salicylic acid on the development and root nodulation of soybean seedlings. Journal of Agronomy and Crop Science 185, 187–192.

Libault, M., Brechenmacher, L., Cheng, J., Xu, D., Stacey, G., 2010. Root hair systems biology. Trends in Plant Science 15, 641–650.

Liu, X., Jin, J., Wang, G., Herbert, S., 2008. Soybean yield physiology and development of high-yielding practices in Northeast China. Field Crops Research 105, 157–171.

Long, S.R., 1996. Rhizobium symbiosis: nod factors in perspective. Plant Cell 8, 1885–1898.

Lynch, J.P., Brown, K.M., 2008. Root strategies for phosphorus acquisition. In: The Ecophysiology of Plant-Phosphorus Interactions. Springer, Netherlands, pp. 83–116.

Mabood, F., Smith, D.L., 2005. Pre-incubation of *Bradyrhizobium japonicum* with jasmonates accelerates nodulation and nitrogen fixation in soybean (*Glycine max*) at optimal and suboptimal root zone temperatures. Physiologia Plantarum 125, 311–323.

Mabood, F., Zhou, X., Smith, D.L., 2014. Microbial signaling and plant growth promotion. Canadian Journal of Plant Science 94, 1051–1063.

Martin, A.C., del Pozo, J.C., Iglesias, J., Rubio, V., Solano, R., de la Pena, A., Leyva, A., Paz-Ares, J., 2000. Influence of cytokinins on the expression of phosphate starvation responsive genes in *Arabidopsis*. Plant Journal 24, 559–567.

Mathesius, U., 2008. Auxin: at the root of nodule development? Functional Plant Biology 35, 651–668.

Méndez, C., Baginsky, C., Hedden, P., Gong, F., Carú, M., Rojas, M.C., 2014. Gibberellin oxidase activities in *Bradyrhizobium japonicum* bacteroids. Phytochemistry 98, 101–109.

Miransari, M., Smith, D., 2009. Rhizobial lipo-chitooligosaccharides and gibberellins enhance barley (*Hordeum vulgare* L.) seed germination. Biotechnology 8, 270–275.

Miransari, M., Smith, D.L., 2014. Plant hormones and seed germination. Environmental and Experimental Botany 99, 110–121.

Miransari, M., 2016. Abiotic and Biotic Stresses in Soybean Production. Elsevier, Academic Press, p. 348. ISBN: 9780128015360.

Mishra, R.P.N., Singh, R.K., Jaiswal, H.K., Kumar, V., Maurya, S., 2006. Rhizobium mediated induction of phenolics and plant growth promotion in rice (*Oryza sativa* L.). Current Microbiology 52, 383–389.

Murphy, A., Peer, W.A., Taiz, L., 2000. Regulation of auxin transport by aminopeptidases and endogenous flavonoids. Planta 211, 315–324.

Nagata, M., Suzuki, A., 2014. Effects of phytohormones on nodulation and nitrogen fixation in leguminous plants. In: Ohyama, T. (Ed.), Advances in Biology and Ecology of Nitrogen Fixation. InTech Open Science. ISBN: 978-953-51-1216-7.

Oldroyd, G.E., Engstrom, E.M., Long, S.R., 2001. Ethylene inhibits the Nod factor signal transduction pathway of *Medicago truncatula*. Plant Cell 13, 1835–1849.

Peer, W., Murphy, A., 2007. Flavonoids and auxin transport: modulators or regulators? Trends Plant Science 12, 556–563.

Penmetsa, R.V., Cook, D.R., 1997. A legume ethylene-insensitive mutant hyperinfected by its rhizobial symbiont. Science 275, 527–530.

Péret, B., Li, G., Zhao, J., Band, L.R., Voss, U., Postaire, O., et al., 2012. Auxin regulates aquaporin function to facilitate lateral root emergence. Nature Cell Biology 14, 991–998.

Pieterse, C.M., Van Pelt, J.A., Ton, J., Parchmann, S., Mueller, M.J., Buchala, A.J., Metraux, J., Van Loon, L., 2000. Rhizobacteria-mediated induced systemic resistance (ISR) in *Arabidopsis* requires sensitivity to jasmonate and ethylene but is not accompanied by an increase in their production. Physiological and Molecular Plant Pathology 57, 123–134.

Puppo, A., Groten, K., Bastian, F., Carzaniga, R., Soussi, M., Lucas, M.M., et al., 2005. Legume nodule senescence: roles for redox and hormone signalling in the orchestration of the natural aging process. New Phytologist 165, 683–701.

Qiao, W., Fan, L.M., 2008. Nitric oxide signaling in plant responses to abiotic stresses. Journal of Integrative Plant Biology 50, 1238–1246.

Salama, A., Wareing, P., 1979. Effects of mineral nutrition on endogenous cytokinins in plants of sunflower (*Helianthus Annuus* L.). Journal of Experimental Botany 30, 971–981.

Sato, T., Fujikake, H., Ohtake, N., Sueyoshi, K., Takahashi, T., Sato, A., Ohyama, T., 2002. Effect of exogenous salicylic acid supply on nodule formation of hypernodulating mutant and wild type of soybean. Soil Science and Plant Nutrition 48, 413–420.

Scheler, C., Durner, J., Astier, J., 2013. Nitric oxide and reactive oxygen species in plant biotic interactions. Current Opinion in Plant Biology 16, 534–539.

Schenk, P.M., Carvalhais, L.C., Kazan, K., 2012. Unraveling plant–microbe interactions: can multi-species transcriptomics help? Trends in Biotechnology 30, 177–184.

Seo, H., Li, J., Lee, S., Yu, J., Kim, K., Lee, S., Lee, I., Paek, N., 2007. The hypernodulating nts mutation induces jasmonate synthetic pathway in soybean leaves. Molecules and Cells 24, 185.

Spaepen, S., Vanderleyden, J., 2011. Auxin and plant-microbe interactions. Cold Spring Harbor Perspectives in Biology 3 a001438.

Sun, J., Cardoza, V., Mitchell, D.M., Bright, L., Oldroyd, G., et al., 2006. Crosstalk between jasmonic acid, ethylene and Nod factor signaling allows integration of diverse inputs for regulation of nodulation. Plant Journal 46, 961–970.

Tanabata, S., Ohyama, T., 2014. Autoregulation of nodulation in soybean plants. In: Ohyama, T. (Ed.), Advances in Biology and Ecology of Nitrogen Fixation. InTech Open Publisher.

Thimann, K.V., 1936. On the physiology of the formation of nodules on legumes roots. Proceedings of the National Academy of Sciences of the United States of America 22, 511–513.

Tirichine, L., Sandal, N., Madsen, L.H., Radutoiu, S., Albrektsen, A.S., Sato, S., Asamizu, E., Tabata, S., Stougaard, J., 2007. A gain-of-function mutation in a cytokinin receptor triggers spontaneous root nodule organogenesis. Science 315, 104–107.

Tsvetkova, G.E., Georgiev, G.I., 2007. Changes in phosphate fractions extracted from difference organs of phosphorus starved nitrogen fixing pea plants. Journal of Plant Nutrition 30, 2129–2140.

Tuteja, N., 2007. Abscisic acid and abiotic stress signaling. Plant Signaling and Behavior 2, 135–138.

Uchida, A., Jagendorf, A.T., Hibino, T., Takabe, T., Takabe, T., 2002. Effects of hydrogen peroxide and nitric oxide on both salt and heat stress tolerance in rice. Plant Science 163, 515–523.

Umehara, M., Hanada, A., Yoshida, S., Akiyama, K., Arite, T., Takeda-Kamiya, N., Magome, H., Kamiya, Y., Shirasu, K., Yoneyama, K., et al., 2008. Inhibition of shoot branching by new terpenoid plant hormones. Nature 455, 195–200.

Umehara, M., 2011. Strigolactone, a key regulator of nutrient allocation in plants. Plant Biotechnology 28, 429–437.

Upreti, K., Murti, G., 2004. Effects of brassmosteroids on growth, nodulation, phytohormone content and nitrogenase activity in French bean under water stress. Biologia Plantarum 48, 407–411.

Valliyodan, B., Nguyen, H.T., 2008. Genomics of abiotic stress in soybean. In: Genetics and Genomics of Soybean. Springer, New York, pp. 343–372.

Xue, R., Zhang, B., 2007. Increased endogenous methyl jasmonate altered leaf and root development in transgenic soybean plants. Journal of Genetics and Genomics 34, 339–346.

Yang, C.J., Zhang, C., Lu, Y.N., Jin, J.Q., Wang, X.L., 2011. The mechanisms of brassinosteroids' action: from signal transduction to plant development. Molecular Plant 4, 588–600.

Yang, H., Mu, J., Chen, L., Feng, J., Hu, J., Li, L., et al., 2015. S-Nitrosylation positively regulates ascorbate peroxidase activity during plant stress responses. Plant Physiology 167, 1604–1615.

Yashima, Y., Kaihatsu, A., Nakajima, T., Kokubun, M., 2005. Effects of source/sink ratio and cytokinin application on pod set in soybean. Plant Production Science 8, 139–144.

Yoneyama, K., Yoneyama, K., Takeuchi, Y., Sekimoto, H., 2007. Phosphorus deficiency in red clover promotes exudation of orobanchol, the signal for mycorrhizal symbionts and germination stimulant for root parasites. Planta 225, 1031–1038.

Yoneyama, K., Xie, X., Kim, H.I., Kisugi, T., Nomura, T., Sekimoto, H., Yokota, T., Yoneyama, K., 2012. How do nitrogen and phosphorus deficiencies affect strigolactone production and exudation? Planta 235, 1197–1207.

Yoon, J., Hamayun, M., Lee, S., Lee, I., 2009. Methyl jasmonate alleviated salinity stress in soybean. Journal of Crop Science and Biotechnology 12, 63–68.

Zhang, Y., Wang, L., Liu, Y., Zhang, Q., Wei, Q., Zhang, W., 2006. Nitric oxide enhances salt tolerance in maize seedlings through increasing activities of proton-pump and Na^+/H^+ antiport in the tonoplast. Planta 224, 545–555.

Zhao, Z., Chen, G., Zhang, C., 2001. Interaction between reactive oxygen species and nitric oxide in drought-induced abscisic acid synthesis in root tips of wheat seedlings. Australian Journal of Plant Physiology 28, 1055–1061.

Zhao, M.G., Zhao, X., Wu, Y.X., Zhang, L.X., 2007a. Enhanced sensitivity to oxidative stress in *Arabidopsis* nitric oxide synthesis mutant. Journal of Plant Physiology 164, 737–745.

Zhao, M.G., Tian, Q.Y., Zhang, W.H., 2007b. Nitric oxide synthase-dependent nitric oxide production is associated with salt tolerance in *Arabidopsis*. Plant Physiology 144, 206–217.

Zhang, S., Wei, Y., Lu, Y., Wang, X., 2009. Mechanisms of brassinosteroids interacting with multiple hormones. Plant Signaling and Behavior 4, 1117–1120.

Zhang, J., Smith, D.L., Liu, W., Chen, X., Yang, W., 2013. Effects of shade and drought stress on soybean hormones and yield of main-stem and branch. African Journal of Biotechnology 10, 14392–14398.

Zhang, Z., Liao, H., Lucas, W.J., 2014. Molecular mechanisms underlying phosphate sensing, signaling, and adaptation in plants. Journal of Integrative Plant Biology 56, 192–220.

Soybean, Protein, and Oil Production Under Stress

7

M. Miransari
AbtinBerkeh Scientific Ltd. Company, Isfahan, Iran

Introduction

The soybean (*Glycine max* (L.) Merr) is among the most important leguminous plants feeding a large number of people in the world. The soybean is able to develop a symbiotic association with its specific rhizobium, *Bradyrhizobium japonicum*, and acquire most of its essential nitrogen (N) for growth and yield production. The soybean, also a great source of food, contains 200 g oil and 400 g/kg (dry matter) protein in its seed, and as a result is the major oilseed and protein crop in different parts of the world, providing about 60% of the total vegetable protein for human use (Rosenthal et al., 1998; Allen et al., 2009).

The economic value of soybean seeds is determined by the rate of soybean protein and oil (Hurburgh et al., 1990). Based on a dry weight, the soybean contains 40% protein and 20% oil. Soybeans produce about 180 kg oil and 800 kg defatted meal by 1 metric ton (MT). The produced meal is also a coproduct of processed soybeans; it has more than 50% protein (Hettiarachchy and Kalapathy, 1997). However, the final composition of soybean seed protein and oil is determined by the interactions between the genotype and the environment (Fehr et al., 2003; Wilson, 2004).

Recently the demand for soybean meal has increased worldwide, with the present standard of 480 g/kg per year for each person. Soybean prices are determined mostly by protein and oil contents (Brumm and Hurburgh, 1990, 2006). Soybeans are also used as sources of raw material for the industry and pharmaceutical usage for the production of soy foods and biodiesel (Clarke and Wiseman, 2000). The soybean is also a suitable source of vitamin B and folic acid. Although there is a high rate of soybean genotypes with a different rate of protein and oil contents, finding a genotype with a high rate of protein and oil is not easy (Medic et al., 2014).

It is accordingly important to determine the environmental conditions, which are the most suitable for the production of soybean seed protein and oil. However, most of the time the high rate of protein is correlated with a lower rate of yield. Accordingly, it is pertinent to find the physiological properties, which can result in genotype differences and hence make the plant produce a higher yield and rate of protein, resulting in the increased market values of soybean grain yield (Liang et al., 2010).

Soybeans are an important source of protein, oil, nutrients, and natural products including isoflavonoids, affecting human health and nutrition. Soybean products are used as sources of protein and edible vegetable oil for human use and a high protein supplement for the chicken industry. According to SOYSTATS (2014), in 2014, the

Environmental Stresses in Soybean Production. http://dx.doi.org/10.1016/B978-0-12-801535-3.00007-3

United States, with the production of 108 million metric tons, has been the first country of soybean production in the world, followed by Brazil (94.5), Argentina (56.0), China (12.4), India (10.5), Paraguay (8.5), and Canada (6.1).

The soybean is a suitable source of oleic and linoleic acid; even if the soy oil is partially hydrogenated, it has 25% linoleic and 3% linolenic acid. It is, as previously mentioned, a good source of protein, oil, and carbohydrates. The rate of amino acids in soybeans is significant and is a complement to cereals. It is also a source of galacto-oligosaccharides by 5% and starch at less than 1%. Although the soybean is a high source of nutrients, due to the presence of antinutrients, such as phenols, trypsin inhibitors, and phytic acid, the use of nutrients is limited (Liener, 1981; Sharma et al., 2014).

In soybean seeds the presence of phytic acid (myo-inositol hexakisphosphate) represents 65–80% of total phosphorous; due to its ability for chelating nutrients including Ca and micronutrients (Fe, Zn, Cu, and Mn), the availability of such nutrients decreases in soybean seeds. However, it is an important health factor with antioxidant and anticarcinogenic properties (Sharma et al., 2014).

In this chapter some of the most important details, which affect soybean seed protein and oil, are presented and analyzed.

Soybean Seeds and Environmental Parameters

Jin et al. (2010) investigated the agronomical and physiological properties contributing to the improvement of soybean genotypes released from 1950 to 2006 in the northern part of China. Such kind of investigations can be useful for the development of genotypes with the higher rate of yield production and higher resistance under stress. Although there was a variation of 37–45.5% for seed protein and 16.7–22% for seed oil, it was not a function of release year.

Seed number per plant was the most important factor affecting soybean yield from year to year. The stable and enhanced production of soybean yield from year to year was due to pod production under different environments. According to their research, Jin et al. (2010) indicated that the rate of seed protein and concentration did not increase in the 56 years of soybean cultivation, indicating that the improvement of soybean genotypes has been for the production and selection of high-yielding genotypes rather than for the selection of genotypes with a high rate of protein and yield.

Seed filling is an interesting period because the rate of protein, oil, and starch increases in plant seeds. It has been previously indicated that the rate of protein and level of transcripts increases significantly during seed filling (Hajduch et al., 2005, 2010). Menacho et al. (2010) compared the rate of bioactive compounds in soybean genotypes with lower and high rates of proteins. The soybean genotype with the high rate of protein contained a 17% lower rate of carbohydrates related to the genotype with the lower rate of protein and higher rate of methionine (1.2%). The genotype with the higher rate of protein contained a higher rate of phosphorous (30.1%), calcium (15.5%), iron (18.7%), zinc (11.5%), and copper (9%), and a higher concentration of lectin (27.1%) and lunasin (20.3%) compared with genotype with the less rate of protein. The genotype with the lower rate of protein had a 75.4% higher rate of isoflavonoids than the genotype with the higher rate of protein. The authors accordingly indicated that the soybean genotype with the lower rate of protein had a higher rate of isoflavonoids and saponins and a lower rate of bioactive compounds such as lunasin and minerals.

The presence of soybean bioactive compounds is a function of plant species, genotype, climate, and location. The complex matrix of soybeans is a source of negligible rate of starch; as previously mentioned, 20% oil, 40% high-quality protein (Liu, 1997), and some important bioactive compounds including the inhibitors of trypsin, lunasin, saponins, and isoflavonoids. Lunasin is a peptide with the ability to prevent cancer and is originated from soy. The lecithin present in soybeans has anticarcinogenic and antinutritional properties (Menacho et al., 2010).

Lectins are located in the protein storage vacuoles of the cotyledons, and their degradation is during the germination and maturation of seeds. The flavor of soybean products results from lipids. Soybean seeds are an important source of isoflavonoids including genistein and glycitein. Research work has indicated the important role of isoflavonoids in preventing health-associated issues including cancer, menopause, and cardiovascular disease (Velazquez et al., 2010).

During processing soybean seeds, the seed compounds may be changed and their rate decreases (Mandarino et al., 1992). During the processing of soy flour, the main isoflavonoids including genistein, malonylgenistin, daidzin, and malonyldaidzin are converted into forms of aglycone and acetylglycoside. Saponins are from the plant terpenoids, which are conjugated to sugar molecules in soy. For the soy protein to be of health significance a minimum of 25 g soy protein should be absorbed daily (Menacho et al., 2010). The biological sufficiency of a food protein is determined by the composition of its total amino acids, especially the essential amino acids. The standard of amino acids by the WHO, FDA, and UNU has been compared with the soybean genotypes containing lower and high rates of protein (Table 7.1).

A large part of fixed N is used by the soybean leaf because it has a high rate of N. About 50% of fixed N in the soybean leaf is found in the form of ribulose 1,5-bisphosphate (Rubisco) as the important photosynthesis enzyme during the fixation of carbon dioxide. Rubisco is an important source of N, because it is produced in large amounts, and it is a safe and stable form of N until it is catalyzed by proteases. A higher part of protein in legume seeds requires a higher part of photosynthate. There is a high rate of protein and lipids in soybeans and peanuts, with a total percentage of 60–80%. This is one of the most important reasons indicating a considerable difference between the yield of such legumes and cereals. With respect to the importance of legumes, especially as a significant source of protein and oil, they are among the most important contents of cropping strategies (Sinclair and Vadez, 2012).

Under drought conditions (rain-fed conditions) during the period of seed filling, the rate of soybean yield decreased by 20% (1765 kg/ha) related to the nonstress conditions (2195 kg/ha). The most important effect of drought stress on the grain seed of soybeans was by decreasing the seed size. A 24% decrease (801 kg/ha) in seed protein content resulted under rain-fed conditions related to nonstress conditions (1055 kg/ha) (Turner et al., 2005).

It is essential to enhance the rate of assimilate per seed (without decreasing the number of seeds), if the high rate of protein is favorable in high-yielding soybean

Table 7.1 Different Crop Plant Seed Proteins Affected by Different Stresses

Crop Plant	Environmental Stresses	References
Drought		
Soybean (*Glycine max* (L.) Merr)	↑	Rotundo and Westgate (2009)
Corn (*Zea mays* L.)	↑	TWumasi-Afriyie et al. (2011)
Barley (*Hordeum vulgare* L.)	↑	Savin and Nicolas (1999)
Peanut (*Arachis hypogaea* L.)	↑	Dwivedi et al. (1996)
Wheat (*Triticum aestivum* L.)	↑	Saint Pierre et al. (2008); Flagella et al. (2010)
Potato (*Solanum tuberosum* L.)	↑	
Heat		
Soybean (*G. max* (L.) Merr)	↑	Rotundo and Westgate (2009)
Barley (*Hordeum vulgare* L.)	↑	Savin and Nicolas (1999)
Sunflower (*Helianthus annuus* L.)	↑	
Wheat (*Triticum aestivum* L.)	↑	Dupont and Altenbach (2003)
Rice (*Oryza sativa* L.)	↑↓	Lin et al. (2010)
Salinity		
Potato (*Solanum tuberosum* L.)	↑	
Tropospheric Ozone		
Soybean (*G. max* (L.) Merr)	↑–	Kress and Miller (1983)
Wheat (*Triticum aestivum* L.)	↑	Pleijel et al. (1998); Feng et al. (2008); Fuhrer (2009)
Corn (*Zea mays* L.)	–	
Bean (*Phaseolus vulgaris* L.)	↑	
Peanut (*A. hypogaea* L.)	–	

genotypes. Accordingly, if the soybean genotypes are genetically modified for their rate of protein content, a higher rate of assimilate will be available to the plant. It is also possible to acquire a higher rate of protein by decreasing the rate of oil content (mg/seed) and carbohydrate without modifying the rate of protein. The important factors determining the final seed protein content include metabolic efficiency and partitioning resulting from genetic differences (Wesley et al., 1998; Liu et al., 2008).

Interestingly, the soybean seed size of genotypes with a lower rate of protein is more responsive to the rate of assimilate per seed than genotypes with a high rate of protein. Such a conclusion is consistent with the decreased responsiveness of seed size to depoding in soybean genotypes with a high rate of protein, because the rate of protein in such genotypes is almost saturated as such genotypes have grown under a high rate of assimilates (Higgins, 1984; Borrás et al., 2004).

The authors indicated that their results can be a response to the negative correlation between soybean yield and seed content. If the higher rate of protein concentration results from a higher rate of protein content, a higher rate of sucrose and N is essential for each seed. Such a case is achieved by a higher leaf area index per seed in genotypes with high rate of protein. In the evaluated genotype, the genotypes with the high rate were able to acquire such a requirement by decreasing the number of seeds per plant. Accordingly, there must be a positive correlation between the composition and number of seeds, resulting in genotypes with a high rate of yield and seed protein content. Accordingly, increasing the rate of assimilate per seed without decreasing the efficiency of reproduction is essential for increasing seed protein content in high-yielding genotypes (Egli and Bruening, 2007a,b; Weichert et al., 2010).

However, extensive research work has shown that although developing cold-tolerant genotypes of soybeans is among the most useful methods for the production of tolerant soybean plants under cold stress, the use of molecular methods can also improve soybean tolerance under cold conditions (Moura et al., 2010). For example, it has been indicated that during the process of biological N fixation, the initial stages, including the process of signaling communication between rhizobium and the host plant, are more sensitive stages to the stress of cold compared with the other stages of the symbiotic process (Zhang and Smith, 2002; Miransari and Smith, 2008; Mabood et al., 2014).

From 1994 to 2004 the research team of Professor Donald Smith, from McGill University, Canada, showed that it is possible to enhance the tolerance of soybean plants under different types of stresses including cold, salinity, acidity, and high soil N rate using the molecular methods. They found that stress decreases the production of the signal molecules, specifically flavonoids including genistein by the host plant roots, and hence the attraction of rhizobium toward the plant roots decreases adversely, affecting the process of biological N fixation. The inoculums of *B. japonicum* were preincubated with the signal molecule genistein and used for the inoculation of soybean seeds under field and greenhouse conditions. The genistein molecule was able to alleviate the stress and increased soybean grain yield and protein (Zhang and Smith, 1995, 2002; Mabood et al., 2006; Miransari and Smith, 2007, 2008, 2009).

Testa is an important part of the seed developing from integuments surrounding the ovule before fertilization (Moïse et al., 2005). In soybeans, testa is among the first parts of the seed, which is developed. Although testa acts as a physical barrier and protects the seed from the environment, it also has some important metabolic functioning including seed development and dormancy, metabolism and transfer of nutrients from the parent plant, and disease resistance (Bellaloui, 2012). The contribution of testa to the total seed weight is 8–10% including (on the basis of dry weight) 9–12% protein, 10–12% pectin, 14–25% cellulose, 9–12% uronic acid, 14–25% hemicellulose, 7–11% ash, 4–5% lipid, and 3–4% lignin (Miernyk and Johnston, 2013).

The biochemical composition of soybean seeds changes under stress, which also decreases seed yield. Dornbos and Mullen (1992) investigated the effects of drought and high temperature on the contents of soybean seed protein and oil. Accordingly, the objective of the research was to evaluate the protein and oil contents and the fatty acid composition of soybean seeds during the filling period under drought and high-temperature stress. A 2-year experiment was conducted and soybean plants were subjected to drought stress at three levels until the seeds fill.

In experiment I, soybean genotype Gnome was grown at 20 and 26°C, and in experiment II, soybean genotype Hodgson 78 was grown at 27, 29, 33, and 35°C. A high rate of drought stress increased protein rate by 4.4% and decreased oil contents by 2.9%. With increasing the intensity of drought stress, there was a linear increase and decrease in soybean protein rate and oil content, respectively. At 29°C the rate of seed protein increased by 4%, and the oil content decreased by 2.6% during the seed fill, compared with 35°C. Although the effects of drought stress and the rate of fatty acid were not significant, high temperature decreased the rate of unsaturated acid (Dornbos and Mullen, 1992).

In a 5-year experiment, Rose (1988) investigated the effects of drought stress on the soybean yield, seed weight, protein, and oil content of different soybean genotypes by the evaluation of genotype×drought interactions. Compared with the nonstress conditions, the deficiency of rainfall decreased seed grain yield by 16–94%. Between 57% and 68% of the seed rate reduction was related to the seed protein and oil.

Regression equations indicated that there was a significant effect of drought stress on the seed weight and the corresponding protein rate and oil content. Under stress the seed rate of protein increased and the oil content decreased. However, in a high season of rainfall, there were not changes in the rate of protein and oil content. Usually during the period of seed fill the intensity of drought stress increased, and hence the rate of seed protein increased and the oil content decreased (Rose, 1988).

Soybean Seed Protein and Oil

The rate of protein and oil content in soybeans is genetically negatively correlated. Accordingly, it may not be possible to improve their rate simultaneously. The heritability rate of soybean seed oil and protein is in the rage of relatively high and high, respectively. The rate of soybean yield has significantly increased during previous years. For example, from 1986 to 2004, it increased at a rate of almost 10.9 kg/ha/year. The rate of seed protein at present is at a suitable level; however, environmental fluctuations may decrease the rate of soybean seed protein. Water stress decreases the rate of seed protein and increases the rate of seed oil (Specht et al., 2001). The seed oil content also increases under a high rate of temperature (Brumm and Hurburgh, 2006).

Soil nutrients can also affect the production of seed protein. For example, potassium (K^+) can regulate the production of protein in plants. The high rate of K^+ in plants can increase the rate of protein production (Blevins, 1985). Using K^+ fertilization increased the rate of seed protein and amino acid. Using manure for the production of maize decreased the rate of seed protein; however, a balanced N-P-K fertilization increased the rate of seed protein. Similar results have been indicated for wheat grain yield, as the use of potassium fertilization increased the rate of seed protein (Pettigrew, 2008).

The quantitative traits of seed protein, oil, and size are controlled by multiple genes with little or large effects. Accordingly, using molecular genetics can be useful for the determination of quantitative trait loci (QTL) controlling such quantitative traits. Different research work has indicated the QTL for soybeans controlling soybean seed protein, oil, and seed size (Specht et al., 2001; Pathan et al., 2013). New QTL has been found by scientists using mapping populations resulting from the interspecific cross of cultivated soybeans (*Glycine max* (L.) Merr) and wild soybeans (*Gycine soja* Siebold and Zucc.) (Pathan et al., 2013).

The seed protein contains two different parts: (1) the content of total protein and (2) the seed proteome, which is the composition of individual proteins. Similar to most of the soybean seeds the two dominant storage proteins are conglycinin (type of 7S vicilin) and glycine (type of 11S legumine). The soybean seed protein also contains other bioactive moderate proteins, including sucrose binding protein, sucrose, urease, and P34 allergen, and a high number of proteins with little abundance such as enzymes and storage substances. The trait of total amino acid compositions is determined by the abundance of each protein and the particular mix of each protein (Herman, 2014).

Since the time of gene transformation, much effort has been made to increase the rate of seed protein. The main goals of such research have been to enhance the nutritional quality of seeds by increasing the rate of amino acids such as methionine in soybeans (Table 7.2). However, such efforts have not been significant; for example, a protein increase of about 1% has resulted. Such results suggest that there are some mechanisms that control the increased rate of proteins in plants (Wu et al., 2010; Qi et al., 2011).

The converse experiment is related to the inactivation of essential storage proteins and determination of a seed's protein content. It has been indicated that if the conglycinin gene is inactivated, soybean protein content is conserved, resulting in a higher rate of glycinin in the seeds. Conglycinin is about 20% of the total protein content. According to Schmidt and Herman (2008), inserting and activating the gene of a non-familiar protein into the conglycinin–glycinin rebalancing increased the rate of the nonfamiliar protein.

Sharma et al. (2014) analyzed the physical and nutritional attributes of eight new soybean genotypes and found the following. The weight of 100 seeds was in the range of 8.7–11.1 g, and the volume of soybean genotypes was in the range of 8.1–12.0 mL. The percent of water absorption and volume expansion were in the range of 94.3–119.5% and 70.8–159.5%, respectively. Crude protein (39.4–44.4%), starch (4.3–6.7%), oil (14.0–18.7%), reducing sugars (0.21–0.33%), total soluble sugars (5.6–7.9%), and sucrose (5.6–11.8%) were also determined. Triglyceride content and free fatty acids were in the range of 90.1–93.9 g/100 g oil and 31–71 mg/100 g oil, respectively. The soybean genotype also contained phytate (2.3–5.6), TIA (41.5–85.0), flavonols (0.20–0.34), phenols (1.0–1.5), and *ortho*-dihydroxy phenols (0.10–0.21) (Table 7.3).

Liang et al. (2010) presented the available QTL for soybean protein rate, oil content, and seed weight according to their significance, effects, mapping parents, and genetic locations. According to SoyBase (2012) a large number of QTL has been detected for different seed traits, including 113 for seed oil, 115 for seed protein, 120 for seed weight and 5 for protein and oil (Pathan et al., 2013).

Table 7.2 The Composition of Total and Free Amino Acids of the Flours of Soybean Genotypes 1 and 2 Related to the WHO/FAO/UNU Standard (2007)

Amino Acid (g/100 g Protein, Dry Basis)	Requirements, WHO/FAO/UNU			Genotypes	
	Children (Years)		Adults	Genotype 1	Genotype 2
	1–2	3–10			
Methionine (Met)	–	–	1.6	1.22±0.01[a]	1.01±0.02[a]
Phenylalanine (Phe)	–	–	–	4.82±0.03[a]	4.88±0.03[a]
Tryptophane (Trp)	7.4	6.6	0.6	nd	nd
Valine (Val)	–	–	3.9	4.49±0.03[a]	4.31±0.02[a]
Cysteine (Cys)	–	–	0.6	1.93±0.02[a]	1.67±0.01[a]
Lycine (Lys)	5.2	4.8	4.5	6.16±0.04[a]	6.10±0.01[a]
Leucine (Leu)	–	–	5.9	7.57±0.04[a]	7.5±0.01[a]
Histidine (Hys)	–	–	1.5	2.63±0.01[a]	2.75±0.01[a]
Isoleucine (Ile)	–	–	3.0	4.44±0.01[a]	4.49±0.03[a]
Threonine (Thr)	2.7	2.5	2.3	3.93±0.01[a]	3.68±0.04[b]
Tyrosine (Tyr)	–	–	–	3.55±0.04[a]	3.37±0.00[a]
Total amino acid essential	15.3	13.9	23.9	40.47	39.76
Alanine (Ala)	–	–	–	4.3±0.00[a]	4.27±0.01[a]
Arginine (Arg)	–	–	–	8.68±0.01[b]	9.42±0.03[a]
Glutamic acid (Glu)	–	–	–	18.54±0.01[a]	18.97±0.02[a]
Proline (Pro)	–	–	–	4.98±0.00[a]	4.99±0.01[a]
Asparatic acid (Asp)	–	–	–	11.61±0.01[a]	11.59±0.04[a]
Glycine (Gly)	–	–	–	5.11±0.01[a]	4.95±0.07[a]
Serine (Ser)	–	–	–	6.06±0.02[a]	6.04±0.01[a]
Total amino acid				100%	100%
Aromatic (Phe+Tyr)	–	–	3.8	8.37	8.25
Sulfur amino acids (Met+Cys)	2.6	2.4	2.4	3.15	2.68
Total free amino acid				2.36%	2.01%

Means (two duplicates±standard error) with similar letter are not significantly different in a similar row ($P<0.05$). nd, not different.

Among the most important effect of cereals is their role in providing humans with their essential proteins. The two important amino acids, which are mostly in lower availability in cereals and legumes compared with the other amino acids, are lysine and methionine, respectively (Wenefrida et al., 2009). In cereals the rate of lysine is 1.5–4.5 mol% versus 5.5% of the WHO recommendation, and in legumes the rate of lysine and methionine is 1–2% compared with the 3.5 mol% of the WHO recommendation.

Accordingly, the nutritional values of such crop plants decreases by 50–70%, related to a food with a balanced rate of amino acids. Protein deficiency results in an abnormal human growth and the related symptoms (Galili and Amir, 2013).

Table 7.3 The Rate of Proteins, Lipids, and Carbohydrates in Seeds of Different Grain Legumes

Legume	Carbohydrate (%)	Lipid (%)	Protein (%)
Soybean (*Glycine max* (L.) Merr)	14	21	43
Broad bean (*Vicia faba*)	55	1	32
Peanut (*Arachis hypogaea*)	14	50	30
Pea (*Pisum sativa*)	60	3	28
Lentil (*Lens esculenta*)	62	3	30
Cowpea (*Vigna unguiculata*)	62	5	27
Pigeon pea (*Cajanus cajan*)	62	5	27
Bean (*Phaseolus vulgaris*)	68	1	27

According to the above-mentioned details, it is important to use genetic methods, which can enhance the rate of lysine and methionine in cereals and legume seeds. The genetic methods, which have been used for the production of crop plants with a suitable rate of lysine and methionine, have been successful, although in a few cases they have resulted in the production of crop plants with an abnormal morphological phenotype.

However, more research is essential for the production of normal phenotypes with a suitable rate of yield. Also, more knowledge must be obtained on how such amino acids are metabolized, how plant growth and metabolism are affected by such amino acids, how the seed storage proteins are regulated, and how the accumulation of proteins are affected by lysine and methionine. Although different research has combined the two strategies, which increase the level of these amino acids in plants, including enhancing their soluble level and expression of lysine and methionine storage proteins, more research is essential on such a combination affecting seed biology.

Arslanoglu et al. (2013) investigated the effects of different environments on the crude oil and protein content of eight soybean genotypes in a 2-year experiment. The experiment was a completely block design with three replicates. The crude oil and protein content of soybean genotypes were significantly affected by genotype, environment, and their interaction. The highest protein and oil content were equal to 34% and 22%, respectively. The interaction of genotype and environment resulted in the least and the highest protein content of 29.5% and 39%, respectively. The genotype and environment interaction resulted in a highly negative correlation between protein and oil content.

Soybean seed protein is controlled by genetics and physiology. Soybean plants are able to regulate their protein content, affecting their proteomes. Although soybean plants have a relatively standardized proteome (Wilson, 2004), it is possible to modify their proteome largely using mutation or genetic methods. The balance of soybean proteome is regulated by genetic and the posttranscriptional control. Realizing the mechanisms that affect soybean seed proteomes makes the production of new soybean phenotypes with a high rate of protein for food, feed, and industry possible (Herman, 2014).

Although plant growth and crop production is affected by climate change, crop quality is also affected. Crop quality is affected by different parameters including technological, nutritional, and environmental ones. For example, among the factors affecting crop protein is the increased rate of CO_2 (Taub et al., 2008). Accordingly, if crop production with a suitable quality is desirable, crop physiologists must also consider the effects of climate change affecting the crop quality (DaMatta et al., 2010; Woolf et al., 2010).

Using a meta-analysis (228 research works), Taub et al. (2008) investigated the protein concentrations of important food crops affected by CO_2 fluctuations (540–958 ppm). The protein concentrations of wheat, barley, and rice decreased by 10–15% in the CO_2 range of 315–400 ppm. For potatoes the decrease in the protein concentration of tubers was equal to 14%; for soybeans it was a much smaller amount at 1.4%. However, it has been indicated that using N fertilization at a much higher rate, it is possible to increase the rate of grain protein as affected by the elevated level of CO_2 (Bloom, 2006). It should be mentioned that increasing the rate of N fertilization at such a level is not an acceptable method, environmentally and economically (Taub et al., 2008).

The major seed storage proteins from an agricultural point of view are prolamins, globulins, and albumins (Xu and Messing, 2009). All seeds contain albumins; however, monocotyledon seeds contain a higher rate of prolamins and dicotyledon seeds contain a higher rate of globulins. The place of seed storage production is the rough endoplasmic reticulum. The produced seed storage proteins are then located to their site of deposition, which is the protein storage vacuole. There are two different signaling pathways for seed storage proteins for prolamins and for nonprolamins (Miernyka and Hajduch, 2011). Classifying proteins according to their functions determines the following functions for the proteins: (1) cellular structure, (2) cellular metabolism, (3) cellular stress response, (4) hormones and signals, (5) amino acid and protein synthesis, (6) transport across membranes, and (7) unknown functions (Brenner and Cheikh, 1995; Miernyka and Hajduch, 2011; Kranner et al., 2010).

All the details related to the physiological mechanisms, which enable soybean seeds to increase their rate of protein content, is not yet known. The rate of assimilate per seed is among the parameters controlling the rate of soybean seed components including protein and oil. Among the most important factors affecting the rate of assimilate per seed are depoding and shading. Depoding increased the rate of seed protein content, oil, and carbohydrates. However, the most responsive component to the treatment of depoding was the seed protein rate. Shading decreased the rate of all seed components, and the rate of seed oil was the one most affected by shading as the rate of soybean seed oil decreased. The soybean genotypes with a lower rate of protein were highly responsive to rate of assimilate per seed (Wahua, and Miller, 1978; Proulx and Naeve, 2009).

However, the authors indicated the reason for the higher response of seed protein content to the assimilate rate than the other components. Depoding increases the rate of sucrose and amino acids for the use of seed, and as a result, all seed components including protein, oil, and carbohydrates can utilize such a source. It has been suggested that sucrose is able to act as a signal molecule, inducing the activity of storage protein genes. Such a mechanism, with the increased availability of sucrose and amino acids, which is used for the production of protein, indicates the higher response of soybean seed protein related to the other seed components (Staswick, 1994; Smeekens, 2000; Gupta and Kaur, 2005).

The authors suggested two strategies related to the increase of seed protein content: (1) using the genotypes with the increased production of protein and (2) by the production of seed oil and carbohydrates. However, interestingly, such strategies are not related to the special patterns of seed development. Accordingly, seed development and composition may be controlled by different genetic processes. Different soybean genotypes may use different strategies to achieve the high rate of seed protein content. It is possible to evaluate the response of different soybean genotypes under different conditions by using the QTL to determine their rate of different seed components. Accordingly, the related genes, which may determine the rate of different seed components, can be indicated (Liang et al., 2010; Qi et al., 2011).

Soybean Seed Protein and Oil Under Stress

Although the highest rate of protein in crop plants is one-third of the total weight, it has some important functioning in crop plants, determining their quality. The protein quality of crop plants is affected by genetic and environmental factors. In most cases the effects of environmental stresses decrease the rate of seed protein, although in some cases the increased rate of seed protein has also been indicated. Proteins are the important part of gene expression and regulate different cellular activities including metabolism and development.

Using a meta-analysis, the effects of different stresses, including drought, N supply, and heat, were determined on the rate of soybean seed protein and oil. According to the meta-analysis, drought stress decreased the rate of seed protein (mg per seed) and oil. However, the effect of drought stress was more pronounced on the rate of seed oil than on the rate of seed protein. The effects of heat were also similar to the effects of drought stress affecting seed protein and oil. The effects of N supply on the rate of seed protein and oil were more pronounced under hydroponic conditions than under field conditions (Qi et al., 2011).

The authors analyzed 10 years of data related to the differences in seed composition as affected by environmental parameters; there was a high rate of responses among different soybean genotypes. The average for genotypes indicated that the difference between the minimum and maximum value for the seed protein rate and oil content was equal to 18% and 23%, respectively. Water stress during the early reproductive stage resulted in a 16% decrease in seed protein rate (Rotundo and Westgate, 2009).

Soybean seed protein and oil are affected by different parameters, including the genotype and environmental factors. It is possible to minimize the adverse effects of environmental stress on the rate of soybean seed protein and oil by using high-yield protein genotypes. However, the production of high-yielding protein genotypes is sometimes adversely correlated by seed grain yield and oil. Increased heat also decreased seed protein at an average of 9%. N fertilization increased the protein content of soybean seeds by 27%. However, the response under hydroponic conditions was higher than field conditions. The amount of N fertilization has a significant effect on the rate of seed protein content, as at 100 kg/ha, the rate of seed protein content increased just at 2%, while at the rate of 200 kg/ha, the rate of seed protein content increased at 14%.

Soybean oil and proteins are also a function of field properties including soil water content and nitrogen rate. The availability of soil N and N fertilization during the seed fill is also an important parameter affecting the rate of seed protein. The strain of *B. japonicum* is also an important factor determining the rate of seed protein. The more efficient species of *B. japonicum* are able to fix the higher rate of N and hence increase the rate of seed protein related to the soil rhizobium (Luna and Planchon, 1995).

The effect of water stress on the reduction of seed oil was more pronounced than its effect on seed protein rate, which is due to the enhanced rate of amino-N remobilization increasing the rate of seed protein. Water stress decreases the length of the seed filling stage. Under water stress the remobilization of N from the plant leaf increases (Turner et al., 2005). Although the important part of plant N is from the soil uptake and by the process of N fixation, remobilization from the plant leaf can alleviate the stress of N deficiency on seed growth and development. Increased heat rate enhances the rate of seed growth and decreases the time of seed filling (Bhullar and Jenner, 1985). Under a high rate of soil mineral N, the rate of amino-N translocated to the plant aerial part from the process of N fixation decreases. Accordingly, N fertilization may not increase the rate of N content and hence seed protein rate (Rodríguez-Navarro et al., 2011).

Seed germination is the most important and sensitive plant growth stage and is affected by different biotic and abiotic stresses. Using the proteomic technique, the proteins of soybean seeds, with changed concentration under salinity stress, were determined. The downregulation of proteins was different in the sensitive and tolerant genotypes of soybeans under salt stress, indicating that such proteins are among the important factors determining soybean genotype tolerance under stress (Xu et al., 2011; Miransari and Smith, 2014).

Soybean seed protein is a function of climate and with decreasing the temperature the rate of seed protein reduces. This is because in cold climates the rate of biological nitrogen fixation also decreases (Hurburgh, 1994; Tabe et al., 2002; Miransari and Smith, 2008). However, interestingly, researchers have been able to find methods, which can enhance soybean tolerance under cold climates, including the use of genetically modified genotypes and molecular methods (Mabood et al., 2014).

Amirjani (2010) investigated the growth, mineral, and protein content, as well as the antioxidant activity of soybean plants as affected by salinity stress including a control, 50, 100, and 200 mM NaCl. The length and the fresh weight of seedlings were determined under salt stress. Salinity stress decreased plant height. The activity of the N-fixing enzyme and the ammonium content of nodules were determined using the chromatography method. Under the highest level of salinity the activity of the N-fixing enzyme and the ammonium content of nodules decreased by 60% and 100%, respectively. The activity of proline under the salt concentration of 50–200 mM increased. Increasing the salinity level significantly increased the rate of plant sodium and significantly decreased the rate of plant potassium, calcium, and magnesium. Under the concentrations of 100 and 200 mM NaCl the activity of antioxidant enzymes including superoxide dismutase, catalase, and peroxidase increased.

Soybean Protein Signaling and Stress

Under stress, different cellular components are affected, including the proteins of the cell wall. Plant cell walls are interesting and complicated and include carbohydrates, lignin, protein, water, and some substances. Plant growth and development, as well as stressful conditions, can alter such cellular properties. Such alteration also affects the nature and functional properties of a plant cell wall with respect to its growth and development, response under stress, cellular signaling communication, and the behavior of the cellular exchange interfaces (Showalter, 1993).

Plant response under stress is regulated by the activation of different physiological and signaling cascade. Plant response under stress is by the production of antioxidant products and enzymes and activation of signaling cascade such as mitogen-activated protein kinase, plant hormonal signaling, etc. The production of antioxidant enzymes increases under stress so that the produced reactive oxygen is scavenged. The production of reactive oxygen species results in oxidative stress, adversely affecting the cellular structure, including the proteins. Depending on the level of produced antioxidant enzymes the stress can be controlled by the plant (Sajedi et al., 2010, 2011). However, it has been indicated that beneficial microbes including plant growth-promoting rhizobacteria are also able to enhance the production of antioxidant enzymes in plants and hence increase plant tolerance under stress (Miransari, 2011, 2014).

Plants are able to sense and translate the signaling of reactive oxygen species to appropriate cellular responses. For such a process the presence of redox-sensitive proteins, which can be oxidized and reduced during the process, is essential. The reactive oxygen species are able to oxidize the redox-sensitive proteins directly in the presence of the redox-sensitive molecules including reduced glutathione and thioredoxins. The redox-sensitive enzymes are able to regulate the cellular mechanism, and the redox-sensitive proteins are functional by their signaling components including phosphatases, kinases, and transcription factors (Bérczi and Moller, 2000; Zhen et al., 2007).

There are two kinds of molecular mechanisms regulating the activity of the redox-sensitive proteins in the cells: (1) the oxidation of protein thiol parts and (2) the oxidation of iron–sulfur parts in such proteins. The alteration of chemistry in the thiol parts of redox-sensitive proteins affect the protein properties and hence its interactions with the other proteins. The oxidation of iron–sulfur by O_2^- inactivates such a part, affecting the enzyme activity (Nagahara, 2011; Natarajan et al., 2013).

Among the most important effects of flavonoids in legumes is the induction of bacterial nodulation genes during the process of biological N fixation. However, da Silva Batista and Hungria (2012) investigated the other effects of flavonoids on the induction of proteins in the commercial strains of B. japonicum used in commercial soybean inoculants in Brazil and in two genetically modified strains of B. japonicum. The complete set of proteins were isolated from the induced (genistein at 1 µM) and noninduced strains. The proteomic analyses indicated that genistein resulted in the induction of 47 proteins including some hypothetical proteins, the FliG component of cytoplasmic flagellar, the preplasmic proteins, the proteins related to the maintenance of redox-state, and a protein essential for the production of exopolysaccharides. Interestingly, genistein also resulted in the induction of the components essential for the process of symbiotic efficiency regulating *B. japonicum* response under stress.

Li et al. (2005) isolated the three genes including *GmDREBa*, *GmDREBb*, and *GmDREBc* from soybeans (*Glycine max* (L.) Merr). Different stresses, including salinity, drought, and cold, induced the transcriptions of *GmDREBa* and *GmDREBb* in the leaves of soybean seedlings. The authors accordingly indicated that such genes are able to regulate soybean growth under stress.

The production of proline is among the most prevalent responses of plants under stress, including salinity, drought, heavy metal, nutrient deficiency, cold, suboptimal root zone temperature, and biotic stresses such as pathogenic microbes. Salinity and drought result in the induction of oxidative stress. Proline can scavenge the reactive oxygen species, which are produced under stress; as a molecular chaperone, it is also able to stabilize the structure of proteins. It can also regulate plant response under stress as a part of plant-signaling cascade (Verbruggen and Hermans, 2008).

Different signaling pathways in plants are regulated by proteins such as protein kinases. For example, the calcium-signaling pathway is regulated by protein kinases. Such a pathway is activated and is a regulator of plant response to different biotic and abiotic stresses. In *Arabidopsis*, such a pathway is activated by the activation of different stress genes. Different signaling sensors recognize calcium pathways and translate them into plant responses, including the phosphorylation of proteins and expression of genes. There is a set of proteins in plants, which are able to bind calcium including protein kinases, calmodulins, and calcineurin B-like proteins (Cheng et al., 2002). Under different types of stress a set of proteins are produced in plants, helping the plant to tolerate the stress. For example, under stress the induction of proteins (α-tubulin and actin) altering plant morphology increases root growth (Cheng et al., 2002).

There is an important role for seed proteins, which is the germination of seeds. During seed germination the seed proteins must be mobilized and available so that the seed can germinate. During seed germination the following protein regulation results in the usage of seed proteins for the process of seed germination: (1) repair of storage protein, (2) utilization of storage protein, (3) alteration of protein changes (synthesis of protein vs degradation of protein), and (4) modification of storage and synthesized protein (Miransari and Smith, 2014; Rensing, 2014; Valmonte et al., 2014).

During the process of seed germination, ABA induces the activation of late embryogenesis abundant proteins, increasing their rate and their use for seed germination, and controls the adverse effects of desiccation stress on seed germination. The synthesis of proteins is essential for the completion of seed germination. Proteomic analysis has indicated that during the process of seed germination, seed proteins are subjected to a set of modifications affecting protein functioning including localization, production, stabilization, and activity (Yang et al., 2007; Xu et al., 2011).

During the process of seed germination the redox potential of seed proteins is significantly changed. In cereals the seed storage proteins, which are in oxidized form, are reduced during the process of germination. Such a process is done by the protein thioredoxin. The enzymatic reduction of this protein, in the presence of nicotinamide adenine dinucleotide phosphate, results in the actin of thioredoxin as an early signal during seed germination, enhancing the mobilization of seed storage by (1) the reduction of storage proteins and hence their increased solubilization and their vulnerability to proteolysis; (2) the reduction and inactivation of proteins, which are able to inhibit the activity of proteases and amylases and hence facilitate the catalysis of storage proteins and starch; and (3) the activation of enzymes, which are essential for the process of seed germination (Bailly, 2004; El-Maarouf-Bouteau and Bailly, 2008; Li et al., 2009).

It has been indicated that a set of protein phosphates and protein kinases are activated during the process of seed germination, by modifying ABA signaling. It is also essential that the enzymes, which are essential for the repair of DNA and the translation of proteins, become phosphorylated so that the molecular controlling of seed germination is properly conducted. Some other biochemical modifications of proteins, which are related to the alteration of protein structure, are also essential for the process of seed germination (Bewley, 1997; Davies, 2010; Law et al., 2014).

Conclusion and Future Perspectives

The soybean is an important source of protein and oil, feeding a large number of people in the world. However, it is subjected to different types of stress, significantly affecting its growth, yield, and grain ingredients, including protein and oil. Under stress, soybean seed oil and protein is affected, and hence the nutritional quality of soybean seeds are changed. In this chapter some of the most important details related to the effects of stress on the rate of soybean protein and oil have been presented and analyzed. Although more research is essential on the effects of stress affecting soybean seed protein and oil, it has been indicated that under stress, usually the rate of soybean seed protein and oil increases and decreases, respectively. The details presented in this chapter include the effects of environmental parameters affecting soybean seeds under unstressed and stressed conditions, the effects of stress on soybean protein and oil, and the effects of soybean protein regulating plant response under stress. However, more research is essential related to the effects of environmental stresses affecting soybean seed protein and oil and how it is possible to enhance soybean seed protein and oil while soybean seed yield remains unchanged.

References

Allen, D.K., Ohlrogge, J.B., Shachar-Hill, Y., 2009. The role of light in soybean seed filling metabolism. The Plant Journal 58, 220–234.

Amirjani, M.R., 2010. Effect of salinity stress on growth, mineral composition, proline content, antioxidant enzymes of soybean. American Journal of Plant Physiology 5, 350–360.

Arslanoglu, F., Aytac, S., Oner, E., 2013. Effect of genotype and environment interaction on oil and protein content of soybean (Glycine max (L.) Merrill) seed. African Journal of Biotechnology 10, 18409–18417.

Bailly, C., 2004. Active oxygen species and antioxidants in seed biology. Seed Science Research 14, 93–107.

Bellaloui, N., 2012. Soybean seed phenol, lignin, and isoflavones partitioning as affected by seed node position and genotype differences. Food and Nutrition Sciences 3, 447–454.

Bérczi, A., Moller, I.M., 2000. Redox enzymes in the plant plasma membrane and their possible roles. Plant Cell and Environment 23, 1287–1302.

Bewley, J., 1997. Seed germination and dormancy. The Plant Cell 9, 1055.

Bhullar, S.S., Jenner, C.F., 1985. Differential responses to high-temperatures of starch and nitrogen accumulation in the grain of 4 cultivars of wheat. Australian Journal of Plant Physiology 12, 363–375.

Blevins, D.G., 1985. Role of potassium in protein metabolism in plants. In: Munson, R.D. (Ed.), Potassium in Agriculture. ASA, CSSA, and SSSA, Madison, WI, pp. 413–424.

Bloom, A.J., 2006. Rising carbon dioxide concentrations and the future of crop production. Journal of the Science of Food and Agriculture 86, 1289–1291.

Borrás, L., Slafer, G., Otegui, M., 2004. Seed dry weight response to source–sink manipulations in wheat, maize and soybean: a quantitative reappraisal. Field Crops Research 86, 131–146.

Brenner, M., Cheikh, N., 1995. The role of hormones in photosynthate partitioning and seed filling. In: Plant Hormones. Springer Netherlands, pp. 649–670.

Brumm, T.J., Hurburgh, C.R., 1990. Estimating the processed value of soybean. Journal of American Oil Chemists' Society 67, 302–307.

Brumm, T.J., Hurburgh, C.R., 2006. Changes in long-term soybean compositional patterns. Journal of American Oil Chemists' Society 83, 981–982.

Cheng, S., Willmann, M., Chen, H., Sheen, J., 2002. Calcium signaling through protein kinases. The Arabidopsis calcium-dependent protein kinase gene family. Plant Physiology 129, 469–485.

Clarke, F.J., Wiseman, J., 2000. Developments in plant breeding for improved nutritional quality in soybean. I. Protein and amino acids content. Journal of Agricultural Science 134, 111–124.

DaMatta, F., Grandis, A., Arenque, B., Buckeridge, M., 2010. Impacts of climate changes on crop physiology and food quality. Food Research International 43, 1814–1823.

Davies, P., 2010. The plant hormones: their nature, occurrence, and functions. In: Plant Hormones. Springer Netherlands, pp. 1–15.

Dornbos Jr., D.L., Mullen, R.E., 1992. Soybean seed protein and oil contents and fatty acid composition adjustments by drought and temperature. Journal of the American Oil Chemists' Society 69, 228–231.

Dupont, F., Altenbach, S., 2003. Molecular and biochemical impacts of environmental factors on wheat grain development and protein synthesis. Journal of Cereal Science 38, 133–146.

Dwivedi, S., Nigam, S., Rao, R., Singh, U., Rao, K., 1996. Effect of drought on oil, fatty acids and protein contents of groundnut (Arachis hypogaea L) seeds. Field Crops Research 48, 125–133.

Egli, D., Bruening, W., 2007a. Accumulation of nitrogen and dry matter by soybean seeds with genetic differences in protein concentration. Crop Science 47, 359–366.

Egli, D., Bruening, W., 2007b. Nitrogen accumulation and redistribution in soybean genotypes with variation in seed protein concentration. Plant and Soil 301, 165–172.

El-Maarouf-Bouteau, H., Bailly, C., 2008. Oxidative signaling in seed germination and dormancy. Plant Signaling & Behavior 3, 175–182.

Fehr, W.R., Hoeck, J.A., Johnson, S.L., Murphy, P.A., Nott, J.D., Padilla, G.I., Welke, G.A., 2003. Genotype and environment influence on protein components of soybean. Crop Science 43, 511–514.

Feng, Z., Kobayashi, K., Ainsworth, E., 2008. Impact of elevated ozone concentration on growth, physiology, and yield of wheat (Triticum aestivum L.): a meta-analysis. Global Change Biology 14, 2696–2708.

Flagella, Z., Giuliani, M., Giuzio, L., Volpi, C., Masci, S., 2010. Influence of water deficit on durum wheat storage protein composition and technological quality. European Journal of Agronomy 33, 197–207.

Fuhrer, J., 2009. Ozone risk for crops and pastures in present and future climates. Naturwissenschaften 96, 173–194.

Galili, G., Amir, R., 2013. Fortifying plants with the essential amino acids lysine and methionine to improve nutritional quality. Plant Biotechnology Journal 11, 211–222.

Gupta, A., Kaur, N., 2005. Sugar signalling and gene expression in relation to carbohydrate metabolism under abiotic stresses in plants. Journal of Biosciences 30, 761–776.

Hajduch, M., Ganapathy, A., Stein, J.W., Thelen, J.J., 2005. A systematic proteomic study of seed filling in soybean: establishment of high resolution two-dimensional reference maps, expression profiles, and an interactive proteome database. Plant Physiology 137, 1397–1419.

Hajduch, M., Hearne, L., Miernyk, J., Casteel, J., Joshi, T., Agrawal, G., Song, Z., Zhou, M., Xu, D., Thelen, J., 2010. Systems analysis of seed filling in Arabidopsis: using general linear modeling to assess concordance of transcript and protein expression. Plant Physiology 152, 2078–2087.

Herman, E., 2014. Soybean seed proteome rebalancing. In: Frontiers in Plant Science, 5.

Hettiarachchy, N., Kalapathy, U., 1997. Soybean protein products. In: Soybeans. Springer US, pp. 379–411.

Higgins, T., 1984. Synthesis and regulation of major proteins in seeds. Annual Review of Plant Physiology 35, 191–221.

Hurburgh Jr., C., Brumm, T., Guinn, J., Hartwig, R., 1990. Protein and oil patterns in US and world soybean markets. Journal of the American Oil Chemists' Society 67, 966–973.

Hurburgh Jr., C., 1994. Long-term soybean composition patterns and their effect on processing. Journal of American Oil Chemist' Society 71, 1425–1427.

Jin, J., Liu, X., Wang, G., Mi, L., Shen, Z., Chen, X., Herbert, S., 2010. Agronomic and physiological contributions to the yield improvement of soybean cultivars released from 1950 to 2006 in Northeast China. Field Crops Research 115, 116–123.

Kranner, I., Minibayeva, F.V., Beckett, R.P., Seal, C.E., 2010. What is stress? Concepts, definitions and applications in seed science. New Phytologist 188, 655–673.

Kress, L., Miller, J., 1983. Impact of ozone on soybean yield. Journal of Environmental Quality 12, 276–281.

Law, S., Narsai, R., Whelan, J., 2014. Mitochondrial biogenesis in plants during seed germination. Mitochondrion 19, 214–221.

Li, X., Tian, A., Luo, G., Gong, Z., Zhang, J., Chen, S., 2005. Soybean DRE-binding transcription factors that are responsive to abiotic stresses. Theoretical and Applied Genetics 110, 1355–1362.

Li, Y., Ren, J., Cho, M., Zhou, S., Kim, Y., Guo, H., et al., 2009. The level of expression of thioredoxin is linked to fundamental properties and applications of wheat seeds. Molecular Plant 2, 430–441.

Liang, H., Yu, Y., Wang, S., Yun, L., Wang, T., Wei, Y., Gong, P., Liu, X., Fang, X., Zhang, M., 2010. QTL mapping of isoflavone, oil and protein contents in soybean (*Glycine max* L. Merr.). Agricultural Sciences in China 9, 1108–1116.

Liener, I., 1981. Factors affecting the nutritional quality of soya products. Journal of the American Oil Chemists' Society 58, 406–415.

Lin, C., Li, C., Lin, S., Yang, F., Huang, J., Liu, Y., Lur, H., 2010. Influence of high temperature during grain filling on the accumulation of storage proteins and grain quality in rice (*Oryza sativa* L.). Journal of Agriculture and Food Chemistry 58, 10545–10552.

Liu, K., 1997. Soybeans: Chemistry, Technology and Utilization. Chapman and Hall, New York.

Liu, C., Wang, X., Ma, H., Zhang, Z., Gao, W., Xiao, L., 2008. Functional properties of protein isolates from soybeans stored under various conditions. Food Chemistry 111, 29–37.

Luna, R., Planchon, C., 1995. Genotype X *Bradyrhizobium japonicum* strain interactions in dinitrogen fixation and agronomic traits of soybean (*Glycine max* L. Merr.). Euphytica 86, 127–134.

Mabood, F., Zhou, X., Smith, D.L., 2006. Preincubated with methyl jasmonate increases soybean nodulation and nitrogen fixation. Agronomy Journal 98, 289–294.

Mabood, F., Zhou, X., Smith, D., 2014. Microbial signaling and plant growth promotion. Canadian Journal of Plant Science 94, 1051–1063.

Mandarino, J.M.G., Carrao-Panizzi, M.C., Oliveira, M.C.N., 1992. Chemical composition of soybean seed from different production areas of Brazil. Arquivos de Biologia e Tecnologia 35, 647–653.

Medic, J., Atkinson, C., Hurburgh Jr., C.R., 2014. Current knowledge in soybean composition. Journal of the American Oil Chemists' Society 91, 363–384.

Menacho, L., Amaya-Farfan, J., Berhow, M., Mandarino, J., Mejia, E., Chang, Y., 2010. A high-protein soybean cultivar contains lower isoflavones and saponins but higher minerals and bioactive peptides than a low-protein cultivar. Food Chemistry 120, 15–21.

Miernyka, J.A., Hajduch, M., 2011. Seed proteomics. Journal of Proteomics 74, 389–400.

Miernyk, J.A., Johnston, M.L., 2013. Proteomic analysis of the testa from developing soybean seeds. Journal of Proteomics 89, 265–272.

Miransari, M., Smith, D.L., 2007. Overcoming the stressful effects of salinity and acidity on soybean nodulation and yields using signal molecule genistein under field conditions. Journal of Plant Nutrition 30, 1967–1992.

Miransari, M., Smith, D., 2008. Using signal molecule genistein to alleviate the stress of suboptimal root zone temperature on soybean-*Bradyrhizobium* symbiosis under different soil textures. Journal of Plant Interactions 3, 287–295.

Miransari, M., Smith, D.L., 2009. Alleviating salt stress on soybean (*Glycine max* (L.) Merr.)–*Bradyrhizobium japonicum* symbiosis, using signal molecule genistein. European Journal of Soil Biology 45, 146–152.

Miransari, M., 2011. Interactions between arbuscular mycorrhizal fungi and soil bacteria. Applied Microbiology and Biotechnology 89, 917–930.

Miransari, M., 2014. Plant growth promoting rhizobacteria. Journal of Plant Nutrition 37, 2227–2235.

Miransari, M., Smith, D.L., 2014. Plant hormones and seed germination. Environmental and Experimental Botany 99, 110–121.

Moïse, J., Han, S., Gudynaite-Savitch, L., Johnson, D., Miki, B., 2005. Seed coats: structure, development, composition, and biotechnology. In Vitro Cellular & Developmental Biology 41, 620–644.

Moura, J., Bonine, C., De Oliveira Fernandes Viana, J., Dornelas, M., Mazzafera, P., 2010. Abiotic and biotic stresses and changes in the lignin content and composition in plants. Journal of Integrative Plant Biology 52, 360–376.

Nagahara, N., 2011. Intermolecular disulfide bond to modulate protein function as a redox-sensing switch. Amino Acids 41, 59–72.

Natarajan, S., Luthria, D., Bae, H., Lakshman, D., Mitra, A., 2013. Transgenic soybeans and soybean protein analysis: an overview. Journal of Agricultural and Food Chemistry 61, 11736–11743.

Pathan, S.M., Vuong, T., Clark, K., Lee, J.D., Shannon, J.G., Roberts, C.A., et al., 2013. Genetic mapping and confirmation of quantitative trait loci for seed protein and oil contents and seed weight in soybean. Crop Science 53, 765–774.

Pettigrew, W.T., 2008. Potassium influences on yield and quality production for maize, wheat, soybean and cotton. Physiologia Plantarum 133, 670–681.

Pleijel, H., Danielsson, H., Gelang, J., Sild, E., Selldén, G., 1998. Growth stage dependence of the grain yield response to ozone in spring wheat (*Triticum aestivum* L.). Agriculture Ecosystem and Environment 70, 61–68.

Proulx, R., Naeve, S., 2009. Pod removal, shade, and defoliation effects on soybean yield, protein, and oil. Agronomy Journal 101, 971–978.

Qi, Z.M., Wu, Q., Han, X., Sun, Y., Du, X., Liu, C., Jiang, H., Hu, G., Chen, Q., 2011. Soybean oil content QTL mapping and integrating with meta-analysis method for mining genes. Euphytica 179, 499–514.

Rensing, S.A., 2014. Gene duplication as a driver of plant morphogenetic evolution. Current Opinion in Plant Biology 17, 43–48.

Rodríguez-Navarro, D., Oliver, I., Contreras, M., Ruiz-Sainz, J., 2011. Soybean interactions with soil microbes, agronomical and molecular aspects. Agronomy for Sustainable Development 31, 173–190.

Rose, I.A., 1988. Effects of moisture stress on the oil and protein components of soybean seeds. Crop and Pasture Science 39, 163–170.

Rosenthal, A., Pyle, D.L., Niranjan, K., 1998. Simultaneous aqueous extraction of oil and protein from soybean: mechanisms for process design. Food and Bioproducts Processing 76, 224–230.

Rotundo, J.L., Westgate, M.E., 2009. Meta-analysis of environmental effects on soybean seed composition. Field Crops Research 110, 147–156.

Saint Pierre, C., Peterson, C., Ross, A., Ohm, J., Verhoeven, M., Larson, M., Hoefer, B., 2008. Winter wheat genotypes under different levels of nitrogen and water stress: changes in grain protein composition. Journal of Cereal Science 47, 407–416.

Sajedi, N., Ardakani, M., Rejali, F., Mohabbati, F., Miransari, M., 2010. Yield and yield components of hybrid corn (*Zea mays* L.) as affected by mycorrhizal symbiosis and zinc sulfate under drought stress. Physiology and Molecular Biology of Plants 16, 343–351.

Sajedi, N., Ardakani, M., Madani, H., Naderi, A., Miransari, M., 2011. The effects of selenium and other micronutrients on the antioxidant activities and yield of corn (*Zea mays* L.) under drought stress. Physiology and Molecular Biology of Plants 17, 215–222.

Savin, R., Nicolas, M., 1999. Effects of timing of heat stress and drought on growth and quality of barley grains. Australian Journal of Agricultural Research 50, 357–364.

Schmidt, M.A., Herman, E.M., 2008. The collateral protein compensation mechanism can be exploited to enhance foreign protein accumulation in soybean seeds. Plant Biotechnology Journal 6, 832–842.

Sharma, S., Kaur, M., Goyal, R., Gill, B.S., 2014. Physical characteristics and nutritional composition of some new soybean (*Glycine max* (L.) Merrill) genotypes. Journal of Food Science and Technology 51, 551–557.

Showalter, A.M., 1993. Structure and function of plant cell wall proteins. The Plant Cell 5, 9–23.

Sinclair, T., Vadez, V., 2012. The future of grain legumes in cropping systems. Crop and Pasture Science 63, 501–512.

Smeekens, S., 2000. Sugar-induced signal transduction in plants. Annual Review of Pant Biology 51, 49–81.

SOYSTATS, 2014. http://soystats.com/facts.

Specht, J.E., Germann, M., Markwell, J.P., Lark, K.G., Orf, J.H., Macrander, M., Chase, K., Chung, J., Graef, J.L., 2001. Soybean response to water: a QTL analysis of drought tolerance. Crop Science 41, 493–509.

Staswick, P., 1994. Storage proteins of vegetative plant tissues. Annual Review of Plant Biology 45, 303–322.

da Silva Batista, J., Hungria, M., 2012. Proteomics reveals differential expression of proteins related to a variety of metabolic pathways by genistein-induced *Bradyrhizobium japonicum* strains. Journal of Proteomics 75, 1211–1219.

Tabe, L., Hagan, N., Higgins, T.J.V., 2002. Plasticity of seed protein composition in response to nitrogen and sulfur availability. Current Opinion in Plant Biology 5, 212–217.

Taub, D., Miller, B., Allen, H., 2008. Effects of elevated CO_2 on the protein concentration of food crops: a meta-analysis. Global Change Biology 14, 565–575.

Turner, N.C., Davies, S.L., Plummer, J.A., Siddique, K.H.M., 2005. Seed filling in grain legumes under water deficits, with emphasis on chickpeas. Advances in Agronomy 87, 211–250.

TWumasi-Afriyie, S., Friesen, D., Pixley, K., 2011. Quality protein maize: progress and prospects. Plant Breeding Reviews 34, 83.

Valmonte, G., Arthur, K., Higgins, C., MacDiarmid, R., 2014. Calcium-dependent protein kinases in plants: evolution, expression and function. Plant and Cell Physiology 55, 551–569.

Velazquez, E., Silva, L., Peix, Á., 2010. Legumes: a healthy and ecological source of flavonoids. Current Nutrition & Food Science 6, 109–144.

Verbruggen, N., Hermans, C., 2008. Proline accumulation in plants: a review. Amino Acids 35, 753–759.

Wahua, T., Miller, D., 1978. Effects of shading on the N_2-fixation, yield, and plant composition of field-grown soybeans. Agronomy Journal 70, 387–392.

Wenefrida, I., Utomo, H., Blanche, S., Linscombe, S., 2009. Enhancing essential amino acids and health benefit components in grain crops for improved nutritional values. Recent Patents on DNA and Gene Sequences 3, 219–225.

Wesley, T., Lamond, R., Martin, V., Duncan, S., 1998. Effects of late-season nitrogen fertilizer on irrigated soybean yield and composition. Journal of Production Agriculture 11, 331–336.

Weichert, N., Saalbach, I., Weichert, H., Kohl, S., Erban, A., Kopka, J., et al., 2010. Increasing sucrose uptake capacity of wheat grains stimulates storage protein synthesis. Plant Physiology 152, 698–710.

Wilson, R.F., 2004. Seed Composition. Soybeans: Improvement, Production and Users, third ed. ASA, CSSA, SSSA, Madison, WI.

Woolf, D., Amonette, J., Street-Perrott, F., Lehmann, J., Joseph, S., 2010. Sustainable biochar to mitigate global climate change. Nature Communications 1, 56.

Wu, W., Wu, X., Hua, Y., 2010. Structural modification of soy protein by the lipid peroxidation product acrolein. Food Science and Technology 43, 133–140.

Xu, J.H., Messing, J., 2009. Amplification of prolamin storage protein genes in different subfamilies of the Poaceae. Theoretical and Applied Genetics 119, 1397–1412.

Xu, X.Y., Fan, R., Zheng, R., Li, C.M., Yu, D.Y., 2011. Proteomic analysis of seed germination under salt stress in soybeans. Journal of Zhejiang University Science B 12, 507–517.

Yang, P., Li, X., Wang, X., Chen, H., Chen, F., Shen, S., 2007. Proteomic analysis of rice (*Oryza sativa*) seeds during germination. Proteomics 7, 3358–3368.

Zhang, F., Smith, D.L., 1995. Preincubation of *Bradyrhizobium japonicum* with genistein accelerates nodule development of soybean at suboptimal root zone temperatures. Plant Physiology 108, 961–968.

Zhang, F., Smith, D.L., 2002. Interorganismal signaling in suboptimum environments: the legume-rhizobia symbiosis. Advances in Agronomy 76, 125–161.

Zhen, Y., Qi, J.L., Wang, S.S., Su, J., Xu, G.H., Zhang, M.S., Miao, L.V., Peng, X., Tian, D., Yang, Y.H., 2007. Comparative proteome analysis of differentially expressed proteins induced by Al toxicity in soybean. Physiologia Plantarum 131, 542–554.

Soybeans, Stress, and Plant Growth-Promoting Rhizobacteria

M. Miransari
AbtinBerkeh Scientific Ltd. Company, Isfahan, Iran

Introduction

Although use of chemicals such as nitrogen (N) and phosphorus (P) fertilization for crop production is essential, due to the economic and environmental consequences, such methods are not recommended, especially an extra amount. Using chemicals for plant growth and crop production results in the contamination of the environment and interrupts the ecological activities, the cycling of nutrients, and the growth and the activities of biological communities. Accordingly, alternative methods must be used, which can be efficiently combined with such chemicals. The use of plant growth-promoting rhizobacteria (PGPR) is among such efficient methods (Sivasakthi et al., 2014; Maksimov et al., 2015).

Different abiotic and biotic factors adversely affect plant growth and yield production. However, the presence of an active community of soil microbes, including bacteria, fungi, algae, and protozoa, around plant roots in the area called the rhizosphere, feeding on the roots products, can effectively and efficiently affect plant growth and yield production by a wide range of activities. The concept of the rhizosphere was first explained by Hiltner, indicating an active layer of soil microbes in the area surrounding plant roots, which are activated by root products. Such a concept is now indicated by the soil around plant roots with physical, chemical, and biological activities, which is altered by the growth and activities of plant roots (Sivasakthivelan and Saranraj, 2013).

Both favorite microbes, including rhizobia, mycorrhizal fungi, and actinomycetes, and nonfavorite microbes are found in the rhizosphere, affecting plant growth and yield production. PGPR are free-living bacteria, with the ability of colonizing the root rhizosphere and hence increasing plant growth. The dominant species are *Pseudomonas* and *Bacillus* (Podile and Kishore, 2006).

Bacteria are the most abundant microbial community in the rhizosphere. However, the most active and useful bacteria for plant growth are selected by plants through the production of organic compounds. Because the bacteria are the most abundant microbial community in the root rhizosphere, especially with respect to their great ability of competitiveness and root colonizing, it is more likely that plant growth and physiology is more affected by the bacterial activities (Glick, 2014).

The soybean, as an important legume, is able to establish a symbiotic association with its specific rhizobium, *Bradyrhizobium japonicum*, and acquires most of its essential N (Miransari, 2016). However, the plant is also able to develop a nonsymbiotic association with other soil microbes, specifically plant growth-promoting rhizobacteria (PGPR). The rhizosphere is a dynamic and active environment surrounding plant

Environmental Stresses in Soybean Production. http://dx.doi.org/10.1016/B978-0-12-801535-3.00008-5

roots. A high rate of biological activities take place in the rhizosphere by plant roots and soil microbes, affecting their growth and their efficiency. The physical, chemical, and biological properties of the soil microbes in the rhizosphere affect their growth, their interaction, their number, and their diversity (Zhuang et al., 2007; Saharan and Nehra, 2011).

Due to the effects of environmental factors, the effects of PGPR on plant growth are different under greenhouse and field conditions. For example, the establishment and survival of PGPR in the rhizosphere and on plant roots are functions of root exudates and microbial products (Miransari, 2011a,b,c). Such details are also affected by microbial motility and quorum sensing. The other important factor, which affects the nature of root products, is the properties of soil microbes, specifically their pathogenic activities. The damages caused by pathogens results in the leakage of nutrients affecting the activity of soil microbes (Bassler, 1999; Wai-Leung and Bassler, 2009).

The most important parameter affecting the colonization ability of PGPR is their survival in the rhizosphere, which is a function of PGPR interactions with the physical, chemical, and biological properties of the rhizosphere, as well as the properties of bacteria. Although anchoring plants and the uptake of water and nutrients is among the most important functions of plant roots, they have some other important functions including the production of root exudates, affecting the rhizosphere microbes, their growth, and their interactions with plants. Root exudates are a major source of nutrients and they can act as signals for bacterial activity, growth, and survival. Root exudates include free oxygen and water, ions, mucilage, and different carbon products (Bertin et al., 2003; Singh et al., 2014).

The biochemical properties of root exudates may attract some microorganisms and deter the other microbes. Such differences can also increase or decrease the availability of different soil nutrients. Also the competition between the soil microbes affects their response to plant products. The root rhizosphere and the population of soil microbes are affected by the process of root exudation, which acidifies the root rhizosphere. Decreasing the rhizosphere pH to less than 5.5 reduces the availability of macronutrients and increases the availability of micronutrients including iron, manganese, and aluminum, adversely affecting plant growth and activity (Haichar et al., 2008).

Under P-deficient conditions the production of organic acid onions by plant roots decreases the rhizosphere pH and hence increases the availability of micronutrients such as iron, manganese, and zinc under calcareous conditions, affecting the activity of PGPR as well as the structure of soil. The production of antibiotic and hence controlling plant pathogens by PGPR is affected by soil minerals. The presence of phosphorus in the soil adversely affects the production of antibiotics by PGPR and hence, under high concentration of phosphorus, plants are more subjected to biotic stresses as the environment is also negatively affected (Milner et al., 1995).

With respect to the great abilities of PGPR and their effects on the growth and yield of crop plants, specifically soybean as the most important legume plants, the most recent findings and details have been presented and analyzed. Some future perspectives have also been presented in the conclusion, which can make the use of PGPR more interesting and applicable, especially their commercial use under field conditions.

Plant Growth-Promoting Rhizobacteria

PGPR includes a wide range of soil microbes, including the symbiotic N fixers, which are able to fix the atmospheric N in a symbiotic association with their specific host plant rhizobia and *Bradyrhizobium*. Most of the biological N fixation is fixed by such soil microbes. *Rhizobium* and *Bradyrhizobium* are able to develop a specific symbiotic association with their legume host plant and fix the atmospheric N for their use. About 50–60% of the essential N of legumes is fixed biologically, and the remaining can be supplied by using chemical and organic fertilization (Davidson, 2009).

In such an association, the communication between the two symbionts results in the production of nodules on the soybean roots, which are the place of rhizobium residence for N biological fixation. However, it has also been interestingly indicated that the combination of *Rhizobium* and *Bradyrhizobium* with the nonsymbiotic PGPR N fixers, such as *Azospirillum*, can also be a suitable method for providing the legumes with their remaining part of their essential N (Hungria et al., 2013, 2015).

A part of N is biologically fixed by nonsymbiotic N fixers as indicated in the following. Although such microbes are not in an intimate association with their host plant, their presence in the rhizosphere is beneficial to their host plant and can significantly affect its growth. Although the nonsymbiotic fixers PGPR are able to enhance the growth of their host plant by providing the plant with biologically fixed N, they can also increase the growth of their host plant by producing different metabolites, such as plant hormones (Bashan, 1998).

PGPR include a wide range of microbes with different activities, including (1) the enhanced availability of soil nutrients, (2) plant hormone production, (3) alleviation of soil stresses, (4) interaction with the other soil microbes including mycorrhizal fungi, (5) controlling pathogens, (6) affecting root growth and functioning, (7) altering plant morphology and physiology, and (8) improving the properties of soil (Lugtenberg and Kamilova, 2009; Singh et al., 2014).

PGPR include a range of indirect mechanisms under different conditions, including the biotic stresses, which are (1) the induction of plant systemic resistance, (2) competition of soil microbes in the rhizosphere, (3) the production of different stress hormones including jasmonic acid and ethylene, and (4) production of antibiotics, preventing the activities of pathogens on plant growth. However, the direct mechanisms by PGPR include (1) biological N fixation, (2) production of different plant hormones including abscisic acid (ABA), auxins, cytokinins, gibberellins, nitric oxide, and 1-amino cyclopropane-1-carboxylate (ACC)-deaminase, (3) sequestration of iron by siderophores, and (4) solubilizaiton of phosphate (Glick, 2014).

Ahmad et al. (2008) evaluated different species of PGPR including fluorescent *Pseudomonas*, *Azotobacter*, *Bacillus*, and *Mesorhizobium* for their plant growth activities. The bacterial species were isolated from the rhizosphere and root nodules of different plant species. Different plant growth-promoting activities were evaluated, including the production of plant hormone [indole acetic acid (IAA)], production of ammonium, siderophore and hydrogen cyanide (HCN), antifungal activity, and solubilization of phosphate.

The ability of hormone production was found in 80% of different bacterial strains including fluorescent *Pseudomonas, Azotobacter*, and *Mesorhizobium*; however, only 20% of the *Bacillus* strains were able to produce the hormone. The phosphate solubilization ability of the bacterial strains was the highest by *Bacillus* and the lowest by *Mesorhizobium* (16.67%), and *Azotobacter* (74.47%) and fluorescent *Pseudomonas* (55.56%) were in between. Although all the isolates had the ability of producing ammonia, none of them were able to catabolize chitin. With the exception of *Mesorhizobium*, the antifungal activity and siderophore production of the isolates were in the range of 10–12.77%. Just *Bacillus* (50%) and *Pseudomonas* (88.89%) were able to produce HCN.

Some of the most efficient and applicable strains of PGPR are presented in greater detail in the following section.

Azospirillum

Azospirillum, as the most important genera of PGPR, are able to colonize the roots of several plant species and increase their growth by the production of different metabolites, including plant hormones. The isolation and use of *Azospirillum* was initiated from 1970 (Steenhoudt and Vanderleyden, 2000). *Azospirillum* consists of 10 species, including *Azospirillum brasilense, Azospirillum amazonense, Azospirillum irakense, Azospirillum lipoferum, Azospirillum largimobile, Azospirillum halopraeferens, Azospirillum oryzae, Azospirillum canadensis, Azospirillum doebereinerae, and Azospirillum melinis* (Tarrand et al., 1978; Magalhães et al., 1983; Dekhil et al., 1997; Peng et al., 2006; Mehnaz et al., 2007; Saharan and Nehra, 2011).

Although *Azospirillum* was first isolated from cereals and most of its inoculation has been done with these crop plants, the use of *Azospirillum* on the noncereals has also been successful. The association of *Azospirillum* with crop plants is not specific, and the bacteria are able to inoculate almost all crop plants, which have not been previously inoculated with the bacteria. *Azospirillum* species are found in the rhizosphere of most important crop plants and are able to fix N under the conditions of microaerophilic. There was a suggestion at the earlier time that *Azospirillum* is able to provide cereal plants with their essential N (Sen, 1929). Usually the establishment of *Azospirillum* in the rhizosphere results in the promotion of plant growth. Although the bacteria is able to fix the atmospheric N (1–10 kg/ha), the most important mechanism, which increases plant growth, is the enhanced growth of plant roots, and hence the uptake of water and nutrients by plants, due to the production of different substances by the bacteria (Saharan and Nehra, 2011).

Azospirillum affects plant growth under the following stress conditions by the production of different plant hormones including auxins (Spaepen et al., 2007): salinity and drought stress (osmotic and matrix stresses), acidic conditions, the limitation of carbon resources, and the production of chemical compounds produced by plants under stress conditions. Although auxins are produced at different stages of *Azospirillum* growth, the highest rate of the hormone is produced at the stage of stationary phase. Under acidic conditions the expression of the ipdC gene in *Azospirillum* is activated, especially during the stationary phase, resulting in the increased production of IAA.

ABA can regulate plant growth under stress by regulating plant response to water stress, which is due to the modification of stomata activities. The first individuals who indicated the production of ABA by *Azospirillum* in defined culture were Kolb and Martin (1985). Perrig et al. (2007) also indicated that *A. brasilense* is able to produce ABA at 75.0 ng/mL in a defined medium. Under the conditions of salty culture medium, NaCl increased the production of ABA by *A. brasilense* (235 ng/mL) compared with control conditions (73 ng/mL).

According to Cohen et al. (2008), inoculated *Arabidopsis thaliana* with *A. brasilense* had a twofold higher ABA than the noninoculated plant. The authors indicated that *A. brasilense* produced ABA under water stress according to the following details. They treated maize plants with fluridone, which prevents the biosynthesis of ABA in plants, and hence stunted plants are produced even under nonstress conditions. However, *A. brasilense* was able to avoid such an effect. Although plants treated with fluridone and subjected to water stress had a significantly lower rate of water, inoculation with *A. brasilense* alleviated such a stress, indicating that the production of ABA by *A. brasilense* can alleviate water stress (Cohen et al., 2008).

A finding has indicated that *A. brasilense* is able to produce polyamine, which results in the enhancement of root growth and alleviates osmotic stress. Different polyamine products are produced by *A. brasilense*, including spermidine, spermien, and putercine. However, such abilities have also been indicated by *Pseudomonas* and *Azetobacter*, affecting plant growth under different conditions, including stress (Cassán et al., 2014).

Pseudomonas

With respect to the great properties of *Pseudomonas*, they can act as PGPR and increase plant growth and yield production. Fluorescent *pseudomonads* are the most efficient and effective *Pseudomonas*. These bacteria contribute to the health of the soil and are metabolically active and diverse. The isolated strains of *Pseudomonas fluorescens* and *Pseudomonas putida* have significantly increased the growth and yield of different crop plants, including different legumes such as chickpeas and sugarcane (Mehnaz et al., 2009). Such strains are able to colonize the plant roots rapidly and increase plant growth. However, different environmental factors, such as type of soil, salinity, pH, moisture, and nutrients, affect the activity of such bacteria in the soil. Plant properties, including plant species and genotype, plant age, and plant growth stage, can also affect the activity of such bacteria (Berg, 2009).

Bacillus

The most important PGPR genus in the rhizosphere is *Bacillus*, and their plant growth-promoting activities have been indicated for a long time. The bacteria are able to produce a high number of metabolites affecting the rhizosphere, other soil microbes, and plant root growth. The increased availability of soil nutrients is among the most important effects of such metabolites, significantly affecting plant growth and yield production. The use of *Bacillus* strains as effective biofertilization on the growth of different plants have been indicated by different research under field and greenhouse conditions (Han et al., 2006; Spaepen et al., 2009).

Azotobacter

The genera of *Azomonas*, which is a noncyst forming strain, with the species of *Azotobacter macrocytogenes*, and *Azotobacter*, which is a cyst-forming strain, compromises six strains including *Azotobacter paspali*, *Azotobacter chroococcum*, *Azotobacter armeniacus*, *Azotobacter beijerinckii*, *Azotobacter nigricans*, and *Azotobacter vinelandii*. The bacteria are aerobic, free-living N fixers. The strain *A. paspali* is a highly specific bacteria and was first isolated from the rhizosphere of the subtropical grass, *Paspalum notatum*. Research has indicated that the growth of different plants, including the annual and perennial grasses, have been increased by *Azospirillum and Azotobacter*. The bacteria are able to increase seed germination and the growth and yield of different crop plants (Dobereiner and Day, 1975; Bloemberg and Lugtenberg, 2001; Basak and Biswas, 2010).

Acetobacter

Acetobacter is usually used as biofertilization for sugarcane. The bacteria are able to colonize sugar cane roots and hence can fulfill an important part of plant N chemical fertilization similar to organic fertilization. The following genera are from the family of Acetobacteriaceae: *Acetobacter*, *Gluconoacetobacter*, *Acidomonas*, and *Gluconobacter*. The name of *Acetobacter diazotrophicus* has been changed to *G. diazotrophicus* according to the analysis of 16S rRNA sequence (Saharan and Nehra, 2011). With respect to the genetic properties of *G. diazotrophicus*, it is not a diverse strain. Some of the strains of *G. diazotrophicus* have become able to colonize plant roots by themselves or with the help of other soil microbes such as fungi. The plasmids of *G. diazotrophicus* have been found to be between 2 and 170 kb.

Azoarcus

The first, as the aerobic N fixer bacteria, *Azoarcus*, was isolated from the tissue of Kaller grass, which had been surface sterilized with the ability of colonizing rice roots. The grass is a salt-tolerant species and can tolerate a high rate of salt concentration. The bacteria are highly efficient N fixers. The genus of *Azoarcus* consists of two strains: *Azoarcus communis* and *Azoarcus indigens* (Bloemberg and Lugtenberg, 2001; Sivasakthi et al., 2014).

Interactions of Plant Growth-Promoting Rhizobacteria and Other Soil Microbes

The interactions between PGPR and the other soil microbes including arbuscular mycorrhizal (AM) fungi can affect their activities and hence make the root rhizosphere more favorable for their growth and for plant growth. Mycorrhizal fungi are among the most important soil fungi developing symbiotic association with most terrestrial plants. In such a kind of symbiosis the fungi increases the uptake of water and nutrients by the host plant and receives carbon from their host plant for their growth and activities (Kanchiswamy et al., 2015).

The fungal spore, especially in the presence of the host plant, germinates and results in the production of an extensive hyphal network, which can significantly increase the uptake of water and nutrients by the plant. Although the fungi are able to enhance the uptake of different nutrients by the host plant, phosphorus is the most affected nutrient by the fungi, which is due to the production of enzymes, such as phosphates by the fungi. Such enzymes can significantly affect the availability and hence the uptake of P by the host plant. The fungi are also able to interact with other soil microbes, including rhizobium, phosphate-solubilizing bacteria, and pathogens, significantly affecting plant growth and yield production (Miransari, 2011a,b,c).

The fungi can affect the activities of other soil microbes including PGPR by directly providing a niche for their growth or indirectly by affecting the host plant physiology. The altering effects of mycorrhizal fungi on plant physiology is by increasing the rate of photosynthesis and the allocation of photosynthates to different parts of the plants and by increasing the uptake of nutrients. Subsequently the cellular properties including the membrane are changed, affecting the leakage of root products (Bianciotto et al., 2000).

Because PGPR are usually responsive to different specific root products, such types of alteration significantly affect the growth and activities of PGPR. For example, *Azospirillum* is able to affect the root physiology and hence root products. The bacteria are able to produce capsular polysaccharides, affecting the adhesion of bacteria to different surfaces and to the other exopolysaccharides (EP). The production of such products by *Azospirillum* can make the bacteria survive under stressful conditions (Bonfante and Anca, 2009; Miransari, 2011a,b,c).

PGPR can also detoxify the unfavorable products produced by plant roots. This is a way for the selection of PGPR compared with the other soil microbes without such an ability. For example, the resistance of soil PGPR is not dependent on the degradation of unfavorable products, produced by plant roots, but on the changes related to the permeability of the bacterial membrane. Accordingly, it is important for PGPR to have some detoxifying mechanisms and use the unfavorable products as sources of carbon after degradation by such mechanisms (Jaiswal, 2011).

Hence, the role of the plant is more important than the bacterial properties in the determination of the PGPR communities, and it is also important that for the efficient use of PGPR inoculums the compatibility of root exudates and the PGPR be determined. Under unfavorable environments, PGPR may use some mechanisms, including the production of spores and biofilms, which can make the bacteria survive the conditions to avoid the stress conditions (Fester et al., 2014).

The following important factors must be determined before using PGPR as inoculums: (1) the compatibility between plant root products and PGPR, (2) the use of PGPR consortium and their related activities, (3) the combination of soil nutrients and soil properties for the activity of PGPR, (4) the combined use of PGPR with the other soil beneficial microbes including rhizobium and mycorrhizal fungi, and (5) the use of genetically modified PGPR, which are able to survive more efficiently under stress. If such details are determined, it would be possible to indicate the long-term benefits of using PGPR. The use of omic can also result in the fast determination of such details for the use of PGPR for organic farming (Berg, 2009; Miransari, 2011b).

Aung et al. (2013) investigated the effects of coinoculation resulting from PGPR and *B. japonicum* on the growth, nodulation, and the microbial combination of the soybean rhizosphere. The results indicated that under pot conditions, coinoculation of soybeans with different strains of *B. japonicum* and *Azospirillum* sp. resulted in a higher plant growth and nodulation compared with the single inoculation with *Bacillus solisalsi*. Compared with the single use of each strain of *B. japonicum*, coinoculated with *Azospirillum* sp., significantly increased nodulation, nodule dry weight, and seed dry weight. Their results also indicated that under both pot and field conditions, the combination of rhizosphere bacteria was not affected by coinoculation with PGPR, and only the plant growth stage was a determining factor. However, neither coinoculation nor plant growth stage affected the combination of soil fungi in the rhizosphere.

Yadav and Aggarwal (2014) investigated the effects of two different species of mycorrhizal fungi and phosphate-solubilizing bacteria (*P. fluorescens)* using different levels of phosphate fertilization on the growth of soybeans in a pot experiment. Different plant growth parameters, including plant height, the fresh and the dry weight of roots, and the aerial part and root length, were significantly affected by the experimental treatments; however, the combined use of the microbial treatments were the most efficient at the lowest rate of P fertilization. Accordingly, the combined use of the microbial treatments resulted in the highest number of AM spores, the percentage of mycorrhizal colonization, the P content of roots and the aerial part, the activity of acidic (IU/g FW) and alkaline phosphatase, and seed protein and oil content, significantly different from the nonmycorrhizal treatments.

The important point about soybean N nutrition is the phase, which is usually about 20 days, after seed germination and before the initiation of the N biological process, in which soybean seedlings may face N deficiency. In such a phase the seed N source has been used for the seedling growth, and the plant is not fed with N from the process of biological N fixation. Accordingly, it is pertinent to use methods, which may speed up the process of nodule production (Chibeba et al., 2015).

Accordingly, Chibeba et al. (2015) hypothesized that using *Azospirillum* coinoculated with *B. japonicum* may be a suitable method, which can result in a faster formation of nodules on the soybean roots. It is because *Azospirillum* is able to increase root growth and enhance the formation of root hairs, and hence accordingly, more root sites will be created for inoculation and production of nodules by *B. japonicum*. The authors investigated the coinoculating effects of *Azospirillum* and *B. japonicum* on the process of nodulation and biological N fixation by soybeans under field and greenhouse conditions.

The number of nodules and their dry weights as affected by the single use of *B, japonicum* or coinoculated with *Azospirillum* were determined at different days after germination. The coinoculation treatments significantly increased the rate of nodulation in the greenhouse. However, the coinoculation treatment was even more effective under field conditions with significant effects on soybean nodulation and grain yield.

Because of the presence of *Azospirillum* and *B. japonicum* from the previous years in the field, Chibeba et al. (2015) conducted their experiment in the greenhouse. Nodule dry weight is a good indicator of biological N fixation as it is highly correlated with plant N content. Accordingly, the authors suggested that it is possible to indicate plant N conditions with respect to the weight of plant aerial part. During the early growth stages of soybeans, *Azospirillum* resulted in a greatly significantly higher nodule dry weight. These results are in accordance with the previous results indicating the positive effects of *Azospirillum* on nodule number and dry weight as well as on root dry weight. The positive effects of *Azospirillum* on nodulation under field conditions indicate that the bacteria is able to act under stress as fields are usually subjected to the stresses of water and temperature (Hungria and Kaschuk, 2014). The authors indicated that the increased weight of the plant aerial part by *Azospirillum* was due to the enhanced availability of N resulted by the process of N fixation and plant hormones, which increase root growth.

Different genera of PGPR have been isolated from the nodules of roots including *Bacillus*, *Erwinia*, *Pseudomonas*, *Aerobacter*, *Agrobacterium*, and *Flavimonas*. PGPR have been isolated by screening the rhizosphere and the plant tissues. Accordingly, they can be used as inoculants. Interestingly, it has been indicated that although the use of rhizobium inoculums has been successful for soybean production under field conditions, their coinoculation with PGPR can enhance their efficiency and hence increase the rate of biologically fixed N. Such a type of inoculation also includes the benefits resulted by the use of PGPR (Hungria and Kaschuk, 2014).

Azospirillum sp.

Azospirillum sp. is definitely among the most important and used PGPR in the world; however, there is not much related to its interaction with rhizobium. Hungria et al. (2015) tested such a hypothesis by coinoculating soybeans with *B. japonicum* and *Azospirillum* sp. under different soil and climatic conditions in Brazil. Their results indicated that the soybean growth and yield increased by coinoculation without using any chemical N fertilization, even in the presence of soil local rhizobium. Accordingly, the authors indicated that it is a beneficial biotechnological tool, if soybean is coinoculated by *B. japonicum* and *Azospirillum*, as the rate of soybean yield will be at a level comparable with the use of N chemical fertilization and hence is of economic and environmental significance and contributes to the sustainability of agriculture.

Different plant hormones are produced by PGPR, including auxin, cytokinines, gibberellins, and ethylene (Tien et al., 1979; Strzelczyk et al., 1994; Hungria et al., 2015). It is essential that the rate of 300 kg/ha N be biologically fixed by rhizobium so that the successful rate of crop yield results (Hungria et al., 2006). There is another important point related to planting soybeans under stress, which is its increased sensitivity to abiotic stresses under biotic stresses. Accordingly, the seeds must be treated with different chemicals, such as fungicides, before planting and hence such practice, which is widely used in different parts of Brazil, can adversely affect the process of soybean seed inoculation with rhizobia (Hungria et al., 2006; Campo et al., 2009).

Accordingly, Hungria et al. (2015) suggested that with respect to the restrictions of biological fixed N by rhizobium and soybeans and the benefits of using PGPR such as

Azospirillum on the growth and yield of soybean, the coinoculation of *B. japonicum* and *Azospirillum* can significantly contribute to the increased soybean growth and yield without a need for the use of N chemical fertilization, as also confirmed by their results. In another study, Hungria et al. (2013) investigated the combined use of *Rhizobium* and *B. japonicum* for the inoculation of seeds, and *Azospirillum* was used in between the rows. Although the results were promising, the practical use of such a method may not be recommended due to the need for a high rate of labor (Hungria et al., 2013). Accordingly, they tested and suggested the combined use of *B. japonicum* and *Azospirillum* for the inoculation of soybean seeds at seeding (Hungria et al., 2015). Their study (Hungria et al., 2013) indicated that a size of 10 times less *Azospirillum* inoculums can fulfill the requirement for the inoculation of soybean seeds, compared with *B. japonicum*.

It is not yet indicated if the beneficial effects of rhizobium and *Azospirillum* on soybean growth and yield production are due to the increased rate of biological N fixation or due to the other reasons. However, research has indicated that the combined use of soil microbes including rhizobium, PGPR, and mycorrhizal fungi can positively increase soybean growth and yield production by the following mechanisms. (1) Increased nutrient uptake of N and P in the aerial part of plant, (2) solubilization of phosphate, (3) production of plant hormones including auxin, and (4) production of ammonium and siderophore by soil microbes including *Pseudomonas* sp.

Although there is not much detail related to the enhancing effects of *Azospirillum* on soybean growth, it has been previously indicated that *Azospirillum* is able to alleviate the stress of drought and adverse temperatures on plant growth and yield production. This can be really important to legumes, including soybean plants, as the process of biological N fixation is sensitive to stresses such as drought. Their results indicated that the coinoculation of soybean seeds with *B. japonicum* and *Azospirillum* significantly increased soybean yield (388 kg, 15.3% increase) compared with single use of *B. japonicum*. Their positive results were related to the areas that soybeans had not been previously planted and the areas that had been planted by soybeans.

A high rate of N chemical fertilization is consumed in Brazil and 70% is imported with high prices; however, the price of microbial inoculums is much cheaper and can result in a savings of US\$7 billion per year. The other important aspect related to the use of chemical N fertilization is its environmental consequences, as for example, using 100 kg N chemical fertilization results in the production of 950 kg CO_2. Such details indicate the importance of using rhizobium and PGPR inoculums in agriculture for an economic and sustainable production. The economic significance of using microbial inoculums is by decreasing the production of greenhouse gases (Kaymak, 2011; Aung et al., 2013).

Plant Growth-Promoting Rhizobacteria and Biological N Fixation

The biological fixation of N is among the abilities of PGPR. Different bacterial species from the following genera are able to fix N biologically: *Azospirillum*, *Arthrobacter*, *Bacillus*, *Enterobacter*, *Flavobacterium*, *Erwinia*, *Rhizobium*, and *Pseudomonas*.

Plant root products play an important role in attracting the suitable species of bacteria toward its roots. Accordingly, the number and diversity of bacterial species is determined by the quality and quantity of such products and their utilization by the bacteria (Mehnaz et al., 2010; Rana et al., 2011).

Biological N fixation is an important process significantly affecting crop production with economic and environmental significance. Hence, use of bacterial inoculums such as PGPR inoculums including rhizobia, which is not widely used in different parts of the world, is an efficient method of providing crop plants with their essential N. The rate of biological N fixation is 180×10^6 tons/year worldwide, with 80% being fixed by rhizobium and 20% by the free-living associative microbes (Wagner, 2011; Bonner and Varner, 2012).

The nonsymbiotic PGPR, which contribute to the process of N fixation, include *Azospirillum*, *Acetobacter*, *Azoarcus*, *diazotrophicus*, *Azotobacter*, etc. Such a nonsymbiotic fixation of N is of great economic significance; however the important constraint is the limitation of carbon and food for utilization by bacteria. Among the most important species of nonsymbiotic N-fixing bacteria are *G. diazotrophicus*, *Azoarcus* sp., *Azotobacter* sp., *Herbaspirillium* sp., *Azospirillim*, *Arthrobacter*, *Achromobacter*, *Bacillus*, *Acetobacter*, *Azomonas*, *Clostridium*, *Entrobacter*, *Pseudomonas*, *Klebseilla*, *Rhodopseudomonas*, and *Rhodospirilum* (Vessey, 2003; Sivasakthi et al., 2014).

The most abundant genus of PGPR in the rhizosphere is *Bacillus*, with a wide range of plant-promoting activities including the production of different metabolites. Different species of *Bacillus*, including *Bacillus subtilis* and *Bacillus licheniformis*, can be used as inoculums promoting plant growth and crop production. Such *Bacillus* can enhance plant growth by the production of plant hormones, increased solubilization, and hence availability and uptake of different nutrients including N, P, K, and Fe by plants. Phosphorus-solubilizing bacteria are also able to increase plant growth and yield production by (1) increasing plant N uptake and the N-fixing ability of a plant and its symbiotic or nonsymbiotic bacteria, and (2) by enhancing the availability of Fe and Zn by the production of plant-promoting substances (Upadhyay et al., 2009; Hayat et al., 2010).

Enhancement of plant growth and controlling pathogens by *Bacillus* and *Pseudomonas* are some complicated processes and include the following mechanisms: (1) production of different metabolites such as plant hormones (auxin, gibberellins, and cytokinin), 1- amino cyclopropane – 1 carboxylate (ACC)-deaminase, HCN, antibiotics, volatile compounds, and siderophore; (2) increased availability of nutrients; and (3) induction of systemic resistance. *Bacillus* sp. is able to significantly affect root properties including root activities, growth, and dry matter (Glick, 2012).

The interesting traits of *Pseudomonas* sp. make them suitable PGPR to be used as microbial inoculums. Fluorescent *Pseudomonas* sp. is among the most effective strains, resulting in the health of soil. However, different environmental factors, such as the properties of soil, availability of nutrients, moisture, pH, plant age, and species, can affect the activities of PGPR. When used as inoculums, *Pseudomonas* sp. can affect plant growth and yield production by different mechanisms. Such PGPR are among the most abundant species in the rhizosphere of different crop plants and the most important metabolite they produce is the plant growth substances (Berg, 2009; Miransari, 2011b).

Egamberdieva (2010) investigated the growth-promoting effects of different PGPR strains including *Pseudomonas* sp. and *P. fluorescens* on the growth of two different wheat genotypes. The PGPR strains were able to colonize the rhizosphere of both wheat genotypes and increased wheat growth significantly in one of the genotypes. The author accordingly indicated that it is possible to increase wheat growth in a calcareous soil using PGPR inoculums; however, the results are genotype-dependent.

Soybeans, Plant Growth-Promoting Rhizobacteria, and Nutrient Deficiency

Plants require nutrients for their growth and yield production; however, the availability of nutrients is sometimes not high for the use of plants and, accordingly, plant face nutrient deficiency. Different methods are used for providing plants with their essential nutrients, including the use of chemical, organic, and microbial fertilization. Although chemical fertilization is a fast method of providing plants with their essential nutrients, due to the environmental and economic concerns, the use of such a method is not recommended compared with the other methods of fertilization, especially at extra amounts (Miransari, 2011b).

Microbial fertilization is the use of soil microbes, which are able to provide plants with their essential nutrients by different methods, including the process of N fixation and increasing the availability of soil nutrients to the plant and the other soil microbes. The most usual type of microbial fertilization is the use of microbial inoculums, specifically rhizobium inoculums, which are now widely used for the production of legumes, including soybeans, worldwide (Berg, 2009; Miransari, 2011a,b; Hungria et al., 2015).

As previously mentioned, AM fungi are also among the soil microbes with the ability to increase the uptake of water and nutrients by the host plant, and they can also be used as microbial fertilization, although their use is not as high as the use of rhizobium inoculums. Research has indicated the positive effects of AM fungi on the growth and yield of their host plants under different conditions, including stress, which is mostly due to the enhanced uptake of water and nutrients by the host plant (Miransari and Smith, 2007, 2008, 2009).

However, the other type of microbial inoculums is the use of PGPR with the ability to increase the availability of different soil nutrients such as P and Fe to the plant. For example, phosphate-solubilizing bacteria are able to enhance the availability of P for the use of the host plant by producing different enzymes. PGPR have the ability to turn the insoluble products of P into the soluble products such as orthophosphate, enhancing the growth and yield of their host plant (Rodriguez et al., 2006).

Phosphate-solubilizing bacteria can be used as a useful source of P for their host plants, with economic and environmental benefits. Soil microbes, with the ability to increase the P availability in the soil, are mostly bacteria including *Rhizobium, Bacillus*, and *Pseudomonas* and fungi including *Penicillium* and *Aspergillus*. Among the rhizobium species, *Mesorhizobium* are the most efficient phosphate-solubilizing bacteria (Laranjo et al., 2008; Guiñazú et al., 2010).

The activities of soil microbes, which result in the increased solubility of phosphorus, is by the production of organic acids dissolving insoluble phosphate compounds and chelating calcium from phosphate compounds, which result in the increased solubility of P Bacteria with the ability of solubilizing P that have been isolated from the rhizosphere of the soybean (Cattelan et al., 1999).

The availability of iron for plant and microbial use is of great importance; however, due to the chemical properties of iron, most of the time the rate of iron availability is not high in the soil, especially in soils with a pH higher than the neutral. Due to such properties the competition for the uptake of iron is high. However, interestingly under the conditions in which the availability of iron is not high in the soil, PGPR are able to produce compounds, which are not of high molecular weight and are able to acquire iron. Such products have a high affinity for chelating iron produced by different organisms, including bacteria, fungi, and grasses (Zhang et al., 2009; Rasouli-Sadaghiani et al., 2014).

The production of such compounds by the soil microbes results in the formation of soluble iron products, which are absorbable by plant and soil microbes. Siderophores are mostly from nonribosomal peptides, and some of them are produced independently (Challis, 2005). Siderophores can strongly bind iron, and enterobactin is one of the strongest ones. According to the analysis of DNA, the PGPR genera, which are able to produce siderophores, include gram negative bacteria, specifically *Entrobacter* and *Pseudomonas*, and gram-positive bacteria, specifically *Rhodococcus* and *Bacillus*.

Soybeans, Plant Growth-Promoting Rhizobacteria, and Salinity

Salinity is among the most important stresses worldwide, significantly affecting plant growth and yield production. The rate of agricultural fields, which are saline, is about 20%, and each year 1–2% of the fields become saline. Salinity decreases plant growth and yield production by adversely affecting the process of photosynthesis, the synthesis of proteins, and the metabolism of lipids (Paul and Lade, 2014; Munns and Gilliham, 2015).

Plants use different morphological and physiological mechanisms to tolerate the stress of salinity. The use of tolerant species, use of organic matter, and treating the saline soils with water are among the methods that are used for the alleviation of salinity stress on plant growth and yield production. However, soil microbes, including mycorrhizal fungi and PGPR, have also been indicated to be effective on the alleviation of salt stress (Upadhyay and Singh, 2015).

Soil salinity adversely affects soybean nodulation, yield, and the symbiotic *B. japonicum*. The soybean and its symbiotic rhizobium are not tolerant under salinity stress; however, it has been indicated that it is possible to alleviate the stress of salinity on soybean nodulation and yield production using the plant to bacteria signal molecule, genistein. It is because the most sensitive stages of soybean nodulation are the initial stages, including the production of signals by the roots of soybeans (Miransari and Smith, 2007, 2008, 2009).

Hence under stresses such as salinity the reduced production of genistein decreases the rate of nodulation and hence biological N fixation by plant and rhizobium. Accordingly, it was hypothesized, tested, and proven that if the inoculums of *B. japonicum* are pretreated with genistein under field and greenhouse conditions, it is possible to alleviate the stress of salinity on soybean nodulation and yield production (Miransari and Smith, 2007, 2008, 2009).

PGPR are also able to alleviate the stress of salinity on plant growth and yield production. For example, the mechanism of osmoadaptation, including the production of proline, glutamate, trehalose, and glycine betaine, can make *Azospirillum* sp. survive the stress (Tripathi et al., 1998). The other important mechanism used by PGPR to alleviate the salinity stress is the production of ACC-deaminase enzyme. This enzyme is able to catabolize ACC to α–ketobutarate and ammonia, the prerequisite for ethylene production, and hence decreases the production of ethylene, the stress enzyme, which can adversely affect plant growth under stress. Hence, the inculcation of crop plants, including soybeans with PGPR producing ACC-deaminase, is a useful method for the alleviation of salinity stress on plant growth and yield production (Jalili et al., 2009; Glick, 2012).

PGPR can make the plant to survive the salinity stress (Table 8.1) using the following mechanisms (Fig. 8.1): (1) increasing chlorophyll content, plant root, and aerial part growth and nutrient uptake, (2) alleviating biotic stresses, (3) improving plant metabolism and production of plant hormones and ACC-deaminase, (4) increasing the production of antioxidant enzymes, which scavenge the produced reactive oxygen species under salt stress, (5) increasing the rate of K^+/Na^+ in plants, and (6) plant osmoregulation by increasing the rate of osmolytes such as proline (Egamberdieva and Lugtenberg, 2014; Munns and Gilliham, 2015).

Although the soil rhizosphere is a more favorable environment for the growth and activity of soil microbes, compared with the bulk soil, salinity stress adversely affects the number, diversity, and functioning of soil microbes. Under salt stress the attachment of *A. brasilense* to the roots of maize and wheat decreased. The adverse effects of salt stress *A. brasilense* were also by reducing the rate of N fixation due to the decreased expression of nifH and the fixation enzyme. The negative effects of salinity stress on the microbial community of the rhizosphere are also achieved by decreasing the rate of root products and the mineralization rate of organic matter (Paul and Lade, 2014).

Soil microbes can survive the stress of salinity using different mechanisms. For example, the release of Na^+ and Cl^- from the vacuole into the cytoplasm results in an osmotic balance between the cytoplasm and the surrounding environment of the bacterial cell. Increased uptake of K^+ and the enhanced production of osmolytes by the bacteria are also among the mechanisms used by bacteria to tolerate the salinity stress. The osmolytes can also stabilize the structure of proteins, resulting in the suitable folding of proteins under stress (Street et al., 2006).

The other mechanism, used by PGPR to alleviate the stress of salinity, is by the production of intracellular polymeric substances, which are able to enhance the volume of soil macropores and improve the structure of soil. Accordingly, the availability

Table 8.1 **The Alleviating Effects of Different PGPR Species on the Growth and Yield of Different Crop Plants Under Salinity Stress (Paul and Lade, 2014)**

Rhizobacteria	Plant	References
Bacillus safensis, Ochrobactrum pseudogregnonense	Wheat (Triticum aestivum)	Chakraborty et al. (2013)
Pseudomonas putida, Enterobacter cloacae, Serratia ficaria, and Pseudomonas fluorescens	Wheat (T. aestivum)	Nadeem et al. (2013)
Alcaligenes faecalis, Bacillus pumilus, Ochrobactrum sp.,	Rice (Oryza sativa)	Bal et al. (2013)
Pseudomonas pseudoalcaligenes, B. pumilus	Rice (O. sativa)	Jha et al. (2013)
Bacillus subtilis, Arthrobacter sp.	Wheat (T. aestivum)	Upadhyay et al. (2012)
Azospirillum sp.	Wheat (T. aestivum)	Zarea et al. (2012)
Streptomyces sp.	Wheat (T. aestivum)	Sadeghi et al. (2012)
Pseudomonas sp., Bacillus sp., Variovorax sp.	Avocado (Persea gratissima)	Nadeem et al. (2012)
Azotobacter chroococcum	Maize (Zea mays)	Rojas-Tapias et al. (2012)
P. pseudoalcaligenes, P. putida	Chickpea (Cicer arietinum)	Patel et al. (2012)
Brachybacterium saurashtrense, Brevibacterium casei, Haererohalobacter sp.	Peanut (Arachis hypogaea)	Shukla et al. (2012)
Pseudomonas extremorientalis, Pseudomonas chlororaphis	Common bean (Phaseolus vulgaris)	Egamberdieva (2011)
Bacillus, Burkholderia, Enterobacter, Microbacterium, Paenibacillus	Wheat (T. aestivum)	Upadhyay et al. (2011)
P. fluorescens, Pseudomonas aeruginosa, Pseudomonas stutzeri	Tomato (Lycopersicon esculentum)	Tank and Saraf (2010)
Pseudomonas sp.	Eggplant (Solanum melongena)	Fu et al. (2010)
Azospirillum sp.	Durum wheat (Triticum durum)	Nabti et al. (2010)
P. putida	Cotton (Gossypium hirsutum)	Yao et al. (2010)
Bacillus megaterium	Maize (Z. mays L.)	Marulanda et al. (2010)
Agrobacterium rubi, Burkholderia gladii, P. putida, B. subtilis, B. megaterium	Radish (Raphanus sativus L.)	Kaymak et al. (2009)
Azospirillum brasilense	Barley (Hordeum vulgare)	Omar et al. (1994)
Pseudomonas mendocina	Lettuce (Lactuca sativa L. cv. Tafalla)	Kohler et al. (2009, 2010)

Continued

Table 8.1 **The Alleviating Effects of Different PGPR Species on the Growth and Yield of Different Crop Plants Under Salinity Stress (Paul and Lade, 2014)—cont'd**

Rhizobacteria	Plant	References
B. subtilis	*Arabidopsis thaliana*	Zhang et al. (2008)
A. brasilense	Pea (*Phaseolus vulgaris*)	Dardanelli et al. (2008)
Bacillus sp., *Ochrobactrum* sp.	Maize (*Z. mays*)	Principe et al. (2007)
Pseudomonas syringae, P. fluorescens, Enterobacter aerogenes	Maize (*Z. mays*)	Nadeem et al. (2007)
P. fluorescens	Groundnut (*A. hypogaea*)	Saravanakumar and Samiyappan (2007)
Azospirillum	Lettuce (*L. sativa*)	Barassi et al. (2006)
P. fluorescence	Black pepper (*Piper nigrum*)	Paul and Sarma (2006)
P. pseudoalcaligenes	Rice (*O. sativa*)	Diby et al. (2005)
Achromobacter piechaudii	Tomato (*L. esculentum*)	Mayak et al. (2004)
Aeromonas hydrophila, Bacillus insolitus Bacillus sp.	Wheat (*T. aestivum*)	Ashraf et al. (2004)
Azospirillum	Maize (*Z. mays*)	Hamdia et al. (2004)
A. brasilense	Chickpeas (*C. arietinum*), faba beans (*Vicia faba* L.)	Hamaoui et al. (2001)
Azospirillum lipoferum, A. brasilense, Azoarcus, Pseudomonas sp.	Kallar grass *Leptochloa fusca*	Malik et al. (1997)

With kind permission from Springer, License number 3779210030047.

of soil water and nutrients increases to the plant, and hence the plant will be able to manage the stress more efficiently (Alami et al., 2000; Paul and Lade, 2014). Such PGPR are able to bind Na^+ cations and decrease their availability to the plant and hence decrease the adverse effects of salinity on plant growth. Such substances have the ability of holding water and connecting the pores of soil. Hence such substances result in the production of a more stable structure of soil and a more smooth water uptake by plant, which is also due to the production of biofilms (Alami et al., 2000; Paul and Lade, 2014).

Naz et al. (2009) isolated PGPR from the rhizosphere of weed-tolerant species under salinity stress (EC: 2.3 dS/m) and compared their results with nonsaline conditions (EC: 0.2 dS/m). The isolated strains were able to produce different plant hormones, including auxin, gibberellins, and ABA in culture media. The strains, which were used for the inoculation of soybean seedlings under nonstress and stress conditions (20 dS/m NaCl), increased plant growth and the rate of proline production. The authors accordingly indicated that it is possible to use such strains as inoculums for soybean production under salinity stress.

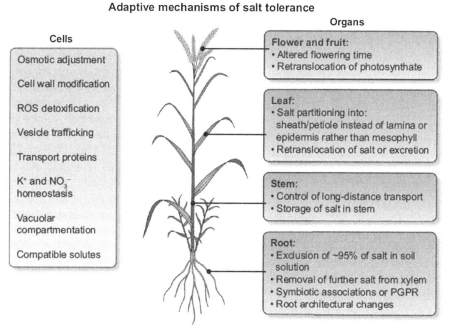

Figure 8.1 Cellular (left) and tissue (right) adaptive mechanisms, used by plants under salinity stress (Munns and Gilliham, 2015).
With kind permission from Wiley, License number 3779210868674.

With respect to the beneficial effects of PGPR on plant growth and yield production, the commercial use of PGPR under different stresses such as salinity is recommended. Accordingly, several strains of PGPR are now being commercially produced using organic and inorganic mediums as well as using the method of solid or liquid fermentation (Bashan et al., 2014; Paul and Lade, 2014).

Soybeans, Plant Growth-Promoting Rhizobacteria, and Drought

Drought is also among the stresses, which adversely affects plant growth and yield production. Under drought stress the rate of water and hence nutrient uptake by plant and soil microbes decrease. Due to the decreased rate of water pressure under drought stress the cellular growth and division and hence the growth of plant and soil microbes decrease. Research has indicated the favorable effects of PGPR on plant growth under drought stress due to the following mechanisms: (1) increased rate of plant water content, (2) a more positive water potential, (3) increased plant growth, (4) enhanced production of osmolytes including proline, (5) increased growth of plant roots and aerial parts, and (5) increased production of plant hormones including IAA, ABA, and ACC-deaminase (Vilchez and Manzanera, 2011; Naseem and Bano, 2014).

The other effective method for the alleviation of drought stress on the growth and yield of crop plants is the combined use of PGPR with mycorrhizal fungi. This method is more efficient under drought stress, compared with the single use of PGPR. Although the single effects of each microbe can positively affect plant growth and yield production, the interactions of PGPR and mycorrhizal fungi can also increase plant growth under stress.

Exopolysaccharides in the soil are an important component of soil organic matter, representing 40–95% of bacterial biomass. The production of EP by bacteria is by (1) slim EP and (2) capsular EP. The important effects of EP are (1) the production of biofilm, (2) protecting the bacteria, (3) aggregation of bacteria, (4) attachment to the surface, (5) bioremediation, and (6) the role in the interaction of plant microbes (Naseem and Bano, 2014). Due to the production of EP, some PGPR such as *Pseudomonas* are able to tolerate the stress of drought. This is mainly due to the presence of 97% water in the polymer matrix of EP, which makes it hydrated and hence can keep the bacteria away from the desiccation stress. EP can also be used for the bioremediation of heavy metal stress due to their enzymatic activities (Naseem and Bano, 2014).

Naseem and Bano (2014) investigated the effects of PGPR producing EP, which had been isolated from the semiarid regions of Pakistan, on the growth of maize under drought stress. Three PGPR strains, including *Proteus pennri* (Pp1), *Pseudomonas aeruginosa* (Pa2), and *Alcaligenes faecalis* (AF3) were selected as the EP producing PGPR. These bacteria are gram negative, with catalase activity, and motile. Strain Pp1 had oxidase activity and the ability of solubilizing phosphate, but not Pa2 and AF3. The sequence of bacteria was indicated using 16SrRNA. Although drought stress significantly decreased the growth of maize seedlings, the PGPR were able to increase plant growth, including the root and the aerial part, by improving the structure of soil. The treated plants also had a higher rate of relative water content, sugar, and protein, which resulted from the increased rate of proline and antioxidant enzymes. *P. aeruginosa*, isolated from the semiarid regions, was the most effective strain under drought stress. The authors accordingly indicated that the use of PGPR consortia is a suitable method for the alleviation of drought stress on maize growth.

Soybeans, Plant Growth-Promoting Rhizobacteria, and Suboptimal Root Zone Temperature

Suboptimal root zone temperature is also an important stress affecting plant growth and yield production in different parts of the world, specifically in places with long winters and cold temperatures. However, the research team of Professor Donald Smith from McGill University, Canada, working with soybeans under the stress of suboptimal root zone temperature from 1995 to 2005, have indicated the positive effects of PGPR on soybean growth and production under such stress conditions. They indicated that if soybean seeds are inoculated with different strains of PGPR, it is possible to alleviate the adverse effects of suboptimal root zone temperature on soybean growth and yield production. The single use of PGPR and their combination with *B. japonicum* were tested under field and greenhouse conditions, and their favorite effects on the alleviation of stress and increased nodulation, growth, and yield of soybeans were indicated (Zhang and Smith, 2002).

Soybeans, Plant Growth-Promoting Rhizobacteria, and Heavy Metals

The presence of heavy metals in the soil is also an important stress adversely affecting plant growth and yield production. Although some heavy meals are essential for plant growth, at higher concentrations, they negatively affect plant growth. Heavy metals decrease plant growth and yield production by their unfavorable effects on plant morphology and physiology. Under the stress of heavy metals the production of reactive oxygen stress decreases the cellular activities and cellular structure. Under such conditions the production of antioxidant enzymes such as proxidase and catalase increases to alleviate the oxidative stress. Different soil microbes such as mycorrhizal fungi and PGPR can help the plant to survive the stress by (1) the production of plant hormones, (2) the production of antioxidant enzymes, (3) increased uptake of water and nutrients by the host plant, and (4) interaction with the other soil microbes (Miransari, 2011a,b,c).

PGPR can alleviate the stress of heavy metals on plant growth and yield production by the process of bioremediation, using the following mechanisms: (1) affecting the availability of heavy metals, (2) producing different chelating compounds, (3) acidifying the rhizosphere, (4) affecting the redox potential in the rhizosphere, and (5) increasing the solubilization of phosphate (He and Yang, 2007; Zhuang et al., 2007). Among the most important mechanisms used by PGPR to alleviate the stress of heavy metals is the production of the ACC-deaminase enzyme, which can decrease the rate of ethylene production in plants and hence decrease the stress of heavy metals (Glick, 2012).

The other important effect of PGPR on plant growth and yield production is by the alleviation of heavy metal stress (Belimov et al., 2005). The use of resistant soil microbes including PGPR, phosphate-solubilizing bacteria, and AM fungi is a suitable method for the bioremediation of heavy metals. The process of bioremediation by soil microbes affects the bioavailability of heavy metals in the soil as well as plant growth. Soil microbes including PGPR produce different biochemicals affecting the bioavailability of heavy metals in the soil and hence decrease their uptake by plants. For example, under the stress of cadmium in the soil, the production of ethylene as the stress hormone increases, which is due to the increased production of ACC. *P. putida* is able to tolerate high concentrations of heavy metals and hence can be used for the process of bioremediation (Zhuang et al., 2007; Lin and Jianlong, 2010).

Soybeans, Plant Growth-Promoting Rhizobacteria, and Pathogens

PGPR can also be used for the alleviation of biotic stresses on plant growth and yield production. Under biotic stresses the systemic resistance is induced in plants, resulting in plant tolerance under biotic stresses. The signaling pathway, which is activated under stress and is related to the induced systemic resistance, is controlled by salicylic acid (SA). When the plant is subjected to biotic stresses the rate of SA in the plant increases, resulting in the increased resistance of plants under the stress. However, the inducing effects of nonpathogenic PGPR on plant growth and the induction of systemic resistance is not by the increased production of SA and the activation of pathogenesis-related gene but by the increased production of ethylene and jasmonic acid (Bloemberg and Lugtenberg, 2001; Saharan and Nehra, 2011).

Conclusion and Future Perspectives

PGPR are beneficial bacteria, which are able to establish a symbiotic or nonsymbiotic association with plants in the rhizosphere. Rhizobium is the most important PGPR, which is able to develop a symbiotic association with its specific host plant and increase its growth and yield by biologically fixing atmospheric N. However, the other PGPR, such as *Pseudomonas* and *Bacillus*, are able to increase plant growth and yield production by colonizing the host plant roots in a nonsymbiotic manner. Research has indicated the enhancing effects of PGPR on soybean growth and yield under different conditions, including stress. The beneficial effects of PGPR on soybean growth have been indicated under different stresses, including nutrient deficiency, salinity, drought, acidity, suboptimal root zone temperature, heavy metals, and biotic stresses. It is time to widely use PGPR in the fields with respect to their economic and environmental significance, although they are being used by just a number of developing nations. However, for a more successful use of PGPR, some important details related to the use of PGPR should be addressed, including their production, formulation, shipping, storage, and their use under field conditions. Using PGPR at a large level requires the education of farmers about the proper use of the PGPR inoculums. The use of genetically modified PGPR strains is a must for the production of more efficient strains; however, the public must be informed that their use is harmless to the environment and human health. It is also important to find if the rhizospheric or the endophytic PGPR are more useful when used as microbial inoculums. The interactions of PGPR and AM fungi must also be modified (Glick, 2012, 2014). The most related details have been reviewed, presented, and analyzed in this chapter. Future research can find the new details, related to the more efficient use of PGPR under field stress conditions. The use of new PGPR products can also make the crop plants, including soybeans, survive the stress more efficiently.

References

Ahmad, F., Ahmad, I., Khan, M., 2008. Screening of free-living rhizospheric bacteria for their multiple plant growth promoting activities. Microbiological Research 163, 173–181.

Alami, Y., Achouak, W., Marol, C., Heulin, T., 2000. Rhizosphere soil aggregation and plant growth promotion of sunflowers by an exopolysaccharide-producing *Rhizobium* sp. strain isolated from sunflower roots. Applied and Environmental Microbiology 66, 3393–3398.

Ashraf, M., Hasnain, S., Berge, O., Mahmood, T., 2004. Inoculating wheat seedlings with exopolysaccharide-producing bacteria restricts sodium uptake and stimulates plant growth under salt stress. Biology and Fertility of Soils 40, 157–162.

Aung, T.T., Buranabanyat, B., Piromyou, P., Longtonglang, A., et al., 2013. Enhanced soybean biomass by co-inoculation of *Bradyrhizobium japonicum* and plant growth promoting rhizobacteria and its effects on microbial community structures. African Journal of Microbiology Research 7, 3858–3873.

Bal, H., Nayak, L., Das, S., Adhya, T., 2013. Isolation of ACC deaminase producing PGPR from rice rhizosphere and evaluating their plant growth promoting activity under salt stress. Plant and Soil 366, 93–105.

Barassi, C., Ayrault, G., Creus, C., Sueldo, R., Sobrero, M., 2006. Seed inoculation with *Azospirillum* mitigates NaCl effects on lettuce. Scientia Horticulturae–Amsterdam 109, 8–14.

Basak, B., Biswas, D., 2010. Co-inoculation of potassium solubilizing and nitrogen fixing bacteria on solubilization of waste mica and their effect on growth promotion and nutrient acquisition by a forage crop. Biology and Fertility of Soils 46, 641–648.

Bashan, Y., 1998. Inoculants of plant growth-promoting bacteria for use in agriculture. Biotechnology Advances 16, 729–770.

Bashan, Y., de-Bashan, L., Prabhu, S., Hernandez, J., 2014. Advances in plant growth-promoting bacterial inoculant technology: formulations and practical perspectives (1998–2013). Plant and Soil 378, 1–33.

Bassler, B., 1999. How bacteria talk to each other, regulation of gene expression by quorum sensing. Current Opinion in Microbiology 2, 582–587.

Belimov, A., Hontzeas, N., Safronova, V., Demchinskaya, S., Piluzza, G., Bullitta, S., Glick, B., 2005. Cadmium-tolerant plant growth-promoting bacteria associated with the roots of Indian mustard (*Brassica juncea* L Czern.). Soil Biology and Biochemesitry 37 (Suppl. 2), 241–250.

Berg, G., 2009. Plant–microbe interactions promoting plant growth and health: perspectives for controlled use of microorganisms in agriculture. Applied Microbiology and Biotechnology 84, 11–18.

Bertin, C., Yang, X., Weston, L., 2003. The role of root exudates and allelochemicals in the rhizosphere. Plant and Soil 256, 67–83.

Bianciotto, V., Lumini, E., Lanfranco, L., Minerdi, D., Bonfante, P., Perotto, S., 2000. Detection and identification of bacterial endosymbionts in arbuscular mycorrhizal fungi belonging to the family Gigasporaceae. Applied and Environmental Microbiology 66, 4503–4509.

Bloemberg, G., Lugtenberg, B., 2001. Molecular basis of plant growth promotion and biocontrol by rhizobacteria. Current Opinion in Plant Biology 4, 343–350.

Bonfante, P., Anca, I., 2009. Plants, mycorrhizal fungi, and bacteria: a network of interactions. Annual Review of Microbiology 63, 363–383.

Bonner, J., Varner, J. (Eds.), 2012. Plant Biochemistry. Elsevier.

Campo, R., Araujo, R., Hungria, M., 2009. Nitrogen fixation with the soybean crop in brazil: compatibility between seed treatment with fungicides and bradyrhizobial inoculants. Symbiosis 48, 154–163.

Challis, G.L., 2005. A widely distributed bacterial pathway for siderophore biosynthesis independent of non ribosomal peptide synthetases. ChemBioChem 6 (Suppl. 4), 601–611.

Cassán, F., Vanderleyden, J., Spaepen, S., 2014. Physiological and agronomical aspects of phytohormone production by model plant-growth-promoting rhizobacteria (PGPR) belonging to the genus *Azospirillum*. Journal of Plant Growth Regulation 33, 440–459.

Cattelan, A., Hartel, P., Fuhrmann, J., 1999. Screening for plant growth–promoting rhizobacteria to promote early soybean growth. Soil Science Society of America Journal 63 (Suppl. 6), 1670–1680.

Chakraborty, N., Ghosh, R., Ghosh, S., Narula, K., Tayal, R., Datta, A., Chakraborty, S., 2013. Reduction of oxalate levels in tomato fruiand consequentmetabolic remodeling following overe xpression of a fungal oxalate decarboxylase. Plant Physiology 162, 364–378.

Chibeba, A., de Fátima Guimarães, M., Brito, O., Nogueira, M., Araujo, R., Hungria, M., 2015. Co-inoculation of soybean with *Bradyrhizobium* and *Azospirillum* promotes early nodulation. American Journal of Plant Sciences 6, 1641–1649.

Cohen, A., Bottini, R., Piccoli, P., 2008. *Azospirillum brasilense* Sp. 245 produces ABA in chemically-defined culture medium and increases ABA content in Arabidopsis plants. Plant Growth Regulation 54, 97–103.

Dardanelli, M., de Cordoba, F., Espuny, M., Carvajal, M., Diaz, M., Serrano, A., et al., 2008. Effect of *Azospirillum brasilense* coinoculated with rhizobium on *Phaseolus vulgaris* flavonoids and nod factor production under salt stress. Soil Biology and Biochemistry 40, 2713–2721.

Davidson, E., 2009. The contribution of manure and fertilizer nitrogen to atmospheric nitrous oxide since 1860. Nature Geoscience 2, 659–662.

Dekhil, S., Cahill, M., Stackebrandt, E., Li, S., 1997. Transfer of *Conglomeromonas largomobilis* subsp. *largomobilis* to the genus *Azospirillum* as *Azopirillum largomobile* comb. nov., and elevation of *Conglomeromonas largomobilis* subsp. *parooensis* to the new type species of *Conglomeromonas, Conglomeromonas parooensis* sp. nov. Systematic and Applied Microbiology 20, 72–77.

Diby, P., Bharathkumar, S., Sudha, N., 2005. Osmotolerance in biocontrol strain of *Pseudomonas pseudoalcaligenes* MSP-538: a study using osmolyte, protein and gene expression profiling. Annals of Microbiology 55, 243–247.

Dobereiner, J., Day, J., 1975. Nitrogen fixation in rhizosphere of grasses. In: Stewart, W. (Ed.), Nitrogen Fixation by Free-Living Microorganisms. Cambridge University Press, Cambridge, pp. 39–56.

Egamberdieva, D., 2010. Growth response of wheat cultivars to bacterial inoculation in calcareous soil. Plant, Soil and Environment 56, 570–573.

Egamberdieva, D., 2011. Survival of *Pseudomonas extremorientalis* TSAU20 and *P. Chlororaphis* TSAU13 in the rhizosphere of common bean (*Phaseolus vulgaris*) under saline conditions. Plant, Soil and Environment 57, 122–127.

Egamberdieva, D., Lugtenberg, B., 2014. Use of plant growth-promoting rhizobacteria to alleviate salinity stress in plants. In: Miransari, M. (Ed.), Use of Microbes for the Alleviation of Soil Stresses, vol. 1. Springer, New York, pp. 73–96.

Fester, T., Giebler, J., Wick, L.Y., Schlosser, D., Kästner, M., 2014. Plant–microbe interactions as drivers of ecosystem functions relevant for the biodegradation of organic contaminants. Current Opinion in Biotechnology 27, 168–175.

Fu, Q., Liu, C., Ding, N., Lin, Y., Guo, B., 2010. Ameliorative effects of inoculation with the plant growth-promoting rhizobacterium *Pseudomonas* sp. DW1 on growth of eggplant (*Solanum melongena* L.) seedlings under salt stress. Agricultural Water Management 97, 1994–2000.

Glick, B., 2012. Plant growth-promoting bacteria: mechanisms and applications. Scientifica 2012.

Glick, B., 2014. Bacteria with ACC deaminase can promote plant growth and help to feed the world. Microbiological Research 169, 30–39.

Guiñazú, L., Andrés, J., Del Papa, M., Pistorio, M., Rosas, S., 2010. Response of alfalfa (*Medicago sativa* L.) to single and mixed inoculation with phosphate-solubilizing bacteria and *Sinorhizobium meliloti*. Biology and Fertility of Soils 46, 185–190.

Haichar, Z.F., Marol, C., Berge, O., Rangel-Castro, J.I., Prosser, J.I., et al., 2008. Plant host habitat and root exudates shape soil bacterial community structure. The ISME Journal 2, 1221–1230.

Hamaoui, B., Abbadi, J., Burdman, S., Rashid, A., Sarig, S., Okon, Y., 2001. Effects of inoculation with *Azospirillum brasilense* on chickpeas (*Cicer arietinum*) and faba beans (*Vicia faba*) under different growth conditions. Agronomie 21, 553–560.

Hamdia, M., Shaddad, M., Doaa, M., 2004. Mechanisms of salt tolerance and interactive effects of *Azospirillum brasilense* inoculation on maize cultivars grown under salt stress conditions. Plant Growth Regulation 44, 165–174.

Han, H., Supanjani, Lee, K., 2006. Effect of co-inoculation with phosphate and potassium solubilizing bacteria on mineral uptake and growth of pepper and cucumber. Plant Soil and Environment 52 (Suppl. 3), 130–136.

Hayat, R., Ali, S., Amara, U., Khalid, R., Ahmed, I., 2010. Soil beneficial bacteria and their role in plant growth promotion: a review. Annals of Microbiology 60, 579–598.

He, Z., Yang, X., 2007. Role of soil rhizobacteria in phytoremediation of heavy metal contaminated soils. Journal of Zhejiang University Science B 8, 192–207.

Hungria, M., Campo, R., Mendes, I., Graham, P., 2006. Contribution of biological nitrogen fixation to the N nutrition of grain crops in the Tropics: the success of soybean (*Glycine max* L. Merr.) in South America. In: Singh, R.P., Shankar, N., Jaiwa, P.K. (Eds.), Nitrogen Nutrition and Sustainable Plant Productivity. Studium Press, Houston, pp. 43–93.

Hungria, M., Nogueira, M., Araujo, R., 2013. Co-inoculation of soybeans and common beans with rhizobia and *Azospirilla*: strategies to improve sustainability. Biology and Fertility of Soils 49, 791–801.

Hungria, M., Kaschuk, G., 2014. Regulation of N_2 fixation and NO_3^-/NH_4^+ assimilation in nodulated and N-fertilized *Phaseolus vulgaris* L. exposed to high-temperature stress. Environmental and Experimental Botany 98, 32–39.

Hungria, M., Nogueira, M., Araujo, R., 2015. Soybean seed co-inoculation with *Bradyrhizobium* spp. and *Azospirillum brasilense*: a new biotechnological tool to improve yield and sustainability. American Journal of Plant Sciences 6, 811–817.

Jaiswal, S., 2011. Role of rhizobacteria in reduction of arsenic uptake by plants: a review. Journal of Bioremediation and Biodegradation 2, 126–130.

Jalili, F., Khavazi, K., Pazira, E., Nejati, A., Rahmani, H., Sadaghiani, H., Miransari, M., 2009. Isolation and characterization of ACC deaminase-producing fluorescent pseudomonads, to alleviate salinity stress on canola (*Brassica napus* L.) growth. Journal of Plant Physiology 166, 667–674.

Jha, M., Chourasia, S., Sinha, S., 2013. Microbial consortium for sustainable rice production. Agroecology and Sustainable Food Systems 37, 340–362.

Kanchiswamy, C., Malnoy, M., Maffei, M., 2015. Chemical diversity of microbial volatiles and their potential for plant growth and productivity. Frontiers in Plant Science 6.

Kaymak, H., Guvenc, I., Yarali, F., Donmez, M., 2009. The effects of biopriming with PGPR on germination of radish (*Raphanus sativus* L.) seeds under saline conditions. Turkish Journal of Agricultural Forestry 33, 173–179.

Kaymak, H., 2011. Potential of PGPR in agricultural innovations. In: Plant Growth and Health Promoting Bacteria. Springer, Berlin/Heidelberg, pp. 45–79.

Kolb, W., Martin, P., 1985. Response of plant roots to inoculation with *Azospirillum brasilense* and to application of indoleacetic acid. In: Klingmuller, W. (Ed.), *Azospirillum* III: Genetics, Physiology, Ecology. Springer, Berlin, pp. 215–221.

Kohler, J., Hernandez, J., Caravaca, F., Roldan, A., 2009. Induction of antioxidant enzymes is involved in the greater effectiveness of a PGPR versus AM fungi with respect to increasing the tolerance of lettuce to severe salt stress. Environmental and Experimental Botany 65, 245–252.

Kohler, J., Caravaca, F., Roldan, A., 2010. An AM fungus and a PGPR intensify the adverse effects of salinity on the stability of rhizosphere soil aggregates of *Lactuca sativa*. Soil Biology and Biochemistry 42, 429–434.

Laranjo, M., Alexandre, A., Rivas, R., Velázquez, E., Young, J., Oliveira, S., 2008. Chickpea rhizobia symbiosis genes are highly conserved across multiple *Mesorhizobium* species. FEMS Microbiology Ecology 66, 391–400.

Lin, Q., Jianlong, W., 2010. Biodegradation characteristics of quinoline by *Pseudomonas putida*. Bioresource Technology 101, 7683–7686.

Lugtenberg, B., Kamilova, F., 2009. Plant-growth-promoting rhizobacteria. Annual Review of Microbiology 63, 541–556.

Malik, K., Rakhshanda, S., Mehnaz, G., et al., 1997. Association of nitrogen-fixing plant-growth-promoting rhizobacteria (PGPR) with kallar grass and rice. Plant and Soil 194, 37–44.

Magalhães, F., Baldani, J., Souto, S., Kuykendall, J., Döbereiner, J., 1983. New acid-tolerant *Azospirillum* species. Anais da Academia Brasileira de Ciências 55 (Suppl. 4), 417–430.

Maksimov, I., Veselova, S., Nuzhnaya, T., Sarvarova, E., Khairullin, R., 2015. Plant growth-promoting bacteria in regulation of plant resistance to stress factors. Russian Journal of Plant Physiology 62, 715–726.

Marulanda, A., Azcon, R., Chaumont, F., Ruiz-Lozano, J., Aroca, R., 2010. Regulation of plasma membrane aquaporins by inoculation with a *Bacillus megaterium* strain in maize (*Zea mays* L.) plants under unstressed and salt-stressed conditions. Planta 232, 533–543.

Mayak, S., Tirosh, T., Glick, B., 2004. Plant growth-promoting bacteria that confer resistance to water stress in tomatoes and peppers. Plant Science 166, 525–530.

Mehnaz, S., Weselowski, B., Lazarovits, G., 2007. *Azospirillum canadense* sp. nov., a nitrogen-fixing bacterium isolated from corn rhizosphere. International Journal of Systematic and Evolutionary Microbiology 57, 620–624.

Mehnaz, S., Weselowski, B., Aftab, F., Zahid, S., Lazarovits, G., Iqbal, J., 2009. Isolation, characterization, and effect of *fluorescent pseudomonads* on micropropagated sugarcane. Canadian Journal of Microbiology 55, 1007–1011.

Mehnaz, S., Baig, D., Lazarovits, G., 2010. Genetic and phenotypic diversity of plant growth promoting rhizobacteria isolated from sugarcane plants growing in Pakistan. Journal of Microbiology and Biotechnology 20, 1614–1623.

Milner, J., Raffel, S., Lethbridge, B., Handelsman, J., 1995. Culture conditions that influence accumulation of zwittermicin A by *Bacillus cereus* UW85. Applied Microbiology and Biotechnology 43, 685–691.

Miransari, M., Smith, D., 2007. Overcoming the stressful effects of salinity and acidity on soybean nodulation and yields using signal molecule genistein under field conditions. Journal of Plant Nutrition 30, 1967–1992.

Miransari, M., Smith, D., 2008. Using signal molecule genistein to alleviate the stress of suboptimal root zone temperature on soybean-*Bradyrhizobium* symbiosis under different soil textures. Journal of Plant Interactions 3, 287–295.

Miransari, M., Smith, D., 2009. Alleviating salt stress on soybean (*Glycine max* (L.) Merr.)–*Bradyrhizobium japonicum* symbiosis, using signal molecule genistein. European Journal of Soil Biology 45, 146–152.

Miransari, M., 2011a. Interactions between arbuscular mycorrhizal fungi and soil bacteria. Applied Microbiology and Biotechnology 89, 917–930.

Miransari, M., 2011b. Soil microbes and plant fertilization. Applied Microbiology and Biotechnology 92, 875–885.

Miransari, M., 2011c. Hyperaccumulators, arbuscular mycorrhizal fungi and stress of heavy metals. Biotechnology Advances 29, 645–653.

Miransari, M., 2016. In: Abiotic and Biotic Stresses in Soybean Production. Academic Press, USA. ISBN: 9780128015360. 344 pp.

Munns, R., Gilliham, M., 2015. Salinity tolerance of crops–what is the cost? New Phytologist 208, 668–673.

Nabti, E., Sahnoune, M., Ghoul, M., Fischer, D., Hofmann, A., Rothballer, M., et al., 2010. Restoration of growth of durum wheat (*Triticum durum* var. waha) under saline conditions due to inoculation with the rhizosphere bacterium *Azospirillum brasilense* NH and extracts of themarine alga *Ulva lactuca*. Journal of Plant Growth Regulation 29, 6–22.

Nadeem, S., Zahir, Z., Naveed, M., Arshad, M., 2007. Preliminary investigations on inducing salt tolerance in maize through inoculation with rhizobacteria containing ACC deaminase activity. Canadian Journal of Microbiology 53, 1141–1149.

Nadeem, S., Shaharoona, B., Arshad, M., Crowley, D., 2012. Population density and functional diversity of plant growth promoting rhizobacteria associated with avocado trees in saline soils. Applied Soil Ecology 62, 147–154.

Nadeem, S., Zahir, Z., Naveed, M., Nawaz, S., 2013. Mitigation of salinity induced negative impact on the growth and yield of wheat by plant growth-promoting rhizobacteria in naturally saline conditions. Annals of Microbiology 63, 225–232.

Naseem, H., Bano, A., 2014. Role of plant growth-promoting rhizobacteria and their exopolysaccharide in drought tolerance of maize. Journal of Plant Interactions 9, 689–701.

Naz, I., Bano, A., Ul-Hassan, T., 2009. Isolation of phytohormones producing plant growth promoting rhizobacteria from weeds growing in Khewra salt range, Pakistan and their implication in providing salt tolerance to *Glycine max* L. African Journal of Biotechnology 8, 5762–5768.

Omar, S., Abdel-Sater, M., Khallil, A., Abdalla, M., 1994. Growth and enzyme activities of fungi and bacteria in soil salinized with sodium chloride. Folia Microbiologica 39, 23–28.

Patel, D., Jha, C., Tank, N., Saraf, M., 2012. Growth enhancement of chickpea in saline soils using plant growth-promoting rhizobacteria. Journal of Plant Growth Regulation 31, 53–62.

Paul, D., Lade, H., 2014. Plant-growth-promoting rhizobacteria to improve crop growth in saline soils: a review. Agronomy for Sustainable Development 34, 737–752.

Paul, D., Sarma, Y., 2006. Plant growth promoting rhizobacteria [PGPR] mediated root proliferation in black pepper (*Piper nigrum* L.) as evidenced through GS root software. Archives of Phytopathology and Plant Protection 39, 311–314.

Peng, G., Wang, H., Zhang, G., Hou, W., Liu, Y., Wang, E., Tan, Z., 2006. *Azospirillum melinis* sp. nov., a group of diazotrophs isolated from tropical molasses grass. International Journal of Systematic and Evolutionary Microbiology 56 (Suppl. 6), 1263–1271.

Perrig, D., Boiero, L., Masciarelli, O., Penna, C., Cassan, F., Luna, V., 2007. Plant growth promoting compounds produced by two agronomically important strains of *Azospirillum brasilense*, and their implications for inoculant formulation. Applied Microbiology and Biotechnology 75, 1143–1150.

Podile, A., Kishore, G., 2006. Plant growth-promoting rhizobacteria. In: Gnanamanickam, S. (Ed.), Plant-Associated Bacteria. Springer, Netherlands, pp. 195–230.

Principe, A., Alvarez, F., Castro, M., Zachi, L., Fischer, S., et al., 2007. Biocontrol and PGPR features in native strains isolated from saline soils of Argentina. Current Microbiology 55, 314–322.

Rana, A., Saharan, B., Joshi, M., Prasanna, R., Kumar, K., Nain, L., 2011. Identification of multi-trait PGPR isolates and evaluating their potential as inoculants for wheat. Annals of Microbiology 61, 893–900.

Rasouli-Sadaghiani, M., Malakouti, M., Khavazi, K., Miransari, M., 2014. Siderophore efficacy of fluorescent Pseudomonades affecting labeled iron (59Fe) uptake by wheat. In: Miransari, M. (Ed.), Use of Microbes for the Alleviation of Soil Stresses. Alleviation of Soil Stress by PGPR and Mycorrhizal Fungi, vol. 2. Springer. ISBN: 978-1-4939-0720-5.

Rodriguez, H., Fraga, R., Gonzalez, T., Bashan, Y., 2006. Genetics of phosphate solubilization and its potential applications for improving plant growth-promoting bacteria. Plant and Soil 287 (Suppl. 1–2), 15–21.

Rojas-Tapias, D., Moreno-Galvan, A., Pardo-Diaz, S., Obando, M., Rivera, D., Bonilla, R., 2012. Effect of inoculation with plant growth-promoting bacteria (PGPB) on amelioration of saline stress in maize (*Zea mays*). Applied Soil Ecology 61, 264–272.

Sadeghi, A., Karimi, E., Dahaji, P., Javid, M., Dalvand, Y., Askari, H., 2012. Plant growth promoting activity of an auxin and siderophore producing isolate of *Streptomyces* under saline soil conditions. World Journal of Microbiology and Biotechnology 28, 1503–1509.

Saharan, B.S., Nehra, V., 2011. Plant growth promoting rhizobacteria: a critical review. Life Sciences and Medicine Research 21, 1–30.

Saravanakumar, D., Samiyappan, R., 2007. ACC deaminase from *Pseudomonas fluorescens* mediated saline resistance in groundnut (*Arachis hypogea*) plants. Journal of Applied Microbiology 102, 1283–1292.

Sen, J., 1929. The role of associated nitrogen-fixing bacteria on nitrogen nutrition of cereal crops. Agricultural Journal of India 24, 967–980.

Shukla, P., Agarwal, P., Jha, B., 2012. Improved salinity tolerance of *Arachis hypogaea* (L.) by the interaction of halotolerant plantgrowth-promoting rhizobacteria. Journal of Plant Growth Regulation 31, 195–206.

Singh, P., Kumar, V., Agrawal, S., 2014. Evaluation of phytase producing bacteria for their plant growth promoting activities. International Journal of Microbiology 2014.

Sivasakthivelan, P., Saranraj, P., 2013. *Azospirillum* and its formulations: a review. International Journal of Microbiological Research 4, 275–287.

Sivasakthi, S., Usharani, G., Saranraj, P., 2014. Biocontrol potentiality of plant growth promoting bacteria (PGPR)-*Pseudomonas fluorescens* and *Bacillus subtilis*: a review. African Journal of Agricultural Research 9, 1265–1277.

Spaepen, S., Vanderleyden, J., Remans, R., 2007. Indole-3-acetic acid in microbial and micro-organism-plant signaling. FEMS Microbiology Reviews 31, 425–448.

Spaepen, S., Vanderleyden, J., Okon, Y., 2009. Plant growth-promoting actions of rhizobacteria. Advances in Botanical Research 51, 283–320.

Street, T., Bolen, D., Rose, G., 2006. A molecular mechanism for osmolyte-induced protein stability. Proceedings of the National Academy of Sciences USA 103, 13997–14002.

Steenhoudt, O., Vanderleyden, J., 2000. *Azospirillum*, a free-living nitrogen-fixing bacterium closely associated with grasses: genetic, biochemical and ecological aspects. FEMS Microbiology Reviews 24 (Suppl. 4), 487–506.

Strzelczyk, E., Kamper, M., Li, C., 1994. Cytokinin-like-substances and ethylene production by *Azospirillum* in media with different carbon sources. Microbiological Research 149, 55–60.

Tank, N., Saraf, M., 2010. Salinity-resistant plant growth promoting rhizobacteria ameliorates sodium chloride stress on tomato plants. Journal of Plant Interactions 5, 51–58.

Tarrand, J., Krieg, N., Döbereiner, J., 1978. A taxonomic study of the *Spirillum lipoferum* group with description of a new genus, *Azospirillum* gen. nov. and two species, *Azospirillum lipoferum* (Beijerinck) comb. nov. And *Azospirillum brasilense* nov. Canadian Journal of Microbiology 24 (Suppl. 8), 967–980.

Tien, T., Gaskins, M., Hubbell, D., 1979. Plant growth substances produced by *Azospirillum brasilense* and their effect on the growth of pearl millet (*Pennisetum americanum* L.). Applied and Environmental Microbiology 37, 1016–1024.

Tripathi, A., Mishra, B., Tripathi, P., 1998. Salinity stress responses in plant growth promoting rhizobacteria. Journal of Biosciences 23 (Suppl. 4), 463–471.

Upadhyay, S., Singh, D., Saikia, R., 2009. Genetic diversity of plant growth promoting rhizobacteria isolated from rhizospheric soil of wheat under saline condition. Current Microbiology 59, 489–496.

Upadhyay, S., Singh, J., Singh, D., 2011. Exopolysaccharide-producing plant growth-promoting rhizobacteria under salinity condition. Pedosphere 21, 214–222.

Upadhyay, S., Singh, J., Saxena, A., Singh, D., 2012. Impact of PGPR inoculation on growth and antioxidant status of wheat under saline conditions. Plant Biology 14, 605–611.

Upadhyay, S., Singh, D., 2015. Effect of salt-tolerant plant growth-promoting rhizobacteria on wheat plants and soil health in a saline environment. Plant Biology 17, 288–293.

Vessey, J., 2003. Plant growth promoting rhizobacteria as biofertilizers. Plant and Soil 255, 571–586.

Vilchez, S., Manzanera, M., 2011. Biotechnological uses of desiccation-tolerant microorganisms for the rhizoremediation of soils subjected to seasonal drought. Applied Microbiology and Biotechnology 91, 1297–1304.

Wagner, S., 2011. Biological nitrogen fixation. Nature Education Knowledge 2, 14.

Wai-Leung, N., Bassler, L., 2009. Bacterial quorum-sensing network architectures. Annual Review of Genetics 43, 197–222.

Yadav, A., Aggarwal, A., 2014. Effect of dual inoculation of AM fungi and *Pseudomonas* with phosphorus fertilizer rates on growth performance, nutrient uptake and yield of soybean. Researcher 6, 5–13.

Yao, L., Wu, Z., Zheng, Y., Kaleem, I., Li, C., 2010. Growth promotion and protection against salt stress by *Pseudomonas putida* Rs-198 on cotton. European Journal of Soil Biology 46, 49–54.

Zarea, M., Hajinia, S., Karimi, N., Goltapeh, E., Rejali, F., Varma, A., 2012. Effect of *Piriformospora indica* and *Azospirillum* strains from saline or non-saline soil on mitigation of the effects of NaCl. Soil Biology and Biochemistyr 45, 139–146.

Zhang, F., Smith, D., 2002. Interorganismal signaling in suboptimum environments: the legume-rhizobia symbiosis. Advances in Agronomy 76, 125–161.

Zhang, H., Kim, M., Sun, Y., Dowd, S., Shi, H., Pare, P., 2008. Soil bacteria confer plant salt tolerance by tissue-specific regulation of the sodium transporter HKT1. Molecular Plant Microbe Interactions 21, 737–744.

Zhang, H., Sun, Y., Xie, X., Kim, M.S., Dowd, S.E., Paré, P., 2009. A soil bacterium regulates plant acquisition of iron via deficiency-inducible mechanisms. The Plant Journal 58, 568–577.

Zhuang, X., Chen, J., Shim, H., Bai, Z., 2007. New advances in plant growth-promoting rhizobacteria for bioremediation. Environment International 33, 406–413.

Role of Genetics and Genomics in Mitigating Abiotic Stresses in Soybeans

G. Raza[1], N. Ahmad[1], M. Hussain[1], Y. Zafar[2], M. Rahman[1]
[1]National Institute for Biotechnology and Genetic Engineering, Pakistan Atomic Energy Commission, Faisalabad, Pakistan; [2]International Atomic Energy Agency, Vienna, Austria

Introduction

The soybean is one of the most important cash crops, contributing 69.74 billion dollars to the economy worldwide, with an annual production of ~280 million metric tons in 2013. The United States is the top producer, with an annual production of 86.58 million metric tons followed by Brazil, Argentina, China, and India, where the annual production is 73.78, 47.74, 13.57, and 12.89 million metric tons, respectively (FAO, 2013). It has been grown for harvesting grains, which are rich in protein (38–40%) and oil (18–20%) content. A total of 95% of the oil has been used for edible purposes, while the remaining is consumed for industrial use, especially in the pharmacological industry. Approximately 98% of the soybean meal is used as a source of nutrition for livestock and aquaculture (Liu, 2008).

Legumes have unique ability of fixing nitrogen in soils, which make them important for multiple cropping patterns. Among these, the soybean is a major contributor (Singh, 2010). Apart from these, biodiesel from soybeans has been accepted as one of the optional alternatives to fossil fuels (Hartman et al., 2011).

Historical Perspective

Accumulating evidence indicates that soybeans originated and domesticated in China between 1100 and 1700 BC (Hymowitz, 1970). Later (from the 1st to 16th centuries), it was introduced to Korea, Japan, Indonesia, Philippines, Vietnam, Thailand, Burma, North India, and Nepal; this whole region is believed to be a second center of soybean origin (Hymowitz, 1990). In the 17th century, it was introduced to Europe and the United States. Currently, it is cultivated in ~60 different countries.

Strength of Genetic Resource

The soybean is grouped in a family *Fabaceae*, and the genus *Glycine*, which is further categorized into two subgenera: *Soja*, containing two important species *Glycine max* L. (cultivated) and *Glycine soja* (wild), and *Glycine*, comprised of 26 perennial species, which are prevalent in Australia or in its surroundings.

Environmental Stresses in Soybean Production. http://dx.doi.org/10.1016/B978-0-12-801535-3.00009-7

In a conventional plant-breeding program, genetic resources play a key role for developing new cultivars. For the genetic improvement of soybeans, a vast collection of soybean germplasm (170,000 accessions) is present in the world, with some duplicated accessions (Nelson, 2009). China has the largest collection of germplasm (>23,000 cultivated and 7000 wild accessions), and this germplasm has been preserved and maintained at the Chinese National Soybean Gene bank (Dong et al., 2004; Wang et al., 2006; Limei et al., 2005). The entire collection was divided into three subcollections, ie, core collection, mini core collection, and integrated applied core collection (Qiu et al., 2013). The USDA-Agricultural Research Service has the second largest soybean germplasm collection, which is comprised of >20,000 accessions of the genus Glycine with a wide range of natural variations (Carter et al., 2004). The details of soybean accessions are available in the International Legume Database and Information Service, USDA-Germplasm Information Resources Network (www.ildis.org, www.ars.grin.gov; Lewis et al., 2005). The USDA has also assembled a large collection of germplasm comprised of 16 perennial species of the genus *Glycine* too. These accessions are also being maintained in Australia. This germplasm has been recognized as world base collection by the International Plant Genetic Resources Institute (Mishra and Verma, 2010). Similarly, The National Institute of Agrobiological Sciences in Japan has a collection of nearly 11,300 accessions including local landraces as well as wild soybean species. This collection contains both the improved cultivars as well as advanced breeding lines in its repository, which were either developed by the regional agriculture research institutes of Japan or introduced from the other countries. In Korea, the Rural Development Administration gene bank has almost 7000 soybean landraces (Yoon et al., 2008).

Challenges to the Sustainability of Soybean Production

Food security is at a potential risk due to fluctuations in environments worldwide and may lead to starvation in regions, which are facing or at potential risk of the changing climate. These climate changes include unusual fluctuations in temperature and rainfall pattern and its frequency, increased salinization, and frequent drought periods. These abiotic factors suppress crop productivities up to 50% (Wang et al., 2003).

Like many other crop species, soybean growth, productivity, and seed quality are severely affected by drought (Mohammadi et al., 2012), temperature (Endo et al., 2009; Sakata and Higashitani, 2008), salinity (Sobhanian et al., 2010), waterlogging (Khatoon et al., 2012; Komatsu et al., 2012), and heavy metals (Hossain et al., 2012a). The soybean has been declared the most drought-sensitive plant (Clement et al., 2008), as drought alone may reduce yield up to 40% (Specht et al., 1999). Although the plant growth is negatively affected at different stages, the seedling stage is the most sensitive to drought and flooding (Tran and Mochida, 2010; Valliyodan and Nguyen, 2006). The soybean plant requires approximately 450–700 mm water from germination to maturity (Dogan et al., 2007).

In soybeans, the potential yield parameters are number of seeds, pods, nodes, and reproductive nodes per unit area, and 100 or 1000 seed weight. These parameters are negatively affected by the onset of drought. For example, drought affects leaf area index, total dry matter (TDM), crop growth rate, and plant height (Meckel et al., 1984; Pandey et al., 1984; Ramseur et al., 1985; Hoogenboom et al., 1987; Cox and Jolliff, 1987; Muchow et al., 1986; Desclaux et al., 2000; Cox and Jolliff, 1986). In another study, a reduction in intermodal length is more obvious than that of the other vegetative growth parameters (Desclaux et al., 2000). If the plant is exposed to reproductive stage one (R1), then the predominant yield losses would be due to a decreased number of pods and seeds while the seed size will be less affected (Ramseur et al., 1984; Pandey et al., 1984; Sionit et al., 1987; Meckel et al., 1984; Constable and Hearn, 1981; Ball et al., 2000). On the contrary, a number of reports have indicated a reduction in pods/m^2 under drought stress during reproductive stages (R1–R6) (Sionit et al., 1987; Ramseur et al., 1984; Pandey et al., 1984; Snyder et al., 1982; Neyshabouri and Hatfield, 1986; Cox and Jolliff, 1986; Westgate and Peterson, 1993; Ball et al., 2000). The water deficiency at the reproductive stage inhibits early expansion of ovaries due to reduced photosynthetic rate (Westgate and Peterson, 1993; Liu et al., 2004). In the reproductive period (R1–R7), initial stages (R1–R5) are more sensitive as compared to the later stage (R6–R7). Two-fold yield reduction was observed when drought stress was imposed on R1–R5 stages than that of R6–R7 (Korte et al., 1983; Brown et al., 1985; Kadhem et al., 1985; Eck et al., 1987; Hoogenboom et al., 1987). In a few reports, it has also been shown that R3–R5 stages are more sensitive within the R1–R6 period (Korte et al., 1983; Kadhem et al., 1985). Nitrogen fixation is a very important physiological process in soybeans. This process is more sensitive to drought as compared to TDM accumulation, photosynthetic rate, and transpiration rate (Purcell and Specht, 2004).

Another important abiotic stress is excessive salt in the soil. In the world, almost one-third of agriculture land is affected by excessive salt in the soil (Zhu, 2001; Munns and Tester, 2008), which induces secondary stresses, ie, toxic metabolites, disarrangement of membranes, weakened nutrient accomplishment, accumulation of reactive oxygen species (ROS), and inhibition of photosynthetic rate. All these factors together disturb the plant growth (Hasegawa et al., 2000). Salinity also causes plant injury by the accumulation of Na^+ and by the osmotic stress (Yeo, 1998; Hasegawa et al., 2000).

Excessive heat, another abiotic factor, is increasing continuously, which impacts the normal functioning of photosynthetic apparatus (electron transport chain, ribulose bisphosphate carboxylase, oxygenase activity, and activity of ribulose bisphosphate activase) and selective phenological stages (ie, development of pollen development, anther opening, pollen germination, fertilization, and grain development) (Sakata and Higashitani, 2008; Endo et al., 2009). Losses due to excessive heat are more pronounced on the crops, which are grown in temperate and subtropical areas (Lobell and Gourdji, 2012; Teixeira et al., 2013). Various studies have shown that temperature more than 35°C affects the germination of pollen and pollen tube growth in soybeans (Koti et al., 2005; Salem et al., 2007). The optimum range of temperature is 25–40°C for the normal canopy photosynthesis (Board and Kahlon, 2011).

In contrary to high temperature environments, when temperature falls to 10–12°C or below, it changes the transition phase of cell membranes from crystalline liquid to gel form (Bramlage et al., 1978). As a result, cell metabolism is interrupted, thus reducing the soybean yield. For example, 24% yield loss has been reported when the temperature in the night drops from 16°C down to 10°C. The effects of cold stress were irreversible when imposed on the flowering to pod formation stage, resulting in huge (~70%) yield losses. However, a much lower loss in yield (25%) was observed at later stages (Board and Kahlon, 2011).

Response of the Soybean Plant to Abiotic Stress

Morphological and Physiological Responses

Under stress conditions, plants use a number of strategies for their survival, including avoidance or tolerance, usually by activating metabolic processes for carrying out normal cell functions (Bita and Gerats, 2013; Hasanuzzaman et al., 2013). For avoidance, plants adapt specialized features such as short duration life cycle, production of specialized morphological structure for the protection of stress sensitive tissue, etc. Under drought stress, the plant reduces the water content by closing the stomata and decreasing transpiration, which results in a decrease of chlorophyll amount, photosynthesis, and CO_2 assimilation. Phenotypically, plants exhibit signs of leaf rolling, wilting, and etiolating during the start of water stress. A first plant part that faces drought and submergence stresses is the root. In response to flooding, plants initially decrease the absorption of water, root permeability, and intake of minerals, which reduce the photosynthetic rate, hormonal imbalance, and development of adventitious roots and parenchyma (Vartapetian and Jackson, 1997).

Response at the Molecular Level

The response of soybeans to stress is linked with the magnitude of stress, time period, and the type of stress. These factors are responsible for a number of changes at the protein level in the plant cell. The nature and degree of response may differ depending upon stress type. However, a few similar response mechanisms have been observed in response to all types of abiotic stresses. For example, an enhanced amount of antioxidant proteins have been observed in response to all types of abiotic stresses (Komatsu et al., 2013). These proteins have been used to scavenge ROS, which ultimately protect important plant cell components from oxidative damage (Hossain et al., 2012b).

Gene expression studies in different crop plants have divided the stress responsive genes into two groups: effectors and regulatory genes (Yamaguchi-Shinozaki and Shinozaki, 2006). The first group (effector) includes the genes, which encode proteins for protecting plants directly, such as the genes responsible for the synthesis of osmolytes, molecular chaperones, membrane channel proteins, antioxidants, late embryogenesis abundant proteins, and enzymes involved in different metabolic pathways. The second group, comprised of regulatory genes, encode products such as localized receptors in membranes, kinases, calcium receptors, and transcription factors (TF). This group of genes is further involved in signal transduction and gene expression. A number of plant drought stress-tolerant TFs, such as dehydration responsible element binding (DREB) protein, are involved in ABA-independent pathway, while TFs (involved in ABA-dependent pathways), ethylene responsive factor, WRKY, MYB, basic leucine zipper domain (bZIP), and NAC have been reported (Tran et al., 2004; Hu et al., 2006; Nakashima et al., 2007; Liao et al., 2008a,b; Zhou et al., 2008; Jeong et al., 2010; Seo et al., 2010; Hao et al., 2010; Niu et al., 2012; Lopes-Caitar et al., 2013; Song et al., 2013).

In plants, the WRKY transcription factor is the largest family. A number of *WRKY* genes have been identified in Arabidopsis (Ulker and Somssich, 2004), rice (Wu et al., 2005), barley (Mangelsen et al., 2008), and wheat (Niu et al., 2012). In soybeans, 233 WRKY members have been identified (http://planttfdb.cbi.pku.edu.cn/family.php? fam=WRKY, Schmutz et al., 2010). Two *WRKY* genes (*GmWRKY21* and *GmWRKY54*) in soybeans were identified, and their role in improving tolerance to cold, salt, and drought has been demonstrated in Arabidopsis. The overexpression of *GmWRKY13* in transgenic Arabidopsis enhanced its tolerance level to salinity and mannitol (Zhou et al., 2008). Recently, the same group characterized the involvement of GmWRKY27 in response to drought and salt stress. Overexpression of GmWRKY27 and GmWRKY27 RNAi in soybeans led to increased tolerance and severe sensitivity to salinity and water-deficit stress, respectively. In the same study, interaction of GmWRKY27 with GmMYB174 was observed, which binds directly to two neighboring *cis*-elements and suppressed GmNAC29 activation, resulting in increased tolerance to abiotic stress (Wang et al., 2015). In another study, novel candidates of *WRKY* genes were found, which elucidated the novel function of WRKY transcription factors under drought stress in soybeans (Tripathi et al., 2015).

Phosphatidylinositol phospholipase C (PI-PLC) belongs to multicellular intracellular enzymes, which is one of the signaling processes involved in plant development and activates in response to abiotic stresses (PEG, NaCl, and saline-alkali). The gene coding for PI-PLC was also identified in soybeans and has been characterized (Shi et al., 1995). Recently, the expression profiling of the *PI-PLC* gene family in response to multiple stresses has been studied in soybeans. A total of 12 putative *PLC* genes were identified. These genes were located on chromosome number Gm2, 11, 14, and 18. It has been demonstrated that PLCs have an important role in imparting the ability to adapt to adverse climatic conditions to plants (Wang et al., 2015).

Heat shock proteins (HSPs) play an important role in combating adverse environmental conditions (Cho and Choi, 2009; Zou et al., 2012; Kim et al., 2014). The role of HSPs have been described in Arabidopsis (Su and Li, 2008), rubber trees (Zhang et al., 2009), wheat (Francki et al., 2002; Duan et al., 2011), pepper (Guo et al., 2014), and cucumber (Li et al., 2014a,b,c). Recently, a total of 61 *HSP70* genes were identified, which were grouped into eight subfamilies (I-VIII). These genes were found to be unevenly distributed on 17 different chromosomes. Out of these, 53 genes differentially expressed in 14 different tissues (Zhang et al., 2015a,b).

The expression profiling studies of the Homeodomain-leucine zipper (HD-Zip) gene family (comprising of 140 *HD-Zip* genes; http://planttfdb.cbi.pku.edu.cn/family.php?fam=HD-ZIP) were carried out in soybeans under water-limited and saline environments. These proteins are homeobox TFs, which are involved in conferring tolerance to different abiotic stresses. Out of the 140, 59 HD-Zip coding genes and three paralogous pairs differentially expressed, while 20 paralogous pairs exhibited similar expression under drought and saline environments in soybeans (Jin et al., 2013).

The other important class of molecules, known as osmoprotectants or compatible solutes such as proline, glycine-betaine, etc., help the plant to counter the extreme stress conditions. In soybeans, a total of 36 differentially expressed genes involved in the synthesis of osmolytes (Proline, Trehalose, Glycine betaine, Myo-inositol) were identified. Out of these, 25 were mapped in the soybean genome (Kido et al., 2013). In another study, a total of 518 and 614 genes differentially expressed in leaves and roots, respectively, of a drought-tolerant soybean cultivar Jindou, as compared to the drought-sensitive cultivar Zhoungdou 33, were identified. While 24 were commonly expressed in the root as well as leaf tissues. A total of seven genes, *Glyma15g03920*, *Glyma05g02470*, *Glyma15g15010*, *Glyma05g09070*, *Glyma06g35630*, *Glyma08g12590*, and *Glyma11g16000*, showed significantly high expression under drought conditions, demonstrating their role in imparting drought tolerance (Chen et al., 2013).

A very comprehensive proteomic study has been conducted in soybean roots grown under submergence and water stress (Oh and Komatsu, 2015). In total, 48 and 97 proteins were induced significantly under water deficit and flooding stresses, respectively. Protein synthesis-related proteins enhanced under drought stress and reduced under flooding environments, while proteins involved in glycolysis enhanced under both stresses. The synthesis of proteins involved in the fermentation process were enhanced under flooding conditions, while the synthesis of proteins involved in cell organization and redox reaction were increased under water stress. Also, three S-adenosyl methionine synthetases, commonly reduced and enhanced under flooding and drought stresses, were identified, demonstrating their role in the regulation of both stress responses (Oh and Komatsu, 2015).

The role of cyclic electron flow (CEF) toward salt tolerance in soybeans was described. It is proposed that salinity stress accelerates the CEF and the overexpression of genes associated with Na^+ transport. In result, the increased CEF raised the ATP contents in the light. The enhanced ATP content together with the genes associated with Na^+ transport assist in Na^+ sequestering into vacuoles, thus protecting the photosynthetic machinery (He et al., 2015).

The circadian clock helps plants maintain better chlorophyll contents, carbon fixation, survival, and faster growth (Dodd et al., 2005), which is under the control of multiple genes. A number of gene paralogues of multiple clocks have been identified in soybeans when exposed to drought and submergence stresses. For example, numerous clocks and *SUB 1* genes were expressed differentially in soybeans under drought and flooding stresses. There are a number of genes (pseudo-response regulators and timing of CAB expression 1—TOC1), discovered to impart tolerance under submergence and drought stresses. These genes can be used to screen the germplasm to genetic resources containing drought and submergence tolerance. It has also been suggested that by editing the clock gene paralogues, one can develop soybean varieties with improved tolerance to drought and flooding (Syed et al., 2015).

Melatonin (*N*-acetyl-5-methoxytryptamine) has an antioxidant role, which protects plants from biotic and abiotic stresses (Tan et al., 2012; Wang et al., 2012; Park et al., 2013; Vitalini et al., 2013; Yin et al., 2013). In soybeans, this chemical (when coated on seed) increased soybean tolerance to salt and drought by upregulating the expression of genes (involved in processes like cell division, photosynthetic rate, metabolism of carbohydrate, etc.) inhibited by the excessive salt (Wei et al., 2015).

Application of Genomic Approaches for Improving Tolerance to Abiotic Stresses

Genetic approaches have revolutionized the traditional ways of improving crops including soybeans. Traditional methods such as reverse genetics and forward genetics have been quite successful in determining gene function for simpler traits. These tools have been quite useful in model plants such as *Arabidopsis thaliana*, rice (*Orzya sativa*), and tobacco (*Nicotiana tabacum*), for which well-established transformation and regeneration protocols have been developed, and genome sequence information has been determined.

Genome Organization

The soybean genome is relatively small compared to the other crops such as maize, sugarcane, and barley (Morrell et al., 2012). Its average size is ~1200 Mb containing 46,000 genes, of these ~78% are located at the chromosomal ends (Schmutz et al., 2010) and 59% belong to the transposable elements (Du et al., 2010; Morrell et al., 2012). Like many other flowering plants, these transposable elements have a key role in plant evolution through recombination, gene expression, *cis*- and *trans*- activation/repression of transcription of other genes, and many unknown mechanisms (Du et al., 2010). The location of the majority of genes in "recombinant zones" suggests that the soybean genome has undergone substantial rearrangements due to domestication, followed by intensive selection (Lam et al., 2010; Li et al., 2014a,b,c).

The soybean genome comprises 20 relatively small and homogenous chromosomes. The information obtained from the sequence of the soybean genome suggests that the genome is paleopolyploid, with large-scale genome duplications (~60%). The hybridization-based mapping showed that 61.4%, 5.63%, or 21.53% of the homologous genes were found in two, three, or even more loci (Chan et al., 2012). One distinguishing feature of the soybean genome, revealed by the resequencing data of wild and cultivated soybean accessions, was its high linkage disequilibrium (LD). For example, the average distance for LD to decay to half of its maximum value both for wild and cultivated species was found to be ~75 and ~150 kb, respectively, much higher than those observed for maize, rice, and *Arabidopsis* (Lam et al., 2010).

This feature is quite attractive from a breeding point of view, as it allows using even a small set of molecular markers in marker-assisted breeding. However, it limits the resolution of genetic maps. That is why the conventional soybean genetic maps suffer from poor resolutions. However, the advances in genomic research are likely to refine the genetic maps.

Another distinguishing feature of the soybean genome is the occurrence of high nonsynonymous to synonymous mutation (Nonsyn/Syn) ratios, which is higher than that of rice and *Arabidopsis* (Chan et al., 2012). This high Nonsyn/Syn ratio could be attributed to high LD values that play a major role in accumulating alleles. It was also observed that the soybean genome exhibited a quite high value (~10%) for large-effect single nucleotide polymorphisms (SNP) (Li et al., 2014a,b,c). These large-effect SNPs together with a high Nonsyn/Syn mutation ratio could cause accumulation of deleterious mutations in the soybean genome. Owing to polyploidization and diploid-ization events, the soybean genome exhibits a mosaic genome structure. For example, a duplication of 1-Mb segment between the soybean chromosomes Gm08 and Gm15 was observed (Lin et al., 2010). Furthermore, a comparison of wild type and cultivated soybean accessions using the cultivated soybean-specific SNPs showed higher Non-syn/Syn ratio among the cultivated species, which is believed to be due to the domes-tication associated with the Hill-Robertson effect (Lam et al., 2010). Further analysis showed a whole array of variations between the cultivated and wild type genomes, suggesting that wild type genomes contain diverse genes/alleles, which can be used for enhancing genetic diversity in elite cultivars (Li et al., 2014a,b,c).

TILLING in Soybeans

The Targeted Induced Local Lesion in Genome (TILLING) method has been exten-sively used in several crop species including soybeans. In 2008, the first soybean TILLING population was developed and established its suitability for high throughput mutation screening. A total of seven genes were screened for mutations in four mutant populations (one developed by exposing Williams-82 with NMU and three with EMS at three different levels). A total of 116 mutations were identified. It has been demon-strated that the NMU-treated population and one with EMS-mutagenized population has similar mutation density (~1/140 kb), while the other EMS-mutagenized popula-tions had shown a mutation density of ~1/250 and ~1/550 kb each (Cooper et al., 2008).

Another TILLING population was developed by bombarding with a fast neutron, aiming to find a dwarf mutant (a desirable trait that protects soybean plants from lodg-ing) in the soybean population. One dwarf mutant was found in ~10,000 M_4 progeny lines. After doing whole genome sequencing followed by making comparisons with the wild type, a total of 13 large deletions were identified. Most of these deletions were positioned in noncoding regions of chromosome 3. While one deletion (803-bp dele-tion) in a mutant allele (*Blyma15g05831*) localized on chromosome 15 was respon-sible for the loss of a start codon that resulted in the complete loss of gene function (*Glyma15g05831*) in the dwarf mutant (Hwang et al., 2015). Thus the aforementioned studies suggest that the TILLING approach can be used to tailor the complex traits conferring resistance to abiotic stresses.

Marker-Assisted Selection

Transferring genes from one plant to another is a useful strategy to engineer traits of interest into crop plants. Several strategies have been developed for transferring genes to a cultivar, which can be grouped into transgenic and nontransgenic approaches. The former requires identification of putative genes, transformation, and then regenerating the transgenic line carrying a gene of interest (discussed in the forthcoming section). However, transgenic approaches are often hampered by the complexity of major agronomic traits. Therefore engineering simpler traits is considered much easier. The nontransgenic requires hybridization of two compatible varieties followed by DNA marker screening, known as marker-assisted selection (MAS). This approach allows combining multiple chromosomal segments containing several QTLs into a single plant. However, it is limited to the same species, and bottlenecks include narrow genetic diversity among hybridizing varieties. Currently, MAS is being applied for QTL introgression, backcrossing, and F_2 enrichment. In soybeans, MAS has been successfully used for developing disease-resistant cultivars (Concibido et al., 2004).

The domestication of crops has resulted in the accumulation of useful alleles. Multiple domestication events yielded a number of landraces in soybeans, which are adapted to a wide range of geographical locations. For the identification of these loci (helped in adaptation) in soybeans, an association-mapping procedure was deployed on 342 landraces and 1062 putative lines. A total of 125 genomic regions harboring several genes were identified, which may have a role in future crop improvement (Wen et al., 2015). These landraces could be a valuable resource for future breeding programs for developing high-yielding soybean cultivars that are resistant to various biotic and abiotic stresses. Efforts are underway to conserve the diversity by looking at the allelic signals of selection before and after domestication (Li et al., 2014a,b,c). Next-generation genomic tools such as genotyping by sequencing (GBS) and whole-genome sequencing (WGS) have greatly accelerated efforts to identify and recover the lost traits into crop plants. For example, a novel ion transporter-coding gene, *GmCHX1*, was identified from a wild soybean using the whole-genome sequencing approach (Qi et al., 2014). The group has also constructed a recombinant inbred population of wild soybeans for the identification of novel QTLs associated with crop improvement. Similarly, a pan-genome of seven accessions of undomesticated soybeans (*Glycine soja* L.) was constructed (Li et al., 2014a,b,c) that would have numerous implications in soybean improvement. In another study, WGS, RNA transcript profiling, and comparative genomics were exploited to reconstruct the evolutionary history of the SWEET gene family in soybeans (Patil et al., 2015). The SWEET gene family plays a crucial role in the development of reproductive organs. The SWEET gene family has been extensively studied in model plants such as Arabidopsis but remain largely unstudied in soybeans. The study could help soybean improvement programs by enhancing carbohydrate delivery to the floral organs to increase yield, for example.

Whole genome sequencing is useful in developing hundreds of simple sequence repeat (SSR) markers (Song et al., 2010) and millions of SNPs (Li et al., 2014a,b,c; Sonah et al., 2013). Thousands of QTLs for different traits spanning across the whole soybean genome have been identified, which are available at www.soybase.org.

Similarly, the genome wide association (GWAS) technique was deployed on 298 soybean accessions to determine QTLs responsible for protein and seed oil content (Hwang et al., 2014).

The availability of a complete genome sequence of soybeans has made MAS a robust tool, as it helps in developing locus-specific markers (Schmutz et al., 2010). With the help of the genome sequence, it is now possible to develop high-density DNA markers with genome-wide coverage that in turn allow haplotype analysis and identification of different alleles for agronomical important traits (Tardivel et al., 2014).

MAS is particularly useful for simpler traits; however, it is not as effective for complex traits such as tolerance against various stresses. The situation becomes further complicated when even major QTLs contribute even a small fraction of a particular phenotype variation and may give an unwanted phenotype in a new genetic background due to epistatic interactions. These challenges can be circumvented using another relatively new and powerful approach known as genomic selection (GS). GS takes into account the information of all available markers (genotypic as well as phenotypic) to compute a prediction model. Thus GS allows identification of "minor" QTLs that account for most phenotypic variations. In soybeans, GS has been employed to generate different models. For example, GS was used to predict primary embryogenic capacity, a highly polygenic trait, using a blend of recombinant inbred lines and SSR markers (Hu et al., 2011). In another study, 288 soybean cultivars and 79 SSR markers were used to calculate genomic-estimated breeding value to be 0.90 (Shu et al., 2012). Although the prediction accuracy was quite high in both studies, these predictions were calculated using a small set of genotypes and markers. For values that are more accurate, GS demands diverse genotypes tested under a range of environmental conditions.

Utilization of Genetically Modified Technology for Improving Tolerance to Abiotic Stresses

Nuclear Approaches

Biotech crops offered a significant increase in yield by protecting crops from the growth-damaging factors such as biotic and abiotic stresses. These are also environment friendly, as a significant reduction of pesticide use was observed on the biotech crops. Since the introduction of genetically modified (GM) soybeans in 1996 in the United States, it is now grown in several countries, including Canada, Argentina, Mexico, Brazil, South Africa, Paraguay, Uruguay, China, and India. Each year, a substantial increase in GM soybean area has been observed. For example, GM soybeans occupied a 47% of the total global area under the GM plantation. To develop drought-tolerant soybean lines, researchers expressed the Δ1-pyrroline-5-carboxylate synthase (*P5CR*) coding gene from Arabidopsis into soybeans. Overexpression of *P5CR* in soybeans resulted in increased levels of free proline, and the transgenic lines showed greater tolerance to drought and heat stresses (De Ronde et al., 2004). Introduction of a *Brassica campestris NTR1* gene coding for jasmonic acid carboxyl methyltransferase in soybeans resulted in high tolerance to dehydration during the germination stage (Xue et al., 2007). In another study, expression of a dehydration-responsive element binding protein, DREB1A, under the drought-inducible promoter Rd29 of Arabidopsis in soybeans was reported. Upregulation of several genes including drought-responsive genes, notably *GmPI-PLC*, *GmSTP*, *GmGRP*, and *GmLEA14*, was found in transgenic soybean lines under drought stress (Polizel et al., 2011). It was also highlighted that the downregulation of strigolactone synthesis enzymes (SSE) could lead to increased tolerance to stress (Quain et al., 2014). For example, downregulation of cysteine proteases, a member of the SSE family, by overexpressing the rice cystatin, oryzacystatin-I, improved stress tolerance in soybeans as well as *Arabidopsis* (Quain et al., 2014). In Argentina, drought tolerant transgenic soybean is very close to its commercial approval (Waltz, 2015).

Like drought, salt-induced toxicity negatively affects the all-growth stages of soybean plants and thus decreases its yield. Expression of an *Arabidopsis* vacuolar Na^+/H^+ antiporter gene (*AtNHX1*) in soybeans improved salt tolerance significantly, rather than that of the nontransformed plants (Li et al., 2010). Constitutive expression of an intrinsic plasma membrane protein 1; 6 (GmPIP1; 6) from a constitutive cauliflower mosaic virus 35S promoter (CaMV 35S) improved soybean root length and Na^+ sequestration (Zhou et al., 2014). Overexpression of a *Solanumtorvum* Δ1-pyrroline-5-carboxylate synthetase gene (*StP5CS*) in soybeans resulted in a higher level of salt tolerance by increasing proline content and reducing membrane peroxidation under salt stress (Zhang et al., 2015a,b).

Although soybean transformation was first reported in 1988 (Christou et al., 1988; Hinchee et al., 1988), it has not become a routine yet. For example, the stable transformation of soybeans is still a challenging task. The success is mainly dependent on the efficient delivery of transforming DNA and the recovery of transgenic lines from a transformed cell. The apical meristem in soybeans consists of three cellular layers, L1, L2, and L3, all of which are involved in the regeneration process of new shoots. The acquisition of stable transgenic lines therefore requires transformation of these layers. Two main methods of delivering transgene at the intended cell location have been reported for soybeans: the particle delivery system and the *Agrobacterium*-mediated method. Particle bombardment has been shown as more efficient compared to *Agrobacterium*-mediated transformation (Wiebke-Strohm et al., 2011). However, its operational cost is high. Since the initial transformation and regeneration of transgenic soybeans from cotyledonary nodes, regeneration from other explants such as hypocotyles, half-seeds, organogenic callus, and immature zygotic cotyledons have been reported as well (see Homrich et al., 2012 for review).

Nonnuclear Approaches

Traditionally, plants have been engineered by inserting transgene in the nuclear genome, known as nuclear insertions. However, nuclear transformation has met with several challenges including variable and poor gene expression, outcrossing of transgenes to weedy relatives (via pollen), and nonallelic interactions, which often results in gene silencing. For example, if the transgenes confer a fitness-enhancing trait such as salt tolerance, drought tolerance, frost tolerance, or insect resistance, transgenes spread to wild relatives could lead to the evolution of unwanted plants. These unwanted plant species will then be able to compete with the crops for nutrients, space, and light resources and would lead to a reduction in yield. Herbicide-resistance traits, if transferred to weeds, would not be killed by the weedicide. For example, the emergence of herbicide-resistant weeds in Canada and the United States is often attributed to the cultivation of transgenic crops in which herbicide resistance genes were introduced (Gilbert, 2013). This shows that cultivation of transgenic crops have a potential threat to drive the evolution of super weeds. This perceived risk has made their cultivation in open fields highly controversial.

The controversy surrounding GM technology necessitates finding alternative means of engineering plants against biotic and abiotic stresses. Plants possess DNA into various other compartments such as mitochondria and plastids that are inherited maternally. This means transgenes from these compartments will not be escaped through natural cross-pollination, and will remain "contained" inside the plant tissue. The advancements in recombinant DNA technology have made it possible to engineer these compartmental genomes. Engineering mitochondrial genomes has not been very promising. However, the engineering of the plastid genome has been quite successful and has emerged as a serious competitor to the conventional plant engineering approaches (Ahmad and Mukhtar, 2013; Oey et al., 2009).

Plastids are cellular organelles, which have arisen through endosymbiosis by engulfing a photosynthetic bacterium. Since then, plastids have evolved to perform a variety of functions in a plant cell, ranging from photosynthesis to store different brightly colored pigments as in flowers and other colored plant parts, accumulation of starch, lipids, and to perform other specialized functions. Apart from carrying essential functions, plastids have their own genome, called plastome, which exhibits a high copy number (10,000 copies per plant cell). Other features of plastome include the arrangement of genes into clusters, higher AT content, and the presence of two inverted repeat regions (in higher plant chloroplasts only). Since plastids are inherited maternally (in most gymnosperms), therefore insertion of transgenes into plastids offer a tight natural gene containment. One of the attractive features of plastid transformation is homologous recombination-based, site-specific integration of transgenes (Fig. 9.1). Plastids do not follow Mendelian laws, and thus insertions of transgenes into plastids always result in uniform gene expression. All these features are quite convincing to make the plastids a choice of genetic transformations.

The transformation of higher plant chloroplasts, or green plastids, often results in an extraordinary expression of foreign proteins (Ahmad et al., 2012; Ahmad and Mukhtar, 2013; Michoux et al., 2011; Oey et al., 2009; Ruhlman et al., 2010). The chloroplast transformation has been deployed to engineer crops for various agronomic traits including different stresses, such as biotic and abiotic (reviewed in Ellstrand et al., 2013).

Initial attempts for establishing plastid transformation in soybeans met with limited success. The development of stable and fertile transplastomic soybean lines were reported for the first time by Dufourmantel et al. (2004). However, this report was limited to the proof-of-concept and expressed only a selection marker, the *aadA* gene coding for an aminoglycoside-3″-adenylyltransferase, which confers resistance against spectinomycin and streptomycin. Dufourmantel et al. (2005) developed transplastomic soybean lines expressing the *Bacillus thuringiensis* insecticidal protoxin Cry1Ab, which

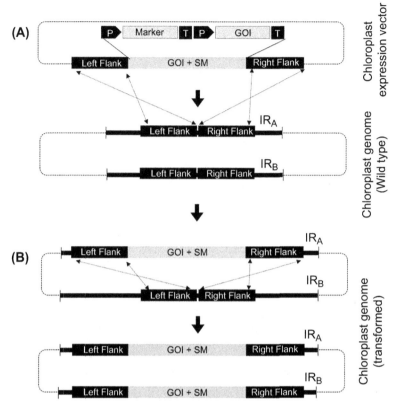

Figure 9.1 Schematic representation of site-specific genetic engineering of the plastid genome. (A) Shows integration of both the expression cassette and selection cassette into the plastid genome at a chosen site through homologous recombination, a hallmark of plastid genome engineering technology. Two flanking regions from the plastid genome at which insertion is required are therefore always incorporated into the chloroplast expression vectors to allow a homologous recombination reaction to occur for the delivery of transgenes at the intended site. The GOI and SM are placed between these flanking regions. The inclusion of these flanking regions into chloroplast expression vectors makes this technology species-specific, which means vectors constructed for one species cannot be used to transform another (when used, both the transformation efficiency as well as expression of recombinant proteins was compromised (Ruhlman et al., 2010)). In addition, the strategy could be employed to study gene functions, either reverse genetics, forward genetics, or the site-directed mutagenesis. (B) Shows the duplication of GOI and SM from IR_A into IR_B through a process known as copy-correction, also mediated by homologous recombination, which takes place between inverted repeat regions. Initially, few copies of the plastome are transformed and therefore the explant contains a mixture of both the transformed as well as untransformed copies, a state known as heteroplastomy. The wild type copies are sorted out gradually by repeating few regeneration cycles against selection to reach homoplastomy, a state when all copies of the plastome are transformed. The dotted arrows show the regions, which will undergo recombination, whereas the thick-black arrows show the progress of the reaction to purify homoplastomic plant lines (see Ahmad and Mukhtar, 2013 for further details). *IR*, inverted repeat; *GOI*, gene of interest; *SM*, selection marker.

was integrated between the *rps12/7* and *trnV* intergenic region of *Glycinemax* plastome. The transplastomic soybean plants showed a strong insecticidal activity (100% mortality in neonates after a three-day feeding) against the velvet bean caterpillar (*Anticarsia gemmatalis*), a major pest of soybeans (Dufourmantel et al., 2005). Since these reports, no other studies were undertaken for engineering the soybean plastome for other traits, in part due to difficult transformation protocols. Due to strong opposition by a large fraction of the scientific as well as public community, the use of antibiotic-based selection markers is discouraged. However, the use of heavy antibiotic-based selection markers in plastids is inevitable. Therefore protocols have been developed to remove selection markers once transplastomic lines are obtained. Excision of marker genes from the soybean plastome has been reported (Dufourmantel et al., 2007). Homology-based direct repeat (repeat length of 367 bp) excision of the *aadA* gene from transplastomic soybeans was quite efficient and was demonstrated by the restoration of a *bar* gene, resulting in a phosphinothricin-resistance phenotype (Dufourmantel et al., 2007).

Like other plant species, transforming the plastid genome of soybeans has appeared quite challenging compared to nuclear transformation. Unlike other plants, such as tobacco, where chloroplasts are targets of the gene delivery process, in soybeans, embryogenic cultures are the most preferred explants due to their regeneration potential. However, the undifferentiated cells in the embryogenic cultures contain smaller and fewer numbers of plastids compared to leaf cells. Therefore transforming soybean plastids is technically challenging, and difficult: one of the major obstacles in the application of this technology in crop plants. However, efforts are underway to develop fast and reliable protocols for the generation of fertile transplastomic soybean plants (Dubald et al., 2014).

Future Prospective

Conventional breeding will not be able to meet the food requirement in 2050, and it would also be difficult for breeders to develop varieties, which can withstand the potential hazards of climate change. Improvement in genomic tools together with the invention of new transformation procedures would be the potential area of research to address the future challenges. The WGS information of soybeans is available and assembled on chromosomes. At present, it is possible to sequence the whole genome of the soybean plant in much less time and in a more cost-effective manner than before. It is imperative to sequence the representative genotypes, which represent the whole genetic diversity available in soybean germplasm. It would help in identifying the function of different genes, and the information generated could also be used in designing new SSRs and SNPs, which would have a potential impact in initiating marker-assisted breeding for addressing the biotic and abiotic stresses. Before sequencing the representative genotypes, characterization of the available germplasm resources, preferably of both species (*Glycine max* and *Glycine soja*), would be an important step. If the resources are limited, the GBS approach can be deployed to explore variations in genomes of multiple representative genotypes, as its reference genome is available.

Studying complex traits is always a difficult task, as it hammers the progress toward initiating marker-assisted breeding (MAB). In multiple reports, a large number of QTLs have been identified. However, it is difficult to use these QTLs in MAB, as these have been detected in different environments and genetic backgrounds. One can overcome this problem by doing a "meta-QTL" analysis for identifying the consensus QTLs associated with the trait of interest. However, there are still many gaps, which are supposed to identify the most reliable QTLs with major effects. There are a number of approaches, including association mapping and nested association mapping strategies, which can be used to develop populations. These populations can be exposed to GWAS for detecting the complex QTLs with much higher resolution. The information generated would help breeders identify DNA markers associated with the traits. Also, the information would be a useful source for geneticists for exploring the genetic mechanisms involved in conferring complex traits.

Like many other important crop species such as wheat, sorghum, pea, rapeseed, etc., it has been demonstrated that TILLING can be used as one of the reverse genetic techniques for identifying the function of different genes. Once the TILLING population has been developed, it should be characterized for variations in morphological traits. If possible, one can explore the desirable mutant plants for physiological as well as biochemical traits (studying proteins, metabolites, etc.) using various high throughput phenotyping and proteomics tools. For studying the variations in nucleotide sequences, it is important to target as many genes as possible. However, one can also sequence the whole genome of the mutants using the available second and or third generation-sequencing methods followed by the detection of putative SNPs. If resources are limited, like in many developing countries, one can carry out exome capturing followed by aligning the sequences with the wild and reference genome sequences. Once the putative SNPs associated with functional diversity are identified, these can be confirmed through crossing the mutant plant with the wild type, followed by studying the segregation of the mutant trait(s) and the associated SNP(s) in F_2 populations. The TILLING populations can also be used to identify the genes involved in different complex traits, ie, seed composition traits. There is a number of new concepts including ZFNs and CRISPR-Cas9 systems, which can precisely edit and or mutate the genes (even present in duplicate forms). This would not only help in studying the function of different genes but would also be helpful in improving the genome of the soybean plant. In the end, conventional approaches, together with marker-assisted selection and genetic engineering approaches, will be the ultimate choice for improving the genetics of soybeans for mitigating the abiotic stresses in the future.

References

Ahmad, N., Michoux, F., McCarthy, J., Nixon, P.J., 2012. Expression of the affinity tags, glutathione-S-transferase and maltose-binding protein, in tobacco chloroplasts. Planta 235, 863–871.

Ahmad, N., Mukhtar, Z., 2013. Green factories: plastids for the production of foreign proteins at high levels. Gene Therapy and Molecular Biology 15, 14–29.

Ball, R.A., Purcell, L.C., Vories, E.D., 2000. Short-season soybean yield compensation in response to population and water regime. Crop Science 40, 1070–1078.

Bita, C.E., Gerats, T., 2013. Plant tolerance to high temperature in a changing environment: scientific fundamentals and production of heat stress-tolerant crops. Frontiers in Plant Science 4, 273.

Board, J.E., Kahlon, C.S., 2011. Soybean yield formation: what controls it and how it can be improved. In: El-Shemy, H. (Ed.), Soybean Physiology and Biochemistry. Intech, Crotia. pp. 1–36.

Bramlage, W.J., Leopold, A.C., Parrish, D.J., 1978. Chilling stress to soybeans during imbibition. Plant Physiology 61, 525–529.

Brown, E., Caviness, C., Brown, D., 1985. Response of selected soybean cultivars to soil moisture deficit. Agronomy Journal 77, 274–278.

Carter, J., Thomas, E., Nelson, R.L., Sneller, C.H., Cui, Z., 2004. Genetic Diversity in Soybean. Soybeans: Improvement, Production, and Uses. , pp. 303–416.

Chan, C., Qi, X., Li, M.-W., Wong, F.-L., Lam, H.-M., 2012. Recent developments of genomic research in soybean. Journal of Genetics and Genomics 39, 317–324.

Chen, L.M., Zhou, X.A., Li, W.B., Chang, W., Zhou, R., Wang, C., Sha, A.H., Shan, Z.H., Zhang, C.J., Qiu, D.Z., Yang, Z.L., Chen, S.L., 2013. Genome-wide transcriptional analysis of two soybean genotypes under dehydration and rehydration conditions. BMC Genomics 14, 687.

Cho, E.K., Choi, Y.J., 2009. A nuclear-localized HSP70 confers thermoprotective activity and drought-stress tolerance on plants. Biotechnology letters 31, 597–606.

Christou, P., McCabe, D.E., Swain, W.F., 1988. Stable transformation of soybean callus by DNA-coated gold particles. Plant Physiology 87, 671–674.

Clement, M., Lambert, A., Herouart, D., Boncompagni, E., 2008. Identification of new up-regulated genes under drought stress in soybean nodules. Gene 426, 15–22.

Concibido, V.C., Diers, B.W., Arelli, P.R., 2004. A decade of QTL mapping for cyst nematode resistance in soybean. Crop Science 44, 1121–1131.

Constable, G., Hearn, A., 1981. Irrigation for crops in a sub-humid environment. Irrigation Science 3, 17–28.

Cooper, J.L., Till, B.J., Laport, R.G., Darlow, M.C., Kleffner, J.M., Jamai, A., El-Mellouki, T., Liu, S., Ritchie, R., Nielsen, N., Bilyeu, K.D., 2008. TILLING to detect induced mutations in soybean. BMC Plant Biology 8, 9–19.

Cox, W., Jolliff, G., 1986. Growth and yield of sunflower and soybean under soil water deficits. Agronomy Journal 78, 226–230.

Cox, W., Jolliff, G., 1987. Crop-water relations of sunflower and soybean under irrigated and dryland conditions. Crop Science 27, 553–557.

De Ronde, J., Cress, W., Krüger, G., Strasser, R., Van Staden, J., 2004. Photosynthetic response of transgenic soybean plants, containing an Arabidopsis P5CR gene, during heat and drought stress. Journal of Plant Physiology 161, 1211–1224.

Desclaux, D., Huynh, T.-T., Roumet, P., 2000. Identification of soybean plant characteristics that indicate the timing of drought stress. Crop Science 40, 716–722.

Dodd, A.N., Salathia, N., Hall, A., Kévei, E., Tóth, R., Nagy, F., Hibberd, J.M., Millar, A.J., Webb, A.A., 2005. Plant circadian clocks increase photosynthesis, growth, survival, and competitive advantage. Science 309, 630–633.

Dogan, E., Kirnak, H., Copur, O., 2007. Deficit irrigations during soybean reproductive stages and CROPGRO-soybean simulations under semi-arid climatic conditions. Field Crops Research 103, 154–159.

Dong, Y., Zhao, L., Liu, B., Wang, Z., Jin, Z., Sun, H., 2004. The genetic diversity of cultivated soybean grown in China. Theoretical and Applied Genetics 108, 931–936.

Du, J., Grant, D., Tian, Z., Nelson, R.T., Zhu, L., Shoemaker, R.C., Ma, J., 2010. SoyTEdb: a comprehensive database of transposable elements in the soybean genome. BMC Genomics 11, 113–119.

Duan, Y.-H., Guo, J., Ding, K., Wang, S.-J., Zhang, H., Dai, X.-W., Chen, Y.-Y., Govers, F., Huang, L.-L., Kang, Z.-S., 2011. Characterization of a wheat HSP70 gene and its expression in response to stripe rust infection and abiotic stresses. Molecular Biology Reports 38, 301–307.

Dubald, M., Tissot, G., Pelissier, B., 2014. Plastid transformation in soybean. In: Maliga, P. (Ed.), Chloroplast Biotechnology: Methods and Protocols. Methods in Molecular Biology, vol. 1132. Springer Science, New York, pp. 345–354.

Dufourmantel, N., Dubald, M., Matringe, M., Canard, H., Garcon, F., Job, C., Kay, E., Wisniewski, J.P., Ferullo, J.M., Pelissier, B., 2007. Generation and characterization of soybean and marker-free tobacco plastid transformants over-expressing a bacterial 4-hydroxyphenylpyruvate dioxygenase which provides strong herbicide tolerance. Plant Biotechnology Journal 5, 118–133.

Dufourmantel, N., Pelissier, B., Garcon, F., Peltier, G., Ferullo, J.M., Tissot, G., 2004. Generation of fertile transplastomic soybean. Plant Molecular Biology 55, 479–489.

Dufourmantel, N., Tissot, G., Goutorbe, F., Garcon, F., Muhr, C., Jansens, S., Pelissier, B., Peltier, G., Dubald, M., 2005. Generation and analysis of soybean plastid transformants expressing *Bacillus thuringiensis* Cry1Ab protoxin. Plant Molecular Biology 58, 659–668.

Eck, H., Mathers, A., Musick, J., 1987. Plant water stress at various growth stages and growth and yield of soybeans. Field Crops Research 17, 1–16.

Ellstrand, N.C., Meirmans, P., Rong, J., Bartsch, D., Ghosh, A., de Jong, T.J., Haccou, P., Lu, B.-R., Snow, A.A., Neal Stewart Jr., C., 2013. Introgression of crop alleles into wild or weedy populations. Annual Review of Ecology, Evolution and Systematics 44, 325–345.

Endo, M., Tsuchiya, T., Hamada, K., Kawamura, S., Yano, K., Ohshima, M., Higashitani, A., Watanabe, M., Kawagishi-Kobayashi, M., 2009. High temperatures cause male sterility in rice plants with transcriptional alterations during pollen development. Plant and Cell Physiology 50, 1911–1922.

FAO, 2013. Statistical Yearbook. Food and Agriculture Organization, United Nations, Rome, Italy.

Francki, M.G., Berzonsky, W.A., Ohm, H.W., Anderson, J.M., 2002. Physical location of a HSP70 gene homologue on the centromere of chromosome 1B of wheat (*Triticum aestivum* L.). Theoretical and Applied Genetics 104, 184–191.

Gilbert, N., 2013. A hard look at GM crops. Nature 497, 24–26.

Guo, M., Zhai, Y.F., Lu, J.P., Chai, L., Chai, W.G., Gong, Z.H., Lu, M.H., 2014. Characterization of CaHsp70-1, a pepper heat-shock protein gene in response to heat stress and some regulation exogenous substances in *Capsicum annuum* L. International Journal of Molecular Sciences 15, 19741–19759.

Hao, Y.J., Song, Q.X., Chen, H.W., Zou, H.F., Wei, W., Kang, X.S., Ma, B., Zhang, W.K., Zhang, J.S., Chen, S.Y., 2010. Plant NAC-type transcription factor proteins contain a NARD domain for repression of transcriptional activation. Planta 232, 1033–1043.

Hartman, G., West, E., Herman, T., 2011. Crops that feed the World 2. Soybean—worldwide production, use, and constraints caused by pathogens and pests. Food Security 3, 5–17.

Hasanuzzaman, M., Nahar, K., Alam, M.M., Roychowdhury, R., Fujita, M., 2013. Physiological, biochemical, and molecular mechanisms of heat stress tolerance in plants. International Journal of Molecular Sciences 14, 9643–9684.

Hasegawa, P.M., Bressan, R.A., Zhu, J.-K., Bohnert, H.J., 2000. Plant cellular and molecular responses to high salinity. Annual Review of Plant Biology 51, 463–499.

He, Y., Fu, J., Yu, C., Wang, X., Jiang, Q., Hong, J., Lu, K., Xue, G., Yan, C., James, A., 2015. Increasing cyclic electron flow is related to Na$^+$ sequestration into vacuoles for salt tolerance in soybean. Journal of Experimental Botany 66, 6877–6889.

Hinchee, M.A., Connor-Ward, D.V., Newell, C.A., McDonnell, R.E., Sato, S.J., Gasser, C.S., Fischhoff, D.A., Re, D.B., Fraley, R.T., Horsch, R.B., 1988. Production of transgenic soybean plants using *Agrobacterium*-mediated DNA transfer. Nature Biotechnology 6, 915–922.

Homrich, M.S., Wiebke-Strohm, B., Weber, R.L.M., Bodanese-Zanettini, M.H., 2012. Soybean genetic transformation: a valuable tool for the functional study of genes and the production of agronomically improved plants. Genetics and Molecular Biology 35, 998–1010.

Hoogenboom, G., Peterson, C.M., Huck, M., 1987. Shoot growth rate of soybean as affected by drought stress. Agronomy Journal 79, 598–607.

Hossain, Z., Hajika, M., Komatsu, S., 2012a. Comparative proteome analysis of high and low cadmium accumulating soybeans under cadmium stress. Amino Acids 43, 2393–2416.

Hossain, Z., Nouri, M.Z., Komatsu, S., 2012b. Plant cell organelle proteomics in response to abiotic stress. Journal of Proteome Research 11, 37–48.

Hu, H., Dai, M., Yao, J., Xiao, B., Li, X., Zhang, Q., Xiong, L., 2006. Overexpressing a NAM, ATAF, and CUC (NAC) transcription factor enhances drought resistance and salt tolerance in rice. Proceedings of the National Academy of Sciences USA 103, 12987–12992.

Hu, Z., Li, Y., Song, X., Han, Y., Cai, X., Xu, S., Li, W., 2011. Genomic value prediction for quantitative traits under the epistatic model. BMC Genetics 12, 1–11.

Hwang, E.-Y., Song, Q., Jia, G., Specht, J.E., Hyten, D.L., Costa, J., Cregan, P.B., 2014. A genome-wide association study of seed protein and oil content in soybean. BMC Genomics 15, 1–12.

Hwang, W.J., Kim, M.Y., Kang, Y.J., Shim, S., Stacey, M.G., Stacey, G., Lee, S.H., 2015. Genome-wide analysis of mutations in a dwarf soybean mutant induced by fast neutron bombardment. Euphytica 203, 399–408.

Hymowitz, T., 1970. On the domestication of the soybean. Economic Botany 24, 408–421.

Hymowitz, T., 1990. Soybeans: the success story. In: Janick, J., Simon, J.E. (Eds.), Advances in New Crops. Timber Press, Portland, pp. 159–163.

Jeong, J.S., Kim, Y.S., Baek, K.H., Jung, H., Ha, S.H., Do Choi, Y., Kim, M., Reuzeau, C., Kim, J.K., 2010. Root-specific expression of OsNAC10 improves drought tolerance and grain yield in rice under field drought conditions. Plant Physiology 153, 185–197.

Jin, J., Zhang, H., Kong, L., Gao, G., Luo, J., 2013. PlantTFDB 3.0: a portal for the functional and evolutionary study of plant transcription factors. Nucleic Acids Research 42, 1182–1187.

Kadhem, F., Specht, J., Williams, J., 1985. Soybean irrigation serially timed during stages R1 to R6. 11. Yield component responses. Agronomy Journal 77, 299–304.

Khatoon, A., Rehman, S., Hiraga, S., Makino, T., Komatsu, S., 2012. Organ-specific proteomics analysis for identification of response mechanism in soybean seedlings under flooding stress. Journal of Proteomics 75, 5706–5723.

Kido, E.A., Ferreira Neto, J.R., Silva, R.L., Belarmino, L.C., Bezerra Neto, J.P., Soares-Cavalcanti, N.M., Pandolfi, V., Silva, M.D., Nepomuceno, A.L., Benko-Iseppon, A.M., 2013. Expression dynamics and genome distribution of osmoprotectants in soybean: identifying important components to face abiotic stress. BMC Bioinformatics 14 (Suppl. 1), S7.

Kim, B.-M., Rhee, J.-S., Jeong, C.-B., Seo, J.S., Park, G.S., Lee, Y.-M., Lee, J.-S., 2014. Heavy metals induce oxidative stress and trigger oxidative stress-mediated heat shock protein (hsp) modulation in the intertidal copepod *Tigriopus japonicus*. Comparative Biochemistry and Physiology Part C: Toxicology & Pharmacology 166, 65–74.

Komatsu, S., Hiraga, S., Yanagawa, Y., 2012. Proteomics techniques for the development of flood tolerant crops. Journal of Proteome Research 11, 68–78.

Komatsu, S., Nanjo, Y., Nishimura, M., 2013. Proteomic analysis of the flooding tolerance mechanism in mutant soybean. Journal of Proteomics 79, 231–250.

Korte, L., Williams, J., Specht, J., Sorensen, R., 1983. Irrigation of soybean genotypes during reproductive ontogeny. I. Agronomic responses. Crop Science 23, 521–527.

Koti, S., Reddy, K.R., Reddy, V., Kakani, V., Zhao, D., 2005. Interactive effects of carbon dioxide, temperature, and ultraviolet-B radiation on soybean (*Glycine max* L.) flower and pollen morphology, pollen production, germination, and tube lengths. Journal of Experimental Botany 56, 725–736.

Lam, H.-M., Xu, X., Liu, X., Chen, W., Yang, G., Wong, F.-L., Li, M.-W., He, W., Qin, N., Wang, B., 2010. Resequencing of 31 wild and cultivated soybean genomes identifies patterns of genetic diversity and selection. Nature Genetics 42, 1053–1059.

Lewis, G.P., Schrire, B., Mackinder, B., Lock, M., 2005. Legumes of the World. Royal Botanic Gardens Kew, Richmond, Surrey, UK.

Li, H., Liu, S.S., Yi, C.Y., Wang, F., Zhou, J., Xia, X.J., Shi, K., Zhou, Y.H., Yu, J.Q., 2014a. Hydrogen peroxide mediates abscisic acid-induced HSP70 accumulation and heat tolerance in grafted cucumber plants. Plant Cell & Environment 37, 2768–2780.

Li, Y.-H., Reif, J.C., Jackson, S.A., Ma, Y.-S., Chang, R.-Z., Qiu, L.-J., 2014b. Detecting SNPs underlying domestication-related traits in soybean. BMC Plant Biology 14, 251–258.

Li, Y.-H., Zhou, G., Ma, J., Jiang, W., Jin, L.-G., Zhang, Z., Guo, Y., Zhang, J., Sui, Y., Zheng, L., 2014c. De novo assembly of soybean wild relatives for pan-genome analysis of diversity and agronomic traits. Nature Biotechnology 32, 1045–1052.

Li, T.Y., Zhang, Y., Liu, H., Wu, Y., Li, W., Zhang, H., 2010. Stable expression of *Arabidopsis* vacuolar Na+/H+ antiporter gene *AtNHX1*, and salt tolerance in transgenic soybean for over six generations. Chinese Science Bulletin 55, 1127–1134.

Liao, Y., Zou, H.F., Wang, H.W., Zhang, W.K., Ma, B., Zhang, J.S., Chen, S.Y., 2008a. Soybean GmMYB76, GmMYB92, and GmMYB177 genes confer stress tolerance in transgenic *Arabidopsis* plants. Cell Research 18, 1047–1060.

Liao, Y., Zou, H.F., Wei, W., Hao, Y.J., Tian, A.G., Huang, J., Liu, Y.F., Zhang, J.S., Chen, S.Y., 2008b. Soybean GmbZIP44, GmbZIP62 and GmbZIP78 genes function as negative regulator of ABA signaling and confer salt and freezing tolerance in transgenic Arabidopsis. Planta 228, 225–240.

Limei, Z., Yingshan, D., Bao, L., Shui, H., Kejing, W., Xianghua, L., 2005. Establishment of a core collection for the Chinese annual wild soybean (*Glycine soja*). Chinese Science Bulletin 50, 989–996.

Lin, J.-Y., Stupar, R.M., Hans, C., Hyten, D.L., Jackson, S.A., 2010. Structural and functional divergence of a 1-Mb duplicated region in the soybean (*Glycine max*) genome and comparison to an orthologous region from *Phaseolus vulgaris*. The Plant Cell 22, 2545–2561.

Liu, F., Jensen, C.R., Andersen, M.N., 2004. Drought stress effect on carbohydrate concentration in soybean leaves and pods during early reproductive development: Its implication in altering pod set. Field Crops Research 86, 1–13.

Liu, K., 2008. Food Use of Whole Soybeans. In: Johnson, L.A., White, P.J., Galloway, R. (Eds.), Soybeans: Chemistry, Production, Processing and Utilization. AOCS Press, Urbana, pp. 441–482.

Lobell, D.B., Gourdji, S.M., 2012. The influence of climate change on global crop productivity. Plant Physiology 160, 1686–1697.

Lopes-Caitar, V.S., de Carvalho, M.C., Darben, L.M., Kuwahara, M.K., Nepomuceno, A.L., Dias, W.P., Abdelnoor, R.V., Marcelino-Guimaraes, F.C., 2013. Genome-wide analysis of the Hsp20 gene family in soybean: comprehensive sequence, genomic organization and expression profile analysis under abiotic and biotic stresses. BMC Genomics 14, 577–593.

Mangelsen, E., Kilian, J., Berendzen, K.W., Kolukisaoglu, Ü.H., Harter, K., Jansson, C., Wanke, D., 2008. Phylogenetic and comparative gene expression analysis of barley (*Hordeum vulgare*) WRKY transcription factor family reveals putatively retained functions between monocots and dicots. BMC Genomics 9, 194–210.

Meckel, L., Egli, D., Phillips, R., Radcliffe, D., Leggett, J., 1984. Effect of moisture stress on seed growth in soybeans. Agronomy Journal 76, 647–650.

Michoux, F., Ahmad, N., McCarthy, J., Nixon, P.J., 2011. Contained and high-level production of recombinant proteins in plant chloroplasts using a temporary immersion bioreactor. Plant Biotechnology Journal 9, 575–584.

Mishra, S., Verma, V., 2010. Soybean genetic resources. In: Singh, D. (Ed.), The Soybean: Botany, Production and Uses. CABI, Walligford, UK. pp. 74–91.

Mohammadi, P.P., Moieni, A., Hiraga, S., Komatsu, S., 2012. Organ-specific proteomic analysis of drought-stressed soybean seedlings. Journal of Proteomics 75, 1906–1923.

Morrell, P.L., Buckler, E.S., Ross-Ibarra, J., 2012. Crop genomics: advances and applications. Nature Reviews Genetics 13, 85–96.

Muchow, R., Sinclair, T., Bennett, J., Hammond, L., 1986. Response of leaf growth, leaf nitrogen, and stomatal conductance to water deficits during vegetative growth of field-grown soybean. Crop Science 26, 1190–1195.

Munns, R., Tester, M., 2008. Mechanisms of salinity tolerance. Annual Review of Plant Biology 59, 651–681.

Nakashima, K., Tran, L.S., Van Nguyen, D., Fujita, M., Maruyama, K., Todaka, D., Ito, Y., Hayashi, N., Shinozaki, K., Yamaguchi-Shinozaki, K., 2007. Functional analysis of a NAC-type transcription factor OsNAC6 involved in abiotic and biotic stress-responsive gene expression in rice. Plant Journal 51, 617–630.

Nelson, R., 2009. Collection, conservation, and evaluation of soybean germplasm. In: The Abstract of Proceedings. The Eight World Soybean Research Conference. Beijing, China.

Neyshabouri, M., Hatfield, J., 1986. Soil water deficit effects on semi-determinate and indeterminate soybean growth and yield. Field Crops Research 15, 73–84.

Niu, C.F., Wei, W., Zhou, Q.Y., Tian, A.G., Hao, Y.J., Zhang, W.K., Ma, B., Lin, Q., Zhang, Z.B., Zhang, J.S., 2012. Wheat WRKY genes TaWRKY2 and TaWRKY19 regulate abiotic stress tolerance in transgenic *Arabidopsis* plants. Plant, Cell & Environment 35, 1156–1170.

Oey, M., Lohse, M., Kreikemeyer, B., Bock, R., 2009. Exhaustion of the chloroplast protein synthesis capacity by massive expression of a highly stable protein antibiotic. Plant Journal 57, 436–445.

Oh, M., Komatsu, S., 2015. Characterization of proteins in soybean roots under flooding and drought stresses. Journal of Proteomics 114, 161–181.

Pandey, R., Herrera, W., Pendleton, J., 1984. Drought response of grain legumes under irrigation gradient: I. Yield and yield components. Agronomy Journal 76, 549–553.

Park, S., Lee, D.E., Jang, H., Byeon, Y., Kim, Y.S., Back, K., 2013. Melatonin-rich transgenic rice plants exhibit resistance to herbicide-induced oxidative stress. Journal of Pineal Research 54, 258–263.

Patil, G., Valliyodan, B., Deshmukh, R., Prince, S., Nicander, B., Zhao, M., Sonah, H., Song, L., Lin, L., Chaudhary, J., 2015. Soybean (*Glycine max*) SWEET gene family: insights through comparative genomics, transcriptome profiling and whole genome re-sequence analysis. BMC Genomics 16, 520–535.

Polizel, A., Medri, M., Nakashima, K., Yamanaka, N., Farias, J., De Oliveira, M., Marin, S., Abdelnoor, R., Marcelino-Guimarães, F., Fuganti, R., 2011. Molecular, anatomical and physiological properties of a genetically modified soybean line transformed with rd29A: AtDREB1A for the improvement of drought tolerance. Genetics and Molecular Research 10, 3641–3656.

Purcell, L.C., Specht, J.E., 2004. Physiological traits for ameliorating drought stress. In: Specht, J.E., Boerma, H.R. (Eds.), Soybeans: Improvement, Production and Uses. ASA-CSSA-SSSA, Madison, WI, pp. 569–620.

Qi, X., Li, M.-W., Xie, M., Liu, X., Ni, M., Shao, G., Song, C., Yim, A.K.-Y., Tao, Y., Wong, F.-L., 2014. Identification of a novel salt tolerance gene in wild soybean by whole-genome sequencing. Nature Communications 5, 4340–4350.

Qiu, L.-J., Xing, L.-L., Guo, Y., Wang, J., Jackson, S., Chang, R.-Z., 2013. A platform for soybean molecular breeding: the utilization of core collections for food security. Plant Molecular Biology 83, 41–50.

Quain, M.D., Makgopa, M.E., Márquez-García, B., Comadira, G., Fernandez-Garcia, N., Olmos, E., Schnaubelt, D., Kunert, K.J., Foyer, C.H., 2014. Ectopic phytocystatin expression leads to enhanced drought stress tolerance in soybean (Glycine max) and Arabidopsis thaliana through effects on strigolactone pathways and can also result in improved seed traits. Plant Biotechnology Journal 12, 903–913.

Ramseur, E., Quisenberry, V., Wallace, S., Palmer, J., 1984. Yield and yield components of 'Braxton'soybeans as influenced by irrigation and intrarow spacing. Agronomy Journal 76, 442–446.

Ramseur, E., Wallace, S., Quisenberry, V., 1985. Growth of 'Braxton'soybeans as influenced by irrigation and intrarow spacing. Agronomy Journal 77, 163–168.

Ruhlman, T., Verma, D., Samson, N., Daniell, H., 2010. The role of heterologous chloroplast sequence elements in transgene integration and expression. Plant Physiology 152, 2088–2104.

Sakata, T., Higashitani, A., 2008. Male sterility accompanied with abnormal anther development in plants–genes and environmental stresses with special reference to high temperature injury. International Journal of Plant Developmental Biology 2, 42–51.

Salem, M.A., Kakani, V.G., Koti, S., Reddy, K.R., 2007. Pollen-based screening of soybean genotypes for high temperatures. Crop Science 47, 219–231.

Schmutz, J., Cannon, S.B., Schlueter, J., Ma, J., Mitros, T., Nelson, W., Hyten, D.L., Song, Q., Thelen, J.J., Cheng, J., 2010. Genome sequence of the palaeopolyploid soybean. Nature 463, 178–183.

Seo, Y.J., Park, J.B., Cho, Y.J., Jung, C., Seo, H.S., Park, S.K., Nahm, B.H., Song, J.T., 2010. Over expression of the ethylene-responsive factor gene BrERF4 from Brassica rapa increases tolerance to salt and drought in Arabidopsis plants. Molecules and Cells 30, 271–277.

Shi, J., Gonzales, R.A., Bhattacharyya, M.K., 1995. Characterization of a plasma membrane-associated phosphoinositide-specific phospholipase C from soybean. Plant Journal 8, 381–390.

Shu, Y., Yu, D., Wang, D., Bai, X., Zhu, Y., Guo, C., 2012. Genomic selection of seed weight based on low-density SCAR markers in soybean. Genetics and Molecular Research 12, 2178–2188.

Singh, G., 2010. The Soybean: Botany, Production and Uses. CABI, Wallingford, UK.

Sionit, N., Strain, B., Flint, E., 1987. Interaction of temperature and CO_2 enrichment on soybean: photosynthesis and seed yield. Canadian Journal of Plant Science 67, 629–636.

Snyder, R., Carlson, R., Shaw, R., 1982. Yield of indeterminate soybeans in response to multiple periods of soil-water stress during reproduction. Agronomy Journal 74, 855–859.

Sobhanian, H., Razavizadeh, R., Nanjo, Y., Ehsanpour, A.A., Jazii, F.R., Motamed, N., Komatsu, S., 2010. Proteome analysis of soybean leaves, hypocotyls and roots under salt stress. Proteome Science 8, 19–33.

Sonah, H., Bastien, M., Iquira, E., Tardivel, A., Légaré, G., Boyle, B., Normandeau, É., Laroche, J., Larose, S., Jean, M., 2013. An improved genotyping by sequencing (GBS) approach offering increased versatility and efficiency of SNP discovery and genotyping. PLoS One 8, e54603.

Song, Q., Jia, G., Zhu, Y., Grant, D., Nelson, R.T., Hwang, E.-Y., Hyten, D.L., Cregan, P.B., 2010. Abundance of SSR motifs and development of candidate polymorphic SSR markers (BARCSOYSSR_1. 0) in soybean. Crop Science 50, 1950–1960.

Song, Q.X., Li, Q.T., Liu, Y.F., Zhang, F.X., Ma, B., Zhang, W.K., Man, W.Q., Du, W.G., Wang, G.D., Chen, S.Y., Zhang, J.S., 2013. Soybean GmbZIP123 gene enhances lipid content in the seeds of transgenic Arabidopsis plants. Journal of Experimental Botany 64, 4329–4341.

Specht, J., Hume, D., Kumudini, S., 1999. Soybean yield potential—a genetic and physiological perspective. Crop Science 39, 1560–1570.

Su, P.-H., Li, H.-M., 2008. Arabidopsis stromal 70-KD heat shock proteins are essential for plant development and important for thermotolerance of germinating seeds. Plant Physiology 146, 1231–1241.

Syed, N.H., Prince, S.J., Mutava, R.N., Patil, G., Li, S., Chen, W., Babu, V., Joshi, T., Khan, S., Nguyen, H.T., 2015. Core clock, SUB1, and ABAR genes mediate flooding and drought responses via alternative splicing in soybean. Journal of Experimental Botany 66, 7129–7149.

Tan, D.-X., Hardeland, R., Manchester, L.C., Korkmaz, A., Ma, S., Rosales-Corral, S., Reiter, R.J., 2012. Functional roles of melatonin in plants, and perspectives in nutritional and agricultural science. Journal of Experimental Botany 63, 577–597.

Tardivel, A., Sonah, H., Belzile, F., O'Donoughue, L.S., 2014. Rapid identification of alleles at the soybean maturity gene E3 using genotyping by sequencing and a haplotype-based approach. The Plant Genome 7, 1–9.

Teixeira, E.I., Fischer, G., van Velthuizen, H., Walter, C., Ewert, F., 2013. Global hot-spots of heat stress on agricultural crops due to climate change. Agricultural and Forest Meteorology 170, 206–215.

Tran, L.-S.P., Mochida, K., 2010. Functional genomics of soybean for improvement of productivity in adverse conditions. Functional & Integrative Genomics 10, 447–462.

Tran, L.-S.P., Nakashima, K., Sakuma, Y., Simpson, S.D., Fujita, Y., Maruyama, K., Fujita, M., Seki, M., Shinozaki, K., Yamaguchi-Shinozaki, K., 2004. Isolation and functional analysis of Arabidopsis stress-inducible NAC transcription factors that bind to a drought-responsive cis-element in the early responsive to dehydration stress 1 promoter. The Plant Cell 16, 2481–2498.

Tripathi, P., Rabara, R.C., Shen, Q.J., Rushton, P.J., 2015. Transcriptomics analyses of soybean leaf and root samples during water-deficit. Genomics Data 5, 164–166.

Ülker, B., Somssich, I.E., 2004. WRKY transcription factors: from DNA binding towards biological function. Current Opinion in Plant Biology 7, 491–498.

Valliyodan, B., Nguyen, H.T., 2006. Understanding regulatory networks and engineering for enhanced drought tolerance in plants. Current Opinion in Plant Biology 9, 189–195.

Vartapetian, B.B., Jackson, M.B., 1997. Plant adaptations to anaerobic stress. Annals of Botany 79, 3–20.

Vitalini, S., Gardana, C., Simonetti, P., Fico, G., Iriti, M., 2013. Melatonin, melatonin isomers and stilbenes in Italian traditional grape products and their antiradical capacity. Journal of Pineal Research 54, 322–333.

Waltz, E., 2015. First stress-tolerant soybean gets go-ahead in Argentina. Nature Biotechnology 33, 682.

Wang, F., Chen, H.W., Li, Q.T., Wei, W., Li, W., Zhang, W.K., Ma, B., Bi, Y.D., Lai, Y.C., Liu, X.L., Man, W.Q., Zhang, J.S., Chen, S.Y., 2015. GmWRKY27 interacts with GmMYB174 to reduce expression of GmNAC29 for stress tolerance in soybean plants. The Plant Journal 83, 224–236.

Wang, L., Guan, Y., Guan, R., Li, Y., Ma, Y., Dong, Z., Liu, X., Zhang, H., Zhang, Y., Liu, Z., Chang, R., Xu, H., Li, L., Lin, F., Luan, W., Yan, Z., Ning, X., Zhu, L., Cui, Y., Piao, R., Liu, Y., Chen, P., Qiu, L., 2006. Establishment of Chinese soybean *Glycine max* core collections with agronomic traits and SSR markers. Euphytica 151, 215–223.

Wang, P., Yin, L., Liang, D., Li, C., Ma, F., Yue, Z., 2012. Delayed senescence of apple leaves by exogenous melatonin treatment: toward regulating the ascorbate–glutathione cycle. Journal of Pineal Research 53, 11–20.

Wang, W., Vinocur, B., Altman, A., 2003. Plant responses to drought, salinity and extreme temperatures: towards genetic engineering for stress tolerance. Planta 218, 1–14.

Wei, W., Li, Q.-T., Chu, Y.-N., Reiter, R.J., Yu, X.-M., Zhu, D.-H., Zhang, W.-K., Ma, B., Lin, Q., Zhang, J.-S., 2015. Melatonin enhances plant growth and abiotic stress tolerance in soybean plants. Journal of Experimental Botany 66, 695–707.

Wen, Z., Boyse, J.F., Song, Q., Cregan, P.B., Wang, D., 2015. Genomic consequences of selection and genome-wide association mapping in soybean. BMC Genomics 16, 671–684.

Westgate, M., Peterson, C., 1993. Flower and pod development in water-deficient soybeans (*Glycine max* L. Merr.). Journal of Experimental Botany 44, 109–117.

Wiebke-Strohm, B., Droste, A., Pasquali, G., Osorio, M., Bücker-Neto, L., Passaglia, L., Bencke, M., Homrich, M., Margis-Pinheiro, M., Bodanese-Zanettini, M., 2011. Transgenic fertile soybean plants derived from somatic embryos transformed via the combined DNA-free particle bombardment and *Agrobacterium* system. Euphytica 177, 343–354.

Wu, K.-L., Guo, Z.-J., Wang, H.-H., Li, J., 2005. The WRKY family of transcription factors in rice and Arabidopsis and their origins. DNA Research 12, 9–26.

Xue, R.-G., Zhang, B., Xie, H.-F., 2007. Overexpression of a NTR1 in transgenic soybean confers tolerance to water stress. Plant Cell Tissue and Organ Culture 89, 177–183.

Yamaguchi-Shinozaki, K., Shinozaki, K., 2006. Transcriptional regulatory networks in cellular responses and tolerance to dehydration and cold stresses. Annual Review of Plant Biology 57, 781–803.

Yeo, A., 1998. Molecular biology of salt tolerance in the context of whole-plant physiology. Journal of Experimental Botany 49, 915–929.

Yin, L., Wang, P., Li, M., Ke, X., Li, C., Liang, D., Wu, S., Ma, X., Li, C., Zou, Y., 2013. Exogenous melatonin improves Malus resistance to Marssonina apple blotch. Journal of Pineal Research 54, 426–434.

Yoon, M.S., Lee, J., Kim, C.Y., Kang, J.H., Cho, E.G., Baek, H.J., 2008. DNA profiling and genetic diversity of Korean soybean (*Glycine max* (L.) Merrill) landraces by SSR markers. Euphytica 165, 69–77.

Zhang, G.-C., Zhu, W.-L., Gai, J.-Y., Zhu, Y.-L., Yang, L.-F., 2015a. Enhanced salt tolerance of transgenic vegetable soybeans resulting from overexpression of a novel Δ1-pyrroline-5-carboxylate synthetase gene from *Solanum torvum* Swartz. Horticulture Environment and Biotechnology 56, 94–104.

Zhang, L., Zhao, H.K., Dong, Q.L., Zhang, Y.Y., Wang, Y.M., Li, H.Y., Xing, G.J., Li, Q.Y., Dong, Y.S., 2015b. Genome-wide analysis and expression profiling under heat and drought treatments of HSP70 gene family in soybean (*Glycine max* L.). Frontiers in Plant Science 6, 773–787.

Zhang, Z.L., Zhu, J.H., Zhang, Q.Q., Cai, Y.B., 2009. Molecular characterization of an ethephon-induced Hsp70 involved in high and low-temperature responses in *Hevea brasiliensis*. Plant Physiology and Biochemistry 47, 954–959.

Zhou, L., Wang, C., Liu, R., Han, Q., Vandeleur, R.K., Du, J., Tyerman, S., Shou, H., 2014. Constitutive overexpression of soybean plasma membrane intrinsic protein GmPIP1; 6 confers salt tolerance. BMC Plant Biology 14, 181–193.

Zhou, Q.Y., Tian, A.G., Zou, H.F., Xie, Z.M., Lei, G., Huang, J., Wang, C.M., Wang, H.W., Zhang, J.S., Chen, S.Y., 2008. Soybean WRKY-type transcription factor genes, GmWRKY13, GmWRKY21, and GmWRKY54, confer differential tolerance to abiotic stresses in transgenic Arabidopsis plants. Plant Biotechnology Journal 6, 486–503.

Zhu, J.-K., 2001. Plant salt tolerance. Trends in Plant Science 6, 66–71.

Zou, J., Liu, C., Liu, A., Zou, D., Chen, X., 2012. Overexpression of OsHsp17.0 and OsHsp23.7 enhances drought and salt tolerance in rice. Journal of Plant Physiology 169, 628–635.

Soybean and Acidity Stress

10

M. Miransari
AbtinBerkeh Scientific Ltd. Company, Isfahan, Iran

Introduction

It is important to provide the increasing world population, which will be approximately 20 billion by year 2035, with their essential food. About 90% of the world population is concentrated in the developing and less developed nations. People in the tropical areas acquire more than 80% of their protein from plant material. At present, 40% of carbon produced by plants through the process of photosynthesis is used by humans (Stewart et al., 2005; Considine, 2012).

The soybean [*Glycine max* (L.) Merr.] is among the most important legumes, with the ability to feed people worldwide. Its nutritious values are due to its high rate of protein and oil. It is also of medicinal use, as some of the products in soybean seeds can enhance human health. Although some nations, such as Brazil, the United States, China, India, and Argentina, have the highest rate of production, the soybean is planted and produced in most parts of the world. One of the most important properties of the soybean is its ability to develop a symbiotic association with its specific rhizobium, *Bradyrhizobium japonicum*, acquiring most of its essential nitrogen by the process of biological N fixation (Jensen et al., 2012; Liu, 2012; Dobereiner, 2013).

The contribution of biologically fixed nitrogen (80%) in agriculture is mainly by the N-fixing bacteria including *Rhizobium*, *Sinorhizobium*, *Bradyrhizobium*, *Azorhizobium*, and *Mezorhizobium*, and the remaining is by actinorrhizal (*Frankia*) and *Anabena-Azolla*. It was approximated in 2005 that about 35% of protein intake is by legumes and the rate of N fixation by these symbionts is 90 Tg N per year, which is of great economic and environmental significance. It is equal to more than 280 Tg of fuel for the production of ammonia by the process of Haber-Bosch. In the United States the reduced use of N fertilizer may save up to $1.0–4.5 billion per year (Hungria et al., 2006; Salvagiotti et al., 2008).

The most important mechanism used by the rhizobium to inoculate the legume roots is via root hairs. Bacterial attachment to the root hairs results in the curling of root hairs 6–18 h after attachment. The bacteria enter the root cortex cells, which have produced the nodules by the process of cell division. The nodules are visible 6–18 days after the rhizobial attachment. The number and size of nodules is determined by (1) plant and bacteria genotype, (2) symbiosis efficiency, and (3) the present nodules and environmental parameters such as soil N and soil water (Long, 2001; Lin et al., 2012).

Biological N fixation is a process in which soybean roots produce some biochemicals, as a signal molecule, specifically called genistein, results in the attraction of *Bradyrhizobium rhizobium* toward the host plant roots. As a result the activated proteins of bacteria produce some biochemicals, which induce some morphological and

Environmental Stresses in Soybean Production. http://dx.doi.org/10.1016/B978-0-12-801535-3.00010-3

physiological changes in the host plant roots, including root hair curling and bulging. The root hairs are the place of entrance for the bacteria, where they reside in the cells of the root cortex, in the produced root nodules, and fix N for the use of the host plant (Miransari and Smith, 2007, 2008, 2009).

Unfavorable soil pH is among the most important stresses adversely affecting different processes of nodulation and N fixation including the survival and growth of rhizobium, infection, nodulation, and N fixation. Under unfavorable soil pH, while the concentration of H^+, Al^{3+}, and Mn^{2+} increases with toxic effects on plant growth and soil microbes, the availability of nutrients such as calcium, molybdenum, and phosphorous decreases (Graham, 1992; Ferguson and Gresshoff, 2015).

Soil acidities, which are not favorable for plant growth and microbial activities, are prevalent in some parts of the world, especially in tropical areas, where there is a high rate of rain. As a result of soil leaching and basic cations and the advanced stages of weathering soil, pH decreases and hence the conditions become unfavorable for plant and soil microbes. The most optimum pH for plant growth and microbial activity is the pH around the neutral zone. However, if the pH increases or decreases both plant growth and soil properties are adversely affected (Von Uexküll and Mutert, 1995).

Soil acidity less than 5.5 can significantly decrease soybean growth and nodulation in different parts of the world. Although decreased soil nutrients and the toxicity of Al are among the most important adverse effects of acidity on plant growth, the high concentration of the H^+ in the soil can also decrease plant growth and yield production by affecting plant morphology and physiology (Marschner, 1991). Under acidic conditions, root growth and the formation of nodules decreases and hence the plant is not able to absorb nutrients for its growth and development (Horst, 1983, 1987).

The soybean is subjected to stress in most parts of the world and hence its growth and yield production decreases. The soybean and its symbiotic rhizobium are not tolerant to stress, and accordingly, their growth and activity decreases. However, it is likely to enhance soybean and bacteria tolerance to stress using different techniques and strategies. Using tolerant plant and bacterial species, improving the properties of soil using biotechnological and biological, etc. are among the most prevalent methods used for the alleviation of soil stresses on the growth and yield of soybeans (Ferguson et al., 2013; Miransari et al., 2013; Miransari, 2014a,b).

Stress parameters are able to affect the host plant and rhizobium, independently and negatively, and in some cases simultaneously affect the bacteria and the plant. Because a large part of the world is subjected to the stress of acidity, a high number of research has evaluated the effects of unsuitable pH on the process of biological N fixation by the host plant and rhizobium. More than 1.5 Gha of the soils are subjected to the stress of acidity and one-fourth of the crop fields are affected by the stress (Munns, 1986).

Plant morphological and physiological properties such as the activity of enzymes may be adversely affected when subjected to pH stress. Nutrient solubility and availability in plants and soil can also be influenced by unfavorable pH. With decreasing soil pH, the availability and solubility of macronutrients decreases. However, the availability and solubility of micronutrients including iron, zinc, copper, and manganese for the use of plant and soil microbes increases, and hence the related symptoms appear (Miransari, 2012).

With respect to the importance of soil acidity as an important stress affecting soybean growth and yield as well as rhizobium activity in different parts of the world, some of the most recent and advanced details related to such a stress is reviewed and analyzed in this chapter. Such details may be used for the more efficient production of tolerant soybean genotypes and bacterial inoculums, especially in areas that are subjected to the stress of acidic conditions.

Soybeans and Acidity

The soybean and its symbiotic rhizobium are not tolerant under acidic conditions, and their growth and activity decreases under such conditions. Under acidic conditions the coexistence of acidic pH (high H^+ concentration), the toxicity of aluminum, and decreased P availability is prevalent. However, the related plant mechanisms to adapt to such conditions are not yet indicated. Accordingly, the research work has been toward the production of tolerant soybean genotypes, which are able to grow and produce yield under acidic conditions. It is because most of the soybean fields in the world are acidic or subjected to the stress of acidity. Under acidic conditions, soybean morphology and physiology is adversely affected, decreasing its growth and yield.

The high concentration of Al under acidic conditions significantly decreases root growth and crop yield in sensitive crop plants by the following mechanisms: (1) decreased water and nutrient uptake and translocation, (2) the disruption of calcium homoeostasis and cellular structure, (3) the deposition of callose affecting the movement of cellular substances in the apoplast, and (4) the production of lipid peroxidation and reactive oxygen species adversely affecting cellular division and growth (Horst, 1995; Yamamoto et al., 2002; Zhou et al., 2009a).

Liang et al. (2013) found that phosphorous is able to alleviate the toxic effects of Al under acidic conditions, especially in P efficient genotypes of soybeans. Using a hydroponic growing medium, the authors indicated that the medium pH, Al, and P are able to affect root growth and malate production. The production of malate by the P efficient genotype was higher than the control treatment, indicating that malate has an important role in the alleviation of Al toxicity and P decreased availability resulting from pH stress in soybeans. *GmALMT1*, as the transporter of malate in soybeans, was obtained from the root tips of the efficient genotype.

The expression of *GmALMT1* was pH dependent, and under acidic conditions, its expression decreased; however, the presence of Al and P in the roots increased its expression. Molecular analyses indicated that *GmALMT1* is able to activate a transporter of a cell plasma membrane, affecting the movement of malate as a function of pH and Al. It was also indicated that the overexpression of *GmALMT1* in the roots of transgenic Arabidopsis and soybeans can be influenced by malate, affecting soybean Al tolerance. Accordingly the authors indicated that under acidic conditions, the exudation of malate is an important factor, affecting soybean tolerance, as affected by pH, Al, and P, and regulated through the activity of *GmALMT1* (Liang et al., 2013).

Shamsi et al. (2008) investigated the effects of Al, Cd, and pH on the photosynthesis, growth, malondialdehyde (MDA), and the activity of antioxidant enzymes in two soybean genotypes with different Al tolerance under hydroponic conditions. According to their results, pH (4) and Al treatment (150 μM) significantly decreased plant growth, chlorophyll content, and the rate of net photosynthesis. The activity of MDA and antioxidant enzymes were higher in plants under Al and Cd (1 μM) treatment. They accordingly indicated that the effects of Al and Cd on the growth of the two soybean genotypes were synergistic.

Indrasumunar et al. (2011) indicated that the number of nodules is controlled by the expression of *NFR1α*. They also found that the overexpression of *GmNFR1a* is able to enhance soybean tolerance under acidity stress for the process of nodulation and for the decreased number of nodules. In other research, Lin et al. (2012) assessed the effects of acidic conditions on the physiological and genetic properties of soybeans using a soybean line, which had been mutated in nodule autoregulation by modifying the receptor kinase *GmNARK*.

They found that a mechanism regulated by the aerial part and *GmNARK* was essential for the enhanced tolerance of soybean nodulation under salinity stress. Interestingly, the inhibition of acidic stress was not correlated with the toxic effects of Al concentration and affected the process of nodulation in 12–96h following the inoculation process with *B. japonicum*. By determining the number of transcripts, which decreased under acidity stress related to the early nodulation genes, the biological effects were indicated. They accordingly expressed that their results are a step further into the determination of processes, regulating the process of nodulation under acidity stress, and can be used for the plantation of soybeans under such conditions.

The cellular components and activities are disrupted by Al due to its high ability of binding to sulfate, phosphate, and carbonyl functional groups of cellular components in the symplast and apoplast. Among plant responses to the adverse effects of Al on plant growth is the production of different compounds by plant roots including organic products, phenolics, phosphates, and polypeptides, which are able to bind Al and detoxify its toxicity in the rhizosphere (Pellet et al., 1996). Accordingly, the most prevalent mechanisms by which plants alleviate Al toxicity is the sequestration of Al by the production of organic products, including malate, citrate, and oxalate (Bianchi-Hall et al., 2000).

Soybean Al tolerance is inherited and physiologically is by the root production of phenolics and citrate. Although the expression of the related genes and the cellular activities are the main causes of plant response to Al stress, more details have yet to be indicated on the related mechanisms. Just a few research works have so far used the method of proteomic for the investigation of Al stress on the growth of plant roots. Such experiments have indicated that the production of organic products and detoxifying enzymes by soybean roots are the most important alleviating mechanisms in the tolerant genotypes of soybeans under acidity stress (Yang et al., 2007; Zhen et al., 2007; Zhou et al., 2009b).

Among the most important enzymes affecting the production of putrescine in plants is arginine decarboxylase. Parameters including the physiological conditions, plant tissue type, and stage of development affect the enzyme activity. Using a cDNA from soybean hypocotyls, Nam et al. (1997) investigated the regulation of this enzyme by the related mechanisms. Accordingly, they examined that under acidic conditions and during tissue development, the enzyme activity and the enzyme mRNA may be related. The results showed that the enzyme and mRNA were positively correlated.

The enzyme activity initiated to increase two days after the germination of seeds and approached its highest level at the fifth day and then decreased. When the five-day-old hypocotyls were incubated with a potassium phosphate solution at pH 3, the enzyme activity increased rapidly, and it approached three times of its initial rate after 2 hours, preceding an increase in mRNA level. Accordingly, the authors indicated that under acidic conditions the increase in mRNA has an important role in the regulation of the enzyme activity, especially during the early growth of seedlings.

Rhizobium and Acidity

The most important mechanism, which results in the acidity tolerance of rhizobium, is maintaining the cytoplasm pH at 7.5–8.0 related to the pH in the environment surrounding the bacterial cells. The pH difference should be at least three units, which results from the exclusion of protons and the uptake of K^+. The cytoplasmic buffering capacity of some bacterial species, the ability of regulating the cellular metabolism and repressing the accumulation of acidic products, are also among the mechanisms used by rhizobium to adjust the cellular pH and acidic conditions (Foster and Hall, 1990). During the establishment of symbiotic association between soybeans and *B. japonicum*, the process of N fixation results in the release of H^+ into the rhizosphere and hence the pH of rhizosphere and nodules decreases (Gibson, 1980). While N fertilization reduces the process of nodulation and N fixation, it increases the rhizosphere pH by producing OH^- (Jimenez et al., 2007).

It has been indicated that the pH genes are placed on the chromosome rather than on the plasmids (Chen et al., 1991; Graham, 1992). Goss et al. (1990) were able to insert the DNA from the acid tolerant *Rhizobium meliloti* WSM419 into pMMB33. Probing of the rhizobium-sensitive mutants indicated the complete insertion of DNA for the maintenance of intercellular pH. Chen et al. (1991) used a practical approach to produce the more effective species of rhizobium by modifying the sym plasmid of an acid-tolerant species and inserting it into the more superior species of N-fixing rhizobium.

It is prevalent that under a pH of 5.0 the legumes are not able to nodulate, which is due to the decreased tolerance and persistency of rhizobium under such conditions (Rice et al., 1977). Among the most sensitive stages of N fixation by legume and rhizobium is the exchange of the signal molecule between the two symbionts. However, the other ones include the processes of rhizobium attachment, infection, and Nod gene induction, and the process of nodulation, which is only affected when the growth and curling of plant root hairs are subjected to the stress (Richardson et al., 1988a,b,c). Richardson et al. (1988b) examined the effects of clover root exudation on the induction of Nod genes in pH ranging from 3.0 to 8.0 and found that the most efficient induction resulted from the range of 5.0–6.0, and in a pH less than 4.5, it significantly decreased.

The presence of *nif* genes in rhizobium regulates the process of N fixation. Although most *nif* genes in rhizobium are plasmid related, in *Bradyrhizobium* the genes are placed on the chromosome. The process of N fixation in the symbionts and nonsymbionts is catabolized by the related enzyme, which is activated by *nifDK* and *nifH* genes. The enzyme by itself consists of two components, including component 1, molybdenum-iron protein (MoFe), and component 2, iron-containing protein (Fe). The subunits of MoFe protein are activated by *nif*K and *nif*D, and for the activation of MoFe protein an FeMo cofactor (MoFeCo) is also essential. However, the activation of the Fe-protein subunit is by *nif*H gene, which is regulated by *nifA* (positive) and *nifL* (negative) regulators (Temme et al., 2012; Shima et al., 2013).

The presence of oxygen and fixed N levels can regulate the expression of the *nif* gene. For example, under an increased level of ammonia (NH_3 and NH_4), NifL can negatively regulate gene expression, which is by preventing the activity of NifA. The increased level of oxygen also prevents FixL from the activation of FixJ, which negatively affects the increase of NifA. Because NifA is an activator for all the other *nif* genes the increased level of oxygen reduces the production of the N fixation enzyme, and hence the rate of biological N fixation is decreased or prevented (Oldroyd and Dixon, 2014; Poza-Carrión et al., 2014).

The *nodD* and plant flavonoid are able to regulate the activity of nodulation genes (Djordjevic et al., 1985; Long, 2001). The production of flavonoid by the legume roots are specific for a special legume–rhizobium interaction and are affected by environmental parameters such as soil fertility, Nod factors, and pH. Nod factors are able to induce the host plant physiology in a way so that more Nod factors are produced, and the morphological changes result in the curling and bulging of root hairs as well as in the cell division of the root cortex (Faucher et al., 1989; Lerouge et al., 1990; Miransari et al., 2006).

The other genes, which are also able to regulate the process of biological N fixation, are *fix* genes. However, mycrosymbionts contain genes, which regulate the uptake of hydrogen, the production of exopolysaccharide and glutamine synthase, the efficiency of nodulation, the transport of dicarboxylate, and the production of β-1,2 glucans, which has been reviewed by different authors (Gresshoff, 2012; Temme et al., 2012). Two types of nodules are produced by rhizobium on the roots of their legume host plant determinant and indeterminant. The latter is formed by the rhizobium on temperate legumes such as alfalfa, pea, and clover; however, the determinant nodules are produced by *B. japonicum* on the roots of soybeans (Friesen, 2012; Biswas and Gresshoff, 2014).

Parameters including pH, heat, and the concentrations of N and P influence the production of Nod factors. Similar to the process of nodulation, the production of Nod factors is adversely affected by the environmental stresses (Wang et al., 2012). Rhizobium are present in the soil in two different forms: if the host plant exists in the soil, they establish a symbiotic association with their host plant and fix the atmospheric nitrogen, and if not, they act as free-living saprophytic heterotrophs. Related to the other soil microbes, a kind of bacterial existence makes the survival and persistence of bacteria possible. The aggregates of soil are safe places for rhizobium, keeping the bacteria away from the environmental stresses. This is also the case for the nodules, although with a higher efficiency, causing the bacteria to multiply and be safe (Liu, 2012).

Soil pH less than 6 resulted in a 10^3 decrease in the number of rhizobium related to the number of rhizobium in pH higher than 7. Similar to the stress of soil drought, some effects result from soil pH on the number, growth, and activities of soil rhizobium, and just a few of them can survive in pH < 4.5–5.0. Although the species of *B. japonicum* are more tolerant than the species of rhizobium under pH stress (Date and Halliday, 1979), some species of *Rhizobium tropici* are really tolerant under the pH stress (Graham, 1992). With respect to the properties of rhizobium, the effects of pH on the growth and activities of rhizobium are different. For example, under natural pH, *Sinorhizobium fredii* was more efficient than *B. japonicum*, while at pH 4.9, *B. japonicum* was the more efficient strain. Although *Rhizobium elti* is a more efficient species than *R. tropici*, decreasing soil pH to less than 7 resulted in the inoculation of beans by *R. tropici* rather than by *R. elti*.

Under pH stress the signaling exchange between the two symbionts, the expression of the nodulation genes, and the production of Nod factors are disrupted (Richardson et al., 1988b; Zhang and Smith, 1995). It is also interesting to mention that nodulated legumes are more sensitive to the toxicity of heavy metals, such as Al and Mn, than to N chemical fertilization. It has been indicated that the following genes are essential for rhizobium to tolerate the acidic pH: *actA*, *actP*, *actR*, *acts*, *exoR*, *lpi*, and *phrR* (Glenn et al., 1999; Reeve et al., 2002).

The response to acidity stress by rhizobium, affecting their survival and competitiveness, is somehow complicated. For example, although the bacterial species, which are isolated from stress conditions, are tolerant under stress, there might be some bacterial species, which are not tolerant to the stress. It is because such sensitive bacterial species are protected from the stress by the soil (Richardson and Simpson, 1989) and not by the development of mechanisms, which result in bacterial tolerance under stress.

The other important reason, which may adversely affect the activity and growth of rhizobium under pH stress, is the increased concentration of Al, Cu, and Mn (Cooper et al., 1983; Reeve et al., 2002). Soil pH can also affect the availability of P and Ca and hence the growth and activities of rhizobium (Munns, 1970; Bell et al., 1989). It has been indicated that rhizobium tolerance under pH stress is determined by the bacterial ability of maintaining the internal cellular pH near the natural pH at 7.2–7.5 (O'Hara et al., 1989).

Bacterial response under pH stress is determined by (1) the exclusion of the proton (Graham, 1992), (2) bacterial response to the stress, (3) the increased buffering capacity of cytoplasm, (4) glutathione presence, (5) increasing and maintaining the concentration of potassium and glutamate, (6) the metabolism of calcium, and (7) the permeability of the membrane (Graham, 1992; Chen et al., 1993). Although the microsymbiont is more sensitive to the pH stress, both the legume host and the rhizobium and their symbiotic processes are negatively affected by the pH stress (Munns, 1986).

Fujihara and Yoneyama (1993) investigated the effects of pH and osmotic stress of the bacterial medium on the growth and polyamine content of *Rhizobium fredii* P220 and *B. japonicum* A1017. Increasing the medium pH adversely affected the growth of *B. japonicum* and slightly increased the bacterial putrescine but not the bacterial homospermidine. However, in *R. fredii* P220 with a good growth in a wide range of pH from 4.0 to 9.5, decreasing medium pH increased the production of homospermidine. Unfavorable pH less than 7 increased the production of Mg^{2+} in the medium. While strain P220 was able to grow in a medium with the salt concentration of 0.4M, strain A1017 was not able to grow in the medium with the salt concentration of 0.15M.

Salinity increased the concentration of glutamate and K^+ in the cells of P220; however, it decreased the concentrations of homospermidine and Mg^{2+}. The contents of other polyamines were not affected by the use of extra salt. According to the osmotic strength, NaCl, KCl, glycerol, and sucrose resulted in a similar decrease in the cellular homospermidine. Accordingly, the authors indicated that the pH and salinity tolerance of P220 may determine the production of homospermidine.

Plant nodules may contain more than 10^{10} rhizobium per gram, which are released into the soil along with the nodule senescence at the end of the growing season (McDermott et al., 1987). Numerous research has indicated that the presence of the host plant is not essential for the survival and persistence (saprophytic competence) of rhizobium in the soil. Although after the return of nodules and rhizobium into the soil, they are subjected to different types of stresses, and they can survive in the soil in the form of free-living and heterotrophic saprophytes for a long time and can become activated when the host plant is present.

The effects of soil acidity on the process of N fixation by alfalfa (*Medicago sativa* L.) and red clover (*Trifolium pratense* L.) was evaluated by Rice et al. (1977) under field conditions in 28 different places and under greenhouse conditions using the field soils. The tested soils (limed and unlimed) had different pHs, ranging from 4.5 to 7.2. The parameters, including the population of rhizobium in the soil, the rate of nodulation, and the relative forage yield (yield without N/yield with N) were determined under both field and greenhouse conditions. At the pH less than 6, the number of *R. meliloti*, the rate of nodulation, and the relative yields of alfalfa decreased quickly. However, in the soils with a pH higher than 6 the effects of pH on the above mentioned parameters were negligible.

The nodulation rate and relative yields of red clover were not affected by the range of pH in the experiments; however, the number of *Rhizobium trifolii* was decreased at a pH less than 4.9. The authors accordingly indicated that the concentration of H^+ is an important factor significantly decreasing the yield of alfalfa under acidic conditions but with less significance for red clover. They also made the conclusion that for the prediction of crop response to liming, soil pH as well as the availability of Al and Mn must be determined on a regular basis.

Wood et al. (1984) investigated the effects of soil acidity-related factors including pH, calcium, Al, and Mn on the symbiotic association of *Trifolium repens* with *R. trifolii* in the lab using a technique of axenic solution culture. Although an Mn concentration at 200 μM increased root length in the pH range of 4.3–5.5, it did not affect the formation of root hair and nodules as well as the rhizobium number in the rhizosphere. Using calcium at 500 and 1000 μM neutralized the effect of pH 4.3 on the root length and formation of root hair; however, a pH range of 4.3–4.7 repressed the multiplication of rhizobium in the rhizosphere as well as the process of nodulation. When Al was not present, Ca at 50–1000 μM did not affect rhizobium multiplication and nodule formation in the pH range of 4.3–5.5.

Root elongation and formation of root hair was suppressed by 50 μM Al in the pH of 4.3 and 4.7; however, the adverse effects of Al on root elongation was decreased by using Ca at the concentration of 50–1000 μM. The multiplication of rhizobium in the rhizosphere was suppressed by 50 μM Al, and the formation of nodules was also decreased by Al at pH 5.5, which resulted in the precipitation of Al from the solution. However, elongation of the root and the formation of root hair were not affected. Such effects of Al at pH 5.5 indicate why white clover is not responsive to inoculation under acidic conditions following liming.

The effects of acidity on the multiplication of two *R. trifolii* strains including HP3 and BELI192 were investigated (Wood and Cooper, 1988). The lowest limit for the multiplication of HP3 was the pH range of 4.8–5.0 and for BELI192 was 4.6–5.6. Although Al precipitates at a pH higher than 4.6, at the concentration of 50 μM, Al suppressed the multiplication of both strains in the pH range of 4.6–5.6. The increasing P concentration from 10–100 μM did not suppress the adverse effects of Al on the multiplication of rhizobium. The multiplication of rhizobium was intensified more significantly in another Al medium, although the rhizobium cells were able to increase their number in a control medium without Al. The results hence indicate the toxic effects of Al on the growth, activity, and multiplication of rhizobium.

Using acid media the growth of 19 strains of *R. trifolli* isolated from different parts of Finland were investigated by Lindström and Myllyniemi (1987). The lowest pH range, which resulted in the growth of tested rhizobial strains, was 4.7–4.9. The effect of Al (50–150 μM) and Mn (200–730 μM) was also tested on the growth and activities of five strains. Al and Mn affected the growth and activity of rhizobium at the least-tested pH and not at the higher level. The symbiotic properties of the five strains with red clover was also tested using axenic medium and a pot experiment with an acidic medium of 4.1, and 5.2 by liming.

In the pot experiment, four strains were used as inoculants, and the fifth one was used as the native bacteria. The acidity of 5.2 did not result in differences in the yields of uninoculated, inoculated, and N control plants, while at pH 4.1, inoculation with the most efficient strain increased the yield by 14 times compared with the uninoculated treatment, and the effect of inoculation was similar to the effect of liming or chemical N fertilization. There was not a correlation between the symbiotic properties of the strains in the acidic medium and in the pot experiment. However, the authors indicated that when using the pH of the field, it is possible to predict the symbiotic activities of the rhizobium strains; the most suitable strain was from the most acidic field.

Watkin et al. (2000) evaluated the acidity tolerance of six different strains of *Rhizobium leguminosarum* bv. *trifolii* (including WU95, NA3001, WSM409, TA1, NA3025, and NA3039) in a 3-year field experiment using a soil with the pH of 4.2. The three strains including WSM409, NA3039, and WU95 were indicated to be more tolerant to the stress of acidity related to the strains NA3025, TA1, and NA3001. The colonization and persistence of the strains WSM409 and NA3039 in the acid soil were greater than TA1 and NA3001. The results indicated that strain WSM409 was the most tolerant and useable rhizobium species under the stress of acidity and can be used for the production of clover under such conditions.

Watkin et al. (2009) also investigated the effects of acidity on the signaling exchange between the two species of *Medicago*, including *Medicago sativa* and *Medicago murex* with *Sinorhizobium*. *Medicago murex* is able to nodulate at a lower pH, related to *M. sativa*. They collected the root products of the two species, grown in nutrient solution at the pH of 4.5, 5.8, and 7.0 and determined the expression of *nodB* in *Sinorhizobium*. The constructed fusion of *nodB-gusA* was placed into the strains of *Sinorhizobium medicae* including CC169 (acid sensitive) and WSM419 (acid tolerant).

The root products of both *Medicago* spp. indicated a higher induction at pH 4.5 than pH 5.8 and 7.0. However, the induction by *M. sativa* was higher than *M. murex* and CC169 was higher than WSM419. The inducing compounds were similar in both *Medicago* spp. including liquiritigenin and isoliquiritigenin at all pH values, and with a higher rate at the lowest pH. The authors accordingly made the conclusion that the lower rate of nodulation in *M. sativa* than *M. murex* is not due to the reduced induction of *Sinorhizobium nodB* resulted by the chemicals in the roots products of both species under acidic conditions.

Nutrients, Acidity, and Biological N Fixation

A high rate of P, ranging from 0.72% to 1.12%, can be absorbed by nodules, which is higher than the P content of other plant parts. Accordingly, in N-fixing plants, higher P is essential for the growth and activity of both plants and rhizobium, related to the plants fed with chemical N. This is due to the reason that different N-fixing activities, including the production of nodules, the process of N fixation, and the activity of nodules, are P dependent and under P-deficient conditions are delayed or suppressed (Cassman et al., 1981; Jakobsen, 1985; Hart, 1989).

According to Beck and Munns (1984), in rhizobium cells with P limitation or in an acidic pH, calcium is essential for the mobilization of phosphorous. Calcium is also essential for the integrity of the rhizobium cell wall. If calcium is not present at the essential level the appearance of nodules is delayed and nodulation will be limited. The attachment of rhizobium to the root hairs of the host plant is affected due to the calcium-dependent components of rhizobium cells. However, Ca at 10 mM enhanced the bacterial *nod* induction by white and subterranean clover at a level of 5–10 times in a pH of 4.5 (Richardson et al., 1988b). Calcium was also able to increase the induction of *nod* genes in a pH range of 4.8–5.2 (Richardson et al., 1988c).

Bradyrhizobium is a more tolerant strain than rhizobium under P-deficient conditions. At high levels of P, rhizobium is able to accumulate extra amounts of P, which may be as high as 2.4% (Beck and Munns, 1984). It has been indicated that under P-deficient conditions, rhizobium is able to absorb P at a rate of 10–180 times higher than P-sufficient conditions. The activity of the alkaline phosphatase enzyme was also evident in the P-limited cells of *Rhizobium* and not in *Bradyrhizobium* under P-deficient conditions.

Plant tolerance is higher to the toxicity of Al and Mn related to the rhizobium and the process of nodulation. However, the tolerance of different plants and rhizobium species differ under different concentrations of Al and Mn. Although Ca is essential for rhizobium growth and activity at a lower amount, under acidic conditions, rhizobium requires higher amounts of Ca for survival. O'Hara et al. (1989) indicated that the Ca concentration of 1–2 mM was essential for maintaining the cellular cytoplasmic pH of the acid-sensitive strains of rhizobium.

The *pho* regulator indicates the microbial response under stress. Using two component products, microbes handle the external stimuli. The components include a protein kinase and a response regulator. Under P-limited conditions the main protein kinase is PhoR and the activity of alkaline phosphatase enzyme is regulated by the PhoB regulator. Although usually under P-deficient conditions the inoculation of soybean roots by *B. japonicum* is significantly delayed; in some cases, P limitation may increase the weight of nodules and hence the content of the aerial part (Al-Niemi et al., 1997; Danhorn and Fuqua, 2007).

The effects of P stress on the growth and development of the young leaf of soybean plants was investigated by Chiera et al. (2002). Soybean seedlings were subjected to the stress of phosphorous for 32 days following seed germination. The significant effects of stress on the growth of seedlings appeared 14 days after the stress. The growth of the first three leaves was severely affected by the P stress. According to the results, decreased cell division was the most important reason for the reduced growth of the soybean leaf under P stress.

Planting soybeans under Fe-deficient conditions, where soybeans have not been previously planted, must be done with care for the proper establishment of symbiosis between *B. japonicum* and soybeans, especially if N fertilization is required. However, N fertilization increases the concentration of OH^- in the rhizosphere, and due to an alkaline pH the availability of Fe decreases too. Accordingly, Wiersma (2010) investigated if the response of soybean genotypes with different Fe tolerance grown under Fe-deficient conditions is different at varying rates of N fertilization.

Using six different genotypes of soybeans to Fe presence, including two efficient, two moderately efficient, two inefficient, and six rates of N fertilization (0, 34, 68, 102, 136, and 170 kg N/ha), Wiersma (2010) evaluated the response of soybean genotypes to N fertilization from 2003 to 2005. The climatic conditions in 2004 were cooler and wetter than 2003 and 2005, while the rate of diethylenetriaminepentaacetic acid extractable Fe was twice as much as 2003 and 2005. However, the results indicated that extractable Fe did not reflect the rate of available Fe, as plant growth and yield in 2004 were significantly less than 2003 and 2005.

Increased N fertilization decreased the rate of nodulation regardless of the year and the properties of soil and genotype with respect to Fe efficiency. Plant growth and yield decreased in response to increased N fertilization for inefficient soybean genotypes, while the response of Fe-efficient and moderately efficient genotypes of soybeans was little. The author accordingly indicated that N fertilization must not be used for the Fe-inefficient genotype of soybeans planted under Fe-deficient conditions.

Under Fe-deficient conditions, plants use different mechanisms to alleviate the stress, including (1) H^+ release by their roots, (2) production of reductants, (3) at the root plasmalemma higher rate of Fe^{3+} is reduced to Fe^{2+}, and (4) higher concentration of organic products, especially citrate in the roots (Marschner and Römheld, 1994; Schmidt, 1999; Hartmann et al., 2009). Accordingly, both Fe stress responses and biological N fixation are reducing processes, while N fertilization increases the rhizosphere pH and hence intensifies Fe deficiency.

The iron-efficient genotypes of soybeans activate the stress mechanism responses in nodulated or nonnodulated genotypes, although higher responses have been observed for the nodulated genotypes. However, the inefficient genotypes activate the stress mechanism responses when nodulated and fixing N. It has been indicated that the mechanisms of the Fe stress response may have little effect on the process of biological N fixation; the activity of the N-fixation enzyme can significantly enhance the mechanisms of Fe stress response in both efficient and inefficient genotypes (Zhang and Li, 2003; Hansen et al., 2006; Miransari, 2013).

The level of response in Fe stress soybeans is determined by the following mechanisms: (1) the genotype Fe efficiency, (2) the rate of nodulation and N fixation, (3) the presence of parameters such as NO^{3-}, which suppress the process of N fixation, and (4) the effectiveness of the nodulating *B. japonicum*. Interestingly, some strains of *B. japonicum* are able to produce siderophores, including citrates, which enhance the chelating and transport of Fe (White et al., 2013; Nasr Esfahani et al., 2014).

The effects of different legumes on the production of H^+ were investigated in an interesting experiment by Liu et al. (1989). They used four legume species, including red clover (*Trifolium prateuse* L.), sweet clover (*Meliotus alba* Medik), vetch (*Vica villosa* Roth), and alfalfa (*Medicago sativa* L.) grown in an N-free nutrient solution. During the experiment the increases in the acidity of the solution were determined. There was a positive correlation between total N and the production of H^+.

The results indicated that the milligrams H^+ produced per g fixed N for the legume species, on the basis of mg, were equal to 49 (red clover), 43 (vetch), 42 (alfalfa), and 37 (sweet clover). The results were calculated according to the amounts of yield in the field, and hence the related values, on the basis of kg, were equal to 15.6 (alfalfa), 9.7 (red clover), 4.6 (sweet clover), and 4.5 (vetch). According to the rate of N fixation, the amounts of acidity produced were equal to: 5.2–14 (alfalfa), 4.2–9.4 (red clover), 3.2–7.1 (sweet clover), and 3.9–6.8 (vetch). The authors accordingly made the conclusion that the process of N fixation by legumes is an important source of acidity in the agricultural fields, and hence in the long term the acidification by legumes can result in the leaching of exchangeable cations and decreasing the pH.

Methods of Acidity Alleviation

The techniques and strategies used for the alleviation of stress are according to the following. The details related to the tolerance of soybeans and the symbiotic rhizobium under acidity stress have been also presented in the previous sections.

1. Tolerant plant species
2. Tolerant rhizobium species
3. Chemical methods
4. Using biochemicals

Tolerant Plant Species

Some plants, especially the tolerant ones, are able to survive under acidity stress using morphological and physiological mechanisms. The most important effects adversely influencing plant growth under acidic conditions are high H^+ concentration, Al toxicity, and P deficiency. Tolerant plants usually have the related gene, which can make the plant survive the stress. Tolerant plants are able to neutralize the acidity of H^+ by producing different compounds in the rhizosphere, and internally, they can alleviate the stress of H^+ in the plant by higher uptake of basic compounds such as Ca and Mg. Plants also have some mechanisms and genes, which can alleviate the toxicity of AL. The deficiency of P can also be mediated by the production of organic compounds by plant roots and by the morphology and architecture of plant roots (Kochian, 1995; Kochian et al., 2004; Li et al., 2012).

Tolerant Rhizobium Species

It is also likely to enhance the efficiency of biological N fixation under acidity stress using the tolerant species of *B. japonicum*. Such tolerant species are able to survive under acidity stress by the production of different products such as exopolysaccharides with a strong cellular membrane, H^+ homeostasis and exclusion, the expression of different stress genes such as the chaperone genes, and the increased buffering capacity of cytoplasm (Correa and Barneix, 1997; Ferreira et al., 2012; Miransari et al., 2013).

Chemical Methods

The use of chemicals may also be applicable for the alleviation of acidity stress on the growth of soybeans and the activity of symbiotic *B. japonicum*. Among the most prevalent chemicals used for the alleviation of acidity stress is "lime," which is able to increase soil pH and hence alleviate the stress of soil acidity. Munns et al. (1981) indicated that the decreased growth of soybean roots under acidity stress is not due to the nodulation failure, which is the cause of reduced root growth in the other legumes. They evaluated such a possibility by mixing lime with the soil and investigating its effects on the process of nodulation, early plant growth, and N concentration in soybean plants inoculated with rhizobium, compared with control soybean plants fertilized with NH_4NO_3.

A two-factorial experiment using lime×N was conducted with acidic soils: (1) under greenhouse conditions with two soybean genotypes and 13 strains of rhizobium for inoculation and (2) under field conditions with nine soybean genotypes and three rates of rhizobium inoculation, resulting in large differences in nodule number and weight. The soils had a high rate of soluble and exchangeable Al. In both experiments, liming pH from 4.4 to 6 (aqueous paste) significantly increased plant growth regardless of plant genotype, *Rhizobium* strain, and number, as well as N source. A high rate of nodulation resulted in inoculated plants with a green color and high N content, even when their growth was severely affected by the acidic soil (Munns et al., 1981).

Solution culture experiments confirmed the toxic effects of Al on soybean growth by resembling field conditions using the similar pH, Al, and Ca concentrations in extracts of the soil solution. A low rate of calcium ($200\,\mu M$) and acidic pH (4.5) did not affect the growth of soybeans, although it was adversely affected by Al concentrations of 30 and $65\,\mu M$. The authors accordingly indicated that the efforts to increase soybean tolerance under acidity stress must be concentrated on the plant rather than on the rhizobium (Munns et al., 1981).

Using a long-term experiment, Cifu et al. (2004) examined the effects of different liming rates (0, 3.75, 7.50, 11.25, and 15.00 Mg/ha) on the acidity of a red soil and crop production. Liming rate and time decreased soil acidity and increased the rate of exchangeable Ca in the plow level (0–20 cm). The acidity of subsoil (20–60) started to decrease 4 years following the liming treatment. The subsoil had a higher rate of exchangeable Mg^{2+} than Ca^{2+}, indicating the faster movement of Mg^{2+} downward.

The maximum rate of crop production was significantly increased by liming for mungbean (2.24 times), wheat (57.3%), sesame (53.4%), broad bean (52.8%), potato (44.1%), cotton (32.1%), corn (28.4%), watermelon (18.5%), cowpea (11%), and soybean (8.8%). The highest rate of liming has more effect on the subsoil acidity and increased yield production, especially during the later period of the experiment. The decreased soil acidity was effective for as long as 5, 7, 12, and 14 years using liming at 0.5, 1.0, 1.5, and 2.0 L, respectively, and for yield it was effective for as long as 15 years.

Indrasumunar and Gresshoff (2013) evaluated the properties of vermiculite as a plant-growth medium for acidity research experiments. Vermiculite is the most prevalently used medium for such experiments due to its high water-holding capacity. The authors used two different nutrient solutions, including Broughton and Dilworth, and modified Herridge nutrient solutions with or without a MES [2-(N-morpholino)ethanesulfonic acid] buffer to irrigate soybeans planted on a vermiculite growth medium. Using 5N HCl and 5N NaOH the acidity of the nutrient solutions was adjusted to 4.0 and 7.0, respectively, prior to the experiments. Three or four weeks following inoculation, soybean growth and nodulation were determined. Because soybean growth and nodulation were not affected by the acidity of the solution at 4.0 or 7.0, the authors indicated that the reason is the high buffering capacity of vermiculite. Accordingly, they made the conclusion that it is not a suitable medium for evaluating soybean growth and nodulation under acidic conditions.

It is important to find how Ca^{2+} may affect the process of biological N fixation. Macció et al. (2002) examined the effects of acidity and calcium concentration on the *Bradyrhizobium* sp., which is a symbiont to the peanut. At a pH of 5.0 and the calcium concentration of 0.05 the growth and viability of *Bradyrhizobium* sp. significantly decreased; however, increasing calcium concentration considerably improved the growth and activity of rhizobium under acidity. The concentrations of molecules related to the plant-rhizobium recognition such as exopolysaccharides and lipopolysaccharides were affected by pH values and calcium concentrations. Under stress, cellular permeability was also affected. The examined attachment of rhizobium to the root hairs showed that it was a function of pH and calcium concentration.

Using Biochemicals

Among the most important biochemical methods, which have been proved successful so far, are the use of signal molecules including genistein during the process of biological N fixation between the soybean and its symbiotic rhizobium. Miransari and Smith (2007, 2008, 2009) hypothesized, tested, and proved that using the signal molecule, genistein, is a suitable method for the alleviation of stresses such as salinity, acidity, and suboptimal root zone temperature on the growth, nodulation, and yield of soybeans under greenhouse and field conditions. It is because the most sensitive stages of biological N fixation are the initial ones, especially the process of signaling exchange between the two symbionts. Hence the process of nodulation and biological N fixation decreases. Using the following experiments the authors approved that genistein is able to alleviate the stress on the growth, nodulation, and yield of soybean plants under greenhouse and field conditions.

Greenhouse Experiments

Under greenhouse conditions with specific heat, light, and moisture the following stages were conducted to examine the effects of acidity stress on the growth and nodulation of soybean plants to find if using genistein can alleviate the stress. The inoculums of *B. japonicum* were prepared in the lab and were preincubated with genistein at different concentrations ranging from 0–20 μM. Soybean seeds were sterilized and planted in pots containing pasteurized vermiculite. Using HCl 1 N the pH of Hoagland nutrient solution, used for watering the posts, was adjusted to the favorable pH, ranging from 4 to 7 (control). Soybean seeds were planted in the pots and inoculated with 1 mL of bacterial inoculums.

Plants were harvested 20, 40, and 60 days after inoculation and were analyzed for different parameters including soybean growth and nodulation. Interestingly, although in some cases genistein was able to enhance plant growth and nodulation under the stress, in some other cases the effects of acidity on plant growth and nodulation were not evident (Miransari and Smith, unpublished data). Although we used turface (inert medium) and sand as the experimental medium, this might have been the reason for the variable results. For example, Indrasumunar and Gresshoff (2013) indicated that due to the strong buffering capacity of vermiculite, it may not be a suitable medium for research on soybean plants under acidity stress.

In other research, using two different experiments, Miransari et al. (2006) evaluated the effects of produced lipochitooligosaccharide (LCO) in the lab on the root hair responses of two different soybean genotypes, including AC Bravor and Maple Glen under acidity stress. In the first experiment, two concentrations of LCO (10^{-7} and 10^{-6} M) were used, and in the second experiment, three concentrations of LCO (10^{-7}, 10^{-6}, and 10^{-5} M) were used. The LCO molecule was prepared in the lab using the medium of *B. japonicum*, which had been preincubated with genistein and high performance liquid chromatography.

The roots of soybean seedlings grown on Petri dishes (autoclaved 1.5% agar medium) were treated with the acidity treatments (4.0, 5.0, 6.0, and 7.0) and different concentrations of LCO for 24 h. The response of Maple Glen roots was similar to different concentrations of LCO; however, the response of AC Bravor was different. LCO at 10^{-5} was able to alleviate the stress of acidity (pH 4.0) on the processes of root hair curling and bulging. The significant interactions between LCO and acidity indicated that LCO might be able to alleviate the stress of acidity on root hair curling and bulging under a higher level of acidity.

Field Experiments

A 2-year field experiment (1998–1999) related to the effects of acidity on soybean growth, nodulation, and yield was conducted at the research farm of McGill University on the Macdonald Campus in Ste-Anne-de-Bellevue, Quebec, Canada. Because increased concentrations of H^+ are among the most important effects of acidity on plant growth and yield production, using the safe method of elemental sulfur at the essential rates, the soil pH was adjusted to the favorite ones, ranging from 4.39–6.45. Soil acidity was determined on a regular basis during the growing season in both years.

Soybean seeds were planted at 50 in each row and inoculated with B. japonicum, which had been previously incubated with genistein at 0, 5, and 20 µM. Plant and soil samples were collected at different growth stages, including harvest, and plant and bacterial parameters, including yield and nodulation, were determined. The results indicated the favorable effects of genistein on different plant growth, nodulation, and yield. There was a 23% yield increase resulted by genistein related to the control treatment under acidity stress (Miransari and Smith, 2007).

Conclusion and Future Perspectives

The soybean and its symbiotic B. japonicum are not tolerant plants and bacterial species under acidic conditions; however, it may be possible to enhance their tolerance under such conditions using biochemical and biotechnological techniques and strategies. Most of the world's soybeans are produced under acidic conditions, and hence it is important to find methods, techniques, and strategies, which are able to alleviate the stress of acidity on soybean growth and production. Different techniques have been tested and used so far, including the use of genetically modified soybean and rhizobium species, which are able to tolerate the stress of acidity. The tolerant species of soybeans are produced genetically, and the tolerant species of rhizobium are produced genetically or are isolated from the stress areas, as the presence of tolerant species under such conditions is prevalent. The other method, which is widely used for the alleviation of acidity stress, is the use of lime, increasing the rate of Ca^{2+} and hence soil pH. The use of biochemical methods has also been tried, tested, and proven to be effective when the plant and bacteria are subjected to the acidity stress. The use of signal molecules, especially genistein, which is produced by the host plant during the initial stages of symbiosis, is among such methods. The production of such a molecule decreases under acidity stress. However, research has indicated that if B. japonicum is pretreated with the signal molecule genistein, it is possible to alleviate the stress of acidity on the process of biological N fixation and hence increase soybean growth and yield under the stress. Future research may focus on the production of more efficient species of soybeans and rhizobium, which are able to develop a symbiotic association under acidic conditions. The production of biochemicals such as the signal molecule genistein at the commercial rate, to be used by the farmers, can also be a suitable method of stress alleviation under acidic conditions.

References

Al-Niemi, T., Summers, M., Elkins, J., Kahn, M., McDermott, T., 1997. Regulation of the phosphate stress response in *Rhizobium meliloti* by PhoB. Applied and Environmental Microbiology 63, 4978–4981.

Beck, D., Munns, D., 1984. Phosphate nutrition of *Rhizobium* spp. Journal of Applied and Environmental Microbiology 47, 278–282.

Bell, W., Edwards, D., Asher, C., 1989. External calcium requirements for growth and nodulation of six tropical food legumes grown in flowing solution culture. Australian Journal of Agricultural Research 40, 85–96.

Bianchi-Hall, C., Carter, T., Bailey, M., Mian, M., Rufty, T., Ashley, D., Boerma, H., Arellano, C., Hussey, R., Parrott, W., 2000. Aluminum tolerance associated with quantitative trait loci derived from soybean PI 416937 in hydroponics. Crop Science 40, 538–545.

Biswas, B., Gresshoff, P.M., 2014. The role of symbiotic nitrogen fixation in sustainable production of biofuels. International Journal of Molecular Sciences 15, 7380–7397.

Cassman, K., Whitney, A., Fox, R., 1981. Phosphorus requirements of soybean and cowpea as affected by mode of N nutrition. Agronomy Journal 73, 17–22.

Chen, W., Li, G., Qi, Y., Wang, E., Yuan, H., Li, J., 1991. *Rhizobium huakuii* sp. nov. isolated from the root nodules of *Astragalus sinicus*. International Journal of Systematic Bacteriology 41, 275–280.

Chen, H., Richardson, A., Rolfe, B., 1993. Studies on the physiological and genetic basis of acid tolerance in *Rhizobium leguminosarum* bv. *Trifolii*. Applied and Environmental Microbiology 59, 1798–1804.

Chiera, J., Thomas, J., Rufty, T., 2002. Leaf initiation and development in soybean under phosphorus stress. Journal of Experimental Botany 53, 473–481.

Cifu, M., Xiaonan, L., Zhihong, C., Zhengyi, H., Wanzhu, M., 2004. Long-term effects of lime application on soil acidity and crop yields on a red soil in Central Zhejiang. Plant and Soil 265, 101–109.

Considine, D.M., 2012. Foods and Food Production Encyclopedia. Springer Science & Business Media.

Cooper, J., Wood, M., Holding, A., 1983. The influence of soil acidity factors on rhizobia. In: Jones, D., Davies, D. (Eds.), Temperate Legumes. Physiology, Genetics and Nodulation. Pittman, London, pp. 319–335.

Correa, O., Barneix, A., 1997. Cellular mechanisms of pH tolerance in *Rhizobium loti*. World Journal of Microbiology and Biotechnology 13, 153–157.

Danhorn, T., Fuqua, C., 2007. Biofilm formation by plant-associated bacteria. Annual Review of Microbiology 61, 401–422.

Date, R., Halliday, J., 1979. Selecting *Rhizobium* for acid, infertile soils of the tropics. Nature 277, 62–64.

Djordjevic, M.A., Schofield, P.R., Rolfe, B.G., 1985. *Tn5* mutagenesis of *Rhizobium trifolii* host specific nodulation genes results in mutants with altered host-range ability. Molecular and General Genetics 200, 463–471.

Dobereiner, J., 2013. Limitations and Potentials for Biological Nitrogen Fixation in the Tropics, vol. 10. Springer Science & Business Media.

Faucher, C., Camut, S., Denarie, J., Truchet, G., 1989. The *nodH* and *nodQ* host range genes of *Rhizobium meliloti* behave as avirulence genes in *R. legumiosarum* bv *viceae* and determine changes in the production of plant-specific extracellular signals. Molecular Plant-Microbe Interaction 2, 291–300.

Ferguson, B., Lin, M.H., Gresshoff, P.M., 2013. Regulation of legume nodulation by acidic growth conditions. Plant Signaling and Behavior 8, e23426.

Ferguson, B.J., Gresshoff, P.M., 2015. Physiological implications of legume nodules associated with soil acidity. In: Legume Nitrogen Fixation in a Changing Environment. Springer International Publishing, pp. 113–125.

Foster, J.W., Hall, H.K., 1990. Adaptive acidification tolerance response of *Salmonella typhimurium*. Journal of Bacteriology 172, 771–778.

Ferreira, P., Bomfeti, C., Soares, B., Moreira, F., 2012. Efficient nitrogen-fixing *Rhizobium* strains isolated from amazonian soils are highly tolerant to acidity and aluminium. World Journal of Microbiology and Biotechnology 28 (5), 1947–1959.

Friesen, M.L., 2012. Widespread fitness alignment in the legume–rhizobium symbiosis. New Phytologist 194, 1096–1111.

Fujihara, S., Yoneyama, T., 1993. Effects of pH and osmotic stress on cellular polyamine contents in the soybean rhizobia *Rhizobium fredii* P220 and *Bradyrhizobium japonicum* A1017. Applied and Environmental Microbiology 59, 1104–1109.

Gibson, A., 1980. Methods for legumes in glasshouses and controlled environment cabinets. In: Bergersen, F.J. (Ed.), Methods for Evaluating Biological Nitrogen Fixation. John Wiley & Sons, New York, pp. 139–148.

Glenn, A.R., Reeve, W.G., Tiwari, R.P., Dilworth, M.J., 1999. Acid tolerance in root nodule bacteria. In: Chadwick, D.J., Cardew, G. (Eds.), Bacterial Response to pH. Novartis Foundation Symposium No. 221, Wiley Publishing, London, pp. 112–126.

Goss, T.G., O'Hara, G.W., Dilworth, M.J., Glenn, A.R., 1990. Cloning, characterization and complementation of lesions causing acid sensitivity in *Tn5*-induced mutants of *Rhizobium meliloti* WSM419. Journal of Bacteriology 172, 5173–5179.

Graham, P.H., 1992. Stress tolerance in *Rhizobium* and *Bradyrhizobium* and nodulation under adverse soil conditions. Canadian Journal of Microbiology 38, 475–484.

Gresshoff, P.M., 2012. Nitrogen Fixation: Achievements and Objectives. Springer Science & Business Media.

Hansen, N.C., Hopkins, B.G., Ellsworth, J.W., Jolley, V.D., 2006. Iron nutrition in field crops. In: Iron Nutrition in Plants and Rhizospheric Microorganisms. Springer Netherlands, pp. 23–59.

Hart, A.L., 1989. Nodule phosphorus and nodule activity in white clover. New Zealand Journal of Agricultural Research 32, 145–149.

Hartmann, A., Schmid, M., Van Tuinen, D., Berg, G., 2009. Plant-driven selection of microbes. Plant and Soil 321, 235–257.

Horst, W., 1983. Factors responsible for genotypic manganese tolerance in cowpea (*Vigna unguiculata*). Plant and Soil 72, 213–218.

Horst, W., 1987. Aluminum tolerance and calcium efficiency of cowpea genotypes. Journal of Plant Nutrition 10, 1121–1129.

Horst, W.J., 1995. The role of the apoplast in aluminum toxicity and resistance of higher plants. Zeitschrift für Pflanzenernährung und Bodenkunde 158, 419–428.

Hungria, M., Franchini, J., Campo, R., Crispino, C., Moraes, J., Sibaldelli, R., Mendes, I., Arihara, J., 2006. Nitrogen nutrition of soybean in Brazil: contributions of biological N_2 fixation and N fertilizer to grain yield. Canadian Journal of Plant Science 86, 927–939.

Indrasumunar, A., Searle, I., Lin, M.H., Kereszt, A., Men, A., Carroll, B.J., Gresshoff, P.M., 2011. Nodulation factor receptor kinase 1α controls nodule organ number in soybean (*Glycine max* L. Merr.). The Plant Journal 65, 39–50.

Indrasumunar, A., Gresshoff, P.M., 2013. Vermiculite's strong buffer capacity renders it unsuitable for studies of acidity on soybean (*Glycine max* L.) nodulation and growth. BMC Research Notes 6, 465.

Jakobsen, I., 1985. The role of phosphorus in nitrogen fixation by young pea plants (*Pisum sativum*). Physiologia Plantarum 64, 190–196.

Jensen, E., Peoples, M., Boddey, R., Gresshoff, P., Hauggaard-Nielsen, H., Alves, B., Morrison, M., 2012. Legumes for mitigation of climate change and the provision of feedstock for biofuels and biorefineries. A review. Agronomy for Sustainable Development 32, 329–364.

Jimenez, S., Gogorcena, Y., Hevin, C., Rombola, A.D., Ollat, N., 2007. Nitrogen nutrition influences some biochemical responses of iron deficiency in tolerant and sensitive genotypes of *Vitis*. Plant and Soil 290, 343–355.

Kochian, L., 1995. Cellular mechanisms of aluminum toxicity and resistance in plants. Annual Review of Plant Biology 46, 237–260.

Kochian, L., Hoekenga, O., Piñeros, M., 2004. How do crop plants tolerate acid soils? Mechanisms of aluminum tolerance and phosphorous efficiency. Annual Review of Plant Biology 55, 459–493.

Lerouge, P., Roche, P., Faucher, C., Maillet, F., Truchet, G., Prome, J.C., Denarie, J., 1990. Symbiotic host specificity of *Rhizobium meliloti* is determined by a sulphated and acylated glucosamine oligosaccharide signals. Nature 344, 781–784.

Li, Y., Yang, T., Zhang, P., Zou, A., Peng, X., Wang, L., Yang, R., Qi, J., Yang, Y., 2012. Differential responses of the diazotrophic community to aluminum-tolerant and aluminum-sensitive soybean genotypes in acidic soil. European Journal of Soil Biology 53, 76–85.

Liang, C., Piñeros, M., Tian, J., Yao, Z., Sun, L., Liu, J., Shaff, J., Coluccio, A., Kochian, L., Liao, H., 2013. Low pH, aluminum, and phosphorus coordinately regulate malate exudation through GmALMT1 to improve soybean adaptation to acid soils. Plant Physiology 161, 1347–1361.

Lin, M.H., Gresshoff, P.M., Ferguson, B.J., 2012. Systemic regulation of soybean nodulation by acidic growth conditions. Plant Physiology 160, 2028–2039.

Lindström, K., Myllyniemi, H., 1987. Sensitivity of red clover rhizobia to soil acidity factors in pure culture and in symbiosis. Plant and Soil 98, 353–362.

Liu, W., Lund, L., Page, A., 1989. Acidity produced by leguminous plants through symbiotic dinitrogen fixation. Journal of Environmental Quality 18, 529–534.

Liu, K., 2012. Soybeans: Chemistry, Technology, and Utilization. Springer.

Long, S., 2001. Genes and signals in the *Rhizobium*-legume symbiosis. Plant Physiology 125, 69–72.

Macció, D., Fabra, A., Castro, S., 2002. Acidity and calcium interaction affect the growth of *Bradyrhizobium* sp. and the attachment to peanut roots. Soil Biology and Biochemistry 34, 201–208.

Marschner, H., 1991. Mechanisms of adaptation of plants to acid soils. Plant and Soil 134, 1–20.

Marschner, H., Römheld, V., 1994. Strategies of plants for acquisition of iron. Plant and Soil 165, 261–274.

McDermott, T., Graham, P., Brandwein, D., 1987. Viability of *Bradyrhizobium japonicum* bacteroids. Archives of Microbiology 148, 100–106.

Miransari, M., Balakrishnan, P., Smith, D.L., Mackenzie, A.F., Bahrami, H.A., Malakouti, M.J., Rejali, F., 2006. Overcoming the stressful effect of low pH on soybean root hair curling using lipochitooligosaccahrides. Communications in Soil Science and Plant Analysis 37, 1103–1110.

Miransari, M., Smith, D.L., 2007. Overcoming the stressful effects of salinity and acidity on soybean [Glycine max (L.) Merr.] nodulation and yields using signal molecule genistein under field conditions. Journal of Plant Nutrition 30, 1967–1992.

Miransari, M., Smith, D.L., 2008. Using signal molecule genistein to alleviate the stress of suboptimal root zone temperature on soybean-Bradyrhizobium symbiosis under different soil textures. Journal of Plant Interactions 3, 287–295.

Miransari, M., Smith, D., 2009. Alleviating salt stress on soybean (Glycine max (L.) Merr.) – Bradyrhizobium japonicum symbiosis, using signal molecule genistein. European Journal of Soil Biology 45, 146–152.

Miransari, M. (Ed.), 2012. Soil Nutrients. Nova Science Publishers.

Miransari, M., 2013. Soil microbes and the availability of soil nutrients. Acta Physiologiae Plantarum 35, 3075–3084.

Miransari, M., Riahi, H., Eftekhar, F., Minaie, A., Smith, D.L., 2013. Improving soybean (Glycine max L.) N_2 fixation under stress. Journal of Plant Growth Regulation 32, 909–921.

Miransari, M., 2014a. Use of Microbes for the Alleviation of Soil Stresses, vol. 1. Springer.

Miransari, M., 2014b. Use of Microbes for the Alleviation of Soil Stresses, vol. 2. Springer.

Munns, D., 1970. Nodulation of Medicago sativa in solution culture. V. Calcium and pH requirements during infection. Plant and Soil 32, 90–102.

Munns, D., Hohenberg, J., Righetti, T., Lauter, D., 1981. Soil acidity tolerance of symbiotic and nitrogen-fertilized soybeans. Agronomy Journal 73, 407–410.

Munns, D., 1986. Acid soils tolerance in legumes and rhizobia. Advances in Plant Nutrition 2, 63–91.

Nam, H.G., 1997. The molecular genetic analysis of leaf senescence. Current Opinion in Biotechnology 8, 200–207.

Nasr Esfahani, M., Sulieman, S., Schulze, J., Yamaguchi-Shinozaki, K., Shinozaki, K., Tran, L.S., 2014. Approaches for enhancement of N_2 fixation efficiency of chickpea (Cicer arietinum L.) under limiting nitrogen conditions. Plant Biotechnology Journal 12, 387–397.

O'Hara, G., Goss, T., Dilworth, M., Glenn, A., 1989. Maintenance of intracellular pH and acid tolerance in Rhizobium meliloti. Applied and Environmental Microbiology 55, 1870–1876.

Oldroyd, G., Dixon, R., 2014. Biotechnological solutions to the nitrogen problem. Current Opinion in Biotechnology 26, 19–24.

Pellet, D.M., Papernik, L.A., Kochian, L.V., 1996. Multiple aluminum-resistance mechanisms in wheat roles of root apical phosphate and malate exudation. Plant Physiology 112, 591–597.

Poza-Carrión, C., Jiménez-Vicente, E., Navarro-Rodríguez, M., Echavarri-Erasun, C., Rubio, L., 2014. Kinetics of nif gene expression in a nitrogen-fixing bacterium. Journal of Bacteriology 196, 595–603.

Reeve, W., Tiwari, R., Kale, N., Dilworth, M., Glenn, A., 2002. ActP controls copper homeostasis in Rhizobium leguminosarum bv. viciae and Sinorhizobium meliloti preventing low pH induced copper toxicity. Molecular Microbiology 43, 981–991.

Rice, W.A., Penney, D.C., Nyborg, M., 1977. Effects of soil acidity on rhizobia numbers, nodulation and nitrogen fixation by alfalfa and red clover. Canadian Journal of Soil Science 57, 197–203.

Richardson, A.E., Simpson, R.J., Djordjevic, M.A., Rolfe, B.G., 1988a. Expression of nodulation genes in Rhizobium leguminosarum biovar Trifolii is affected by low pH and by Ca and Al ions. Applied and Environmental Microbiology 54, 2541–2548.

Richardson, A.E., Djordjevic, M.A., Rolfe, B.G., Simpson, R.J., 1988b. Effects of pH, Ca and Al on the exudation from clover seedlings of compounds that induce the expression of nodulation genes in *Rhizobium trifolii*. Plant and Soil 109, 37–47.

Richardson, A.E., Simpson, R., Djordjevic, M.A., Rolfe, B.G., 1988c. Expression and induction of nodulation genes in *Rhizobium trifolii* at low pH and in the presence of Ca and Al. Applied Environmental Microbiology 54, 2541–2548.

Richardson, A., Simpson, R., 1989. Acid-tolerance and symbiotic effectiveness of *Rhizobium trifolii* associated with a *Trifolium subterraneaum* L. based pasture growing in an acid soil. Soil Biology and Biochemistry 21, 87–95.

Salvagiotti, F., Cassman, K., Specht, J., Walters, D., Weiss, A., Dobermann, A., 2008. Nitrogen uptake, fixation and response to fertilizer N in soybeans: a review. Field Crops Research 108, 1–13.

Schmidt, W., 1999. Mechanisms and regulation of reduction-based iron uptake in plants. New Phytologist 141, 1–26.

Shamsi, I.H., Wei, K., Zhang, G.P., Jilani, G.H., Hassan, M.J., 2008. Interactive effects of cadmium and aluminum on growth and antioxidative enzymes in soybean. Biologia Plantarum 52, 165–169.

Shima, T., Hu, S., Luo, G., Kang, X., Luo, Y., Hou, Z., 2013. Dinitrogen cleavage and hydrogenation by a trinuclear titanium polyhydride complex. Science 340, 1549–1552.

Stewart, W.M., Dibb, D.W., Johnston, A.E., Smyth, T.J., 2005. The contribution of commercial fertilizer nutrients to food production. Agronomy Journal 97, 1–6.

Temme, K., Zhao, D., Voigt, C., 2012. Refactoring the nitrogen fixation gene cluster from *Klebsiella oxytoca*. Proceedings of the National Academy of Sciences 109, 7085–7090.

Von Uexküll, H., Mutert, E., 1995. Global extent, development and economic impact of acid soils. Plant and Soil 171, 1–15.

Wang, N., Khan, W., Smith, D.L., 2012. Changes in soybean global gene expression after application of lipo-chitooligosaccharide from *Bradyrhizobium japonicum* under sub-optimal temperature. PLoS One 7, e31571.

Watkin, E., O'Hara, G., Howieson, J., Glenn, A., 2000. Identification of tolerance to soil acidity in inoculant strains of *Rhizobium leguminosarum* bv. *trifolii*. Soil Biology and Biochemistry 32, 1393–1403.

Watkin, E., Mutch, L., Rome, S., Reeve, W., Castelli, J., Gruchlik, Y., Best, W., O'Hara, G., Howieson, J., 2009. The effect of acidity on the production of signal molecules by *Medicago* roots and their recognition by *Sinorhizobium*. Soil Biology and Biochemistry 41, 163–169.

White, P.J., George, T.S., Dupuy, L.X., Karley, A.J., Valentine, T.A., Wiesel, L., Wishart, J., 2013. Root traits for infertile soils. Frontiers in Plant Science 4.

Wiersma, J.V., 2010. Nitrate-induced iron deficiency in soybean varieties with varying iron-stress responses. Agronomy Journal 102, 1738–1744.

Wood, M., Cooper, J.E., Holding, A.J., 1984. Soil acidity factors and nodulation of *Trifolium repens*. Plant and Soil 78, 367–379.

Wood, M., Cooper, J., 1988. Acidity, aluminium and multiplication of *Rhizobium tripolii*: possible mechanisms of aluminium toxicity. Soil Biology and Biochemistry 20, 95–99.

Yamamoto, Y., Kobayashi, Y., Devi, S., Rikiishi, S., Matsumoto, H., 2002. Aluminum toxicity is associated with mitochondrial dysfunction and the production of reactive oxygen species in plant cells. Plant Physiology 128, 63–72.

Yang, Q., Wang, Y., Zhang, J., Shi, W., Qian, C., Peng, X., 2007. Identification of aluminum-responsive proteins in rice roots by a proteomic approach: cysteine synthase as a key player in Al response. Proteomics 7, 737–749.

Zhang, F., Smith, D.L., 1995. Preincubation of *Bradyrhizobium japonicum* with genistein accelerates nodule development of soybean at suboptimal root zone temperatures. Plant Physiology 108, 961–968.

Zhang, F., Li, L., 2003. Using competitive and facilitative interactions in intercropping systems enhances crop productivity and nutrient-use efficiency. Plant and Soil 248, 305–312.

Zhen, Y., Qi, J.L., Wang, S., Su, J., Xu, G., Zhang, M., Miao, L., Peng, X., Tian, D., Yang, Y., 2007. Comparative proteome analysis of differentially expressed proteins induced by Al toxicity in soybean. Physiologia Plantarum 131, 542–554.

Zhou, S., Sauve, R., Thannhauser, T.W., 2009a. Proteome changes induced by aluminum stress in tomato roots. Journal of Experimental Botany 57, 4201–4213.

Zhou, S., Sauve, R., Thannhauser, T.W., 2009b. Aluminum induced proteome changes in tomato cotyledons. Plant Signaling and Behavior 4, 769–772.

Soybean Production and Compaction Stress

11

M. Miransari
AbtinBerkeh Scientific Ltd. Company, Isfahan, Iran

Introduction

The soybean [*Glycine max* (Merr.) L.] is one of the most important legumes used as a source of food and oil with nutritional value. The plant is able to establish a symbiotic association with the nitrogen (N) fixing bacteria, *Bradyrhizobium japonicum*, and acquire most of its essential N for growth and yield production. High rates of N are fixed by the bacteria for the use of the host plant. Accordingly, such a process is of environmental and economic significance, and it is important to increase its efficiency under different conditions, including stress (Long, 2001; Salvagiotti et al., 2008).

The increasing world population has resulted in the intensive use of farming and cropping and hence the higher use of agricultural machinery for the production of food crops including soybeans. As a result, the soybean and its symbiotic rhizobium are subjected to compaction stress, which is mostly due to the use of agricultural machinery, especially under a high rate of soil moisture, a low rate of organic matter, unsuitable cropping rotations, intensive crop production, a high rate of grazing, and the use of inappropriate soil practices (Figs. 11.1 and 11.2). Accordingly, the soils become compacted and deteriorated and unsuitable for crop production, especially under arid and semiarid conditions. It is hence important to use suitable methods, which result in the establishment of an optimum medium growth for yield production and microbial activities (Clark et al., 2004; Beutler et al., 2008; Nawaz et al., 2013).

It is significant to use soil fertility practices, which enhance soil biological activities, for crop production and soil health. Accordingly, soil quality is a suitable parameter for evaluating the effects of agricultural practices and can be used as a tool for the enhancement of agricultural sustainability. Soil biodiversity is an important factor affecting the productivity and sustainability of agriculture and is a function of agricultural practices (Xavier et al., 2010). Selecting the most suitable agricultural practices, which maintain the soil biodiversity and enhance its productivity and efficiency, is among the most important strategies and must be achieved by conducting proper agricultural practices.

The compaction stress, which results in the degradation of agricultural fields, is among the most important subjects of global research, as it affects the soil productivity, the environment, food production, and quality of life. The most important reason for the mechanization of agricultural fields is the increased world demand for food production (Ishaq et al., 2001). Vehicular traffic has resulted in compaction

Environmental Stresses in Soybean Production. http://dx.doi.org/10.1016/B978-0-12-801535-3.00011-5

Figure 11.1 Formation of ruts resulted by agricultural machinery, an example of soil compaction (Nawaz et al., 2013).

stress in 68 million ha of agricultural fields worldwide, including Europe, Africa, Asia, and Australia at 33, 18, 10, and 4 million ha, respectively, and some parts of North America (Flowers and Lal, 1998).

The definition of compaction stress is the process, which results in the rearrangement of soil particles decreasing the void space and bringing the soil particles to a closer contact with one another and hence increasing bulk density. Subsequently, the rate of soil aggregates and their size decreases, their shape is altered, and the remaining ones will have a new arrangement, and hence the pore spaces inside and between these aggregates are affected (Defossez and Richard, 2002).

The threshold of compaction stress for a given soil and under given climatic conditions is a function of soil texture, aggregation degree, and soil moisture, specifically matric potential (Horn et al., 1995). Accordingly, different soil physical, chemical, and biological properties are affected (Gupta et al., 1989). The silty soils are more subjected to the compaction stress, at the low rate of moisture, compared with the other soil textures (Horn et al., 1995).

The soil parameters, including soil texture, structure, moisture, and organic matter, in combination with soil strength and the vehicle weight, determine the level of compaction stress. Although the higher number of tractor passing results in the higher level of compaction stress, the first passage is the most effective and up to 10 passes can affect the soil compactness to 50 cm (Hamza and Anderson, 2005). The other important parameter, which is influenced by the stress, is the removal of soil air, affecting the behavior of carbon (C) and nitrogen (N) in the soil (Soane and Van Ouwerkerk, 1995).

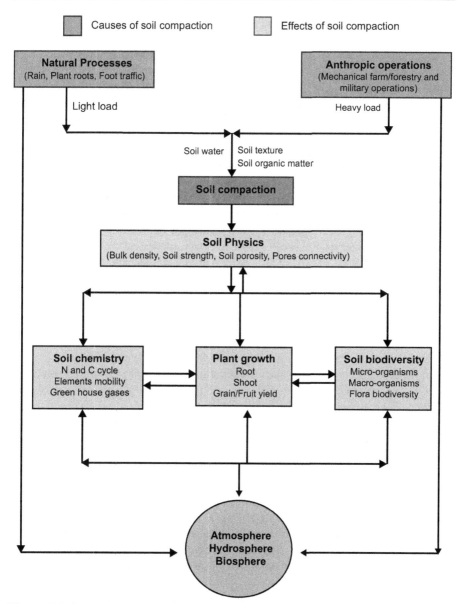

Figure 11.2 Causes of soil compaction and its effects on soil physical, chemical, and physiological properties (Nawaz et al., 2013).

The stress affects CO_2 concentration and the mineralization of C and N in the soil. Under the stress, due to the decreased rate of soil oxygen, denitrification increases, and as a result, soil N decreases and symptoms of N deficiency appear in the plant. Due to anaerobic conditions, the rate of methane production and emission from the soil increases. It is because under such conditions the activity of methanogenic bacteria (producers of methane) increases and the activity of methanotrophic bacteria (oxidizers of methane) decrease (Soane and Van Ouwerkerk, 1995; Miransari et al., 2006).

Although using the agricultural machinery resulted in the compaction of the top soil, with time the subsoil is also subjected to the stress, and due to the difficulty of access, it is hard and expensive to treat the compaction stress in the subsoil, which is considered as an important source of soil degradation by the European Union. A 38% reduction resulted in wheat yield due to the subsoil stress to the 0.15 cm depth at the bulk density of $1.93\,Mg/m^3$ (Ishaq et al., 2001).

Under compaction stress, the soil physical fertility, especially the supply and storage of water and nutrients, is adversely affected, resulting in some unsuitable soil properties such as increased soil bulk density and strength, as well as decreased porosity, soil water infiltration, and water moisture rate. The strength of the soil increases under compaction stress; however, soil fertility decreases, and as a result, a higher rate of chemical fertilization must be used, resulting in greater expenses for crop production. Subsequently, plant growth and crop production decreases, followed by reduced organic matter and the number and activities of soil microbes (Or and Ghezzehei, 2002; Hamza and Anderson, 2005).

Under compaction stress the root growth is adversely affected, as the soil structure will not be suitable for root growth. The rate of macropores decreases under the stress; however, the rate of micropores increases. Accordingly, air circulation including oxygen, which is essential for root metabolic activity, decreases, and hence plant growth and yield production is negatively affected. The activities of soil microbes are also adversely affected. The compaction stress also influences the rate of C and N mineralization, and as a consequence the concentrations of CO_2 and the other greenhouse gases in the soil and their subsequent emission from the soil are also affected (Chen and Weil, 2011; Jensen et al., 2012; Ball, 2013).

The rate of organic matter in the soil is among the most important factors affecting the rate of soil degradation under the compaction stress. Soil organic matter is able to act as a buffer and prevent or lessen the transfer of top soil stress to the subsoil and hence alleviate the stress. Unlike stresses such as soil erosion, which can be easily recognized, the compaction stress is not easily recognizable and hence with time decreases plant growth and yield production, which is often blamed on other causes (West and Post, 2002; Hamza and Anderson, 2005).

Soil bulk density is the most used and precise parameter, which can reflect the state of compaction stress. However, soil strength is also another used parameter indicating soil resistance to the growth of plant roots. Water infiltration can also be used for the determination of compaction stress level; as in a soil with suitable structure, water can enter the soil at a faster rate related to a soil under the compaction stress (West and Post, 2002; Hamza and Anderson, 2005).

Under the stress and due to the above mentioned details, the uptake of nutrients by plant roots decreases. Due to the decreased rate of oxygen in the soil, a part of the soil N is emitted into the atmosphere by the process of denitrification and hence faces N deficiency and it turns pale. Plant height decreases and the growth of roots will be like cluster growth. The activity of soil microbes is also affected similarly to the plant, because the essential parameters for microbial growth and activity will not be accessible by the microbes or will be accessible at a reduced rate (Miransari et al., 2006).

In the following, some of the most important advancements and details related to the effects of soil compaction on plant growth and yield production, specifically soybeans, have been presented, reviewed, and analyzed. Such details can be used for the more efficient production of crop plants under the compaction stress and include methods, techniques, and strategies, which can be used for the alleviation of stress, as a high rate of agricultural fields in the world are subjected to such a stress.

Soybean and Compaction Stress

The unsuitable use of agricultural fields worldwide has resulted in land degradation, pollution of water, and global warming, adversely affecting agricultural sustainability. Due to the negative effects of tillage practices on the soil properties, research has been looking for alternative methods for controlling soil degradation. Using a nontilled method is among the most efficient practices for conserving soil and water. Such practices maintain the soil structure by keeping the highest rate of residue on the soil surface and increasing soil moisture by the infiltration of water into the soil and decreasing the rate of evaporation (Mathew et al., 2012; Toliver et al., 2012).

Similar to the other crop plants, the soybean is also sensitive to the compaction stress and its growth, and yield decreases under such conditions. The other important factor, which must be considered related to soybeans, is its symbiotic association with *B. japonicum*, which is also adversely affected under the compaction stress. Due to the unfavorable effects of compaction, including high bulk density and soil strength and a reduced level of oxygen on plant and bacterial growth and activity, the nodulation of soybeans by *B. japonicum* significantly decreases under the stress (Beutler et al., 2008; Bavin et al., 2009).

The selection and use of a suitable tillage method for soybean production under cool-season conditions is of significance. Accordingly, Carter (2005) investigated the effects of long-term tillage methods on the protein and yield of cool-season soybeans in rotation with barley and on the soil properties, including soil C and soil N using two different experiments under the conditions of Atlantic Canada. In the first experiment (1989–2000) the effects of direct drilling (nontill method), conventional moldboard plowing, and shallow tillage, and in the second experiment (1991–1993) the effects of chisel plowing and moldboard plowing were evaluated. Both experiments were conducted in the spring and fall.

The results indicated that there were not significant differences between different tillage methods related to soybean yield (ranging from 1.5 to 2.1 Mg/ha), grain protein, and nutrients (P and K). In experiment one, direct drilling resulted in the improvement of surface soil properties including microbial biomass C, organic C, and available P, related to moldboard plowing. The high rate of P build up (>200 µg P/kg) by the latter indicated an environmental concern, as the level was higher than the plant requirement. However, for the depth of 0–60 cm the soil C and N were not affected by different tillage methods, although there was a relative increase in the rate of total N for the 0–20 cm depth by direct drilling. With respect to their results, the authors made the conclusion that conservation tillage is a suitable method for planting cool-season soybeans under the conditions of Atlantic Canada.

The effects of compaction stress on soybeans and field bean growth was evaluated by Kahnt et al. (1986) under greenhouse conditions. Using cylindrical tubes of plastic quality with a height of 58 cm and a diameter of 12 cm the pots were filled with a silty soil at three different bulk densities including 1.25, 1.45, and 1.65 g/cm^3, respectively. The most important effect of compaction stress was the decreased plant growth including the root and the aerial part, due to the soil impedance. Compaction stress reduced root growth by 32% and root length up to 60%. However, because the roots became thickened under the stress the root volume was not much affected. The growth of the aerial part decreased up to 31%. A two-year field experiment was conducted by Beutler et al. (2008) to evaluate the effects of compaction stress on the growth of two soybean genotypes using tractor passing. Machine traffic resulted in compaction stress and decreased soybean growth and yield in the soil resistance of 1.64–2.35 MPa and the bulk density of 1.50–1.53 Mg/m^3.

Izumi et al. (2009) investigated the effects of subsoiling on a field regularly cropped with soybeans or wheat, where nontillage practices had been practiced for 5 years, and the crop yield had decreased during this time. Different plant parameters including the growth of the aerial part, root distribution, water uptake, and the yield of both plants were determined. Accordingly, the authors investigated if subsoiling can improve plant growth and yield production by modifying root development and how different the effects of subsoiling in a tilled and nontilled field can be. Under nontillage treatment the wheat roots were less distributed on the surface (0–5 cm) and grew more deeply in the soil (20–25 cm).

Among the most prevalent methods of soil tillage for corn and soybean production is reduced tillage, including the nontill method and rotation. The effects of nontill practice on plant growth and yield production is a function of latitude and cropping methods. Accordingly, Wilhelm and Wortmann (2004) conducted an experiment to investigate the influence of seasonal heat and precipitation on the effects of different conventional tillage methods including plow, chisel, disk, ridge till, subsoil, and nontill on the production of continuous and rotated corn and soybeans under rain-fed conditions in a 16-year period. Higher summer heat decreased corn and soybean grain yield. While both tillage and rotation increased corn yield, just rotation had a positive effect on soybean yield.

The interaction of tillage and year was significant for both crops. A warmer spring diminished the effects of the plow method on both plants. The chisel had the least effect on soybean yield in response to the favorable environmental conditions, related to the other tillage methods. The effect of rotation on the grain yield of both corn (7.10 vs. 5.83 Mg/ha) and soybean (2.57 vs. 2.35 Mg/ha) plants was greater than continuous cropping. For corn the effects of rotation were greater under cool-season conditions, while this was not the case for soybeans. Corn yield under tillage and rotation was a function of both seasonal heat and precipitation, while for soybean; it was just a function of heat affecting the influence of tillage method on grain yield.

Mazzoncini et al. (2008) evaluated the effects of a long-term tillage (nontill (NT) and conventional tillage (CT)) experiment (1990–2005) on the productivity of rain-fed wheat and soybeans under a Mediterranean climate. Under NT the wheat yield was 8.9% less (3.97 vs. 4.36 Mg/ha) than CT; however, for soybeans, NT resulted in significantly less grain yield than CT (2.60 vs. 3.08 Mg/ha). The rate of weeds was higher for NT soybeans than wheat. The tillage method did not affect N concentration in wheat and soybeans. The concentration of P in wheat grain and straw was higher under NT, while in soybeans, P concentration was not affected by tillage method.

Water uptake from the deeper soil was increased by subsoiling under nontilled treatment. Nontilled and subsoil treatments significantly increased plant growth and yield production. The effects of subsoiling might have affected water supply by planting deep roots during spring. Soybean growth and yield was significantly increased by the nontilled treatment; however, this was not the case for subsoiling. However, the distribution of soybean roots was modified by both treatments. Under nontilled treatment, soybean roots were distributed on the soil surface, though subsoiling did not affect root distribution in the deeper soil. Hence subsoiling did not significantly affect the trend of water uptake and plant yield. In the nontilled treatment, subsoiling increased rooting depth and the wheat yield, which can be the case for a long time; however, it did not affect soybean behavior similarly.

Micucci and Taboada (2006) investigated the response of soybean roots to the physical properties of soil under conventional and nontilled tillage in humid Pampas of Argentina. The rate of soil organic matter in the 0–5 cm depth, which had had been recompensated in a 4–11 year period under nontilled conditions, was 53–72% of the pasture soil. The stability of soil aggregates in the conventional tillage was 10.1–46.8% less than the pasture soil and had become similar to the pasture soil in nontilled treatment. The relative level of soil compaction was in the range of 60.8–83.6%, which was less than the threshold level (90%) for crop yield.

The volume rate of soil porosity (>50 μM) ranged from 0.91–5.09%, which was much less than the critical rate (10%) for the elongation and aeration of soybean roots. The threshold level of soil resistance (2–3 MPa) was only passed over in the plow pan of conventional (5.9 MPa) and nontilled (3.7–4.2 MPa) fields of some areas. However, the growth of soybean roots was not suppressed by the decreased rate of soil macroporosity and high soil resistance in any of the sites. There was a negative correlation between soybean root growth and clay particles. The authors accordingly made the conclusion that soybean root growth is a function of subsoil properties and not tillage practices.

Frac et al. (2009) investigated the effects of mulching and compaction stress on the microbial community of the soybean rhizosphere under the light, medium, and heavy levels of compaction stress. The soybean seeds were inoculated with *B. japonicum* and planted in the plot rows with the 0.3 m spacing. The wheat straw was used for mulching the plots when the seeds were planted. The rhizosphere was sampled three times during the growing season. Using microbiological methods, the samples were analyzed three times, and the number of fungi and bacteria, as well as the number of *Bacillus* sp. and *Pseudomonas* sp., were determined. While the light and the medium level of compaction stress resulted in the highest number of soil microbes, the heavy level of stress significantly decreased the number of soil microbes. Wheat mulching had a positive effect on the number and activity of soil microbes under the stress.

Botta et al. (2004) investigated the effects of machinery traffic on the yield of soybeans under direct drilling. Using different tractor passing, the traffic loadings of 0, 60, 120, and 180 Mg km/ha were created in the field using a tractor with the mass of 39.38 kN. The rate of soybean yield was determined 5 months later, as a 9–38% yield decrease resulted. The yield reductions of 38%, 22.6%, and 9.8% were related to the tractor loading of 180, 120, and 60 Mg km/ha, respectively. The level of compaction stress was determined by measuring soil bulk density, soil impedance (resistance to penetrometer), and rut depth.

The effects of compaction stress on the rate of soybean seeds, and its physical quality and size, was evaluated by Mattioni et al. (2012). The research area was at 60.5 ha, and one point was created for each sampling grid. For each sampling grid the resistance to penetrometer, as well as the rate of seed germination and its physical quality, was determined. The results indicated that seed amount and properties were negatively affected by compaction.

In an interesting experiment, Ramos et al. (2010) investigated the correlation between the level of compaction stress and the properties of soybean root hairs. The corresponding bulk density values at different levels of compaction stress were equal to 1.1, 1.3, and 1.5 g m^3, representing the mechanical impedances of 0.1, 0.5, and 3.5 MPa, respectively. The distance of root hairs from the apex and the related diameter at the 1.5 cm distance was determined. There was an inverse correlation between the level of soil impedance and root hair length and diameter. There were also changes in the cells of rhizodermis. The authors accordingly expressed that the morphological properties of root apexes can be used as appropriate indicators of compaction stress.

The effects of soybean intercropping with pigeon pea, as well as subsoiling on the yield of soybeans under rain-fed conditions, was investigated by Ghosh et al. (2006) in a 3-year experiment. Different cropping systems (sole pigeon pea, sole soybean, and their intercropping) and tillage practices (conventional, conventional and subsoiling in alternate years, and yearly) were used as the experimental treatments. They accordingly investigated the effects of subsoiling on the yield production of soybeans/pigeon peas and the frequency of subsoiling, which resulted in the highest and the most profitable soybean yield. A 60% higher soybean yield resulted from soybean/pigeon pea intercropping related to sole soybean cropping.

There was a consistent response of yield to subsoiling as a 20% increase in yield resulted due to the enhanced water storage and root length density. There was a 21–25% yield increase resulting from the interactions of subsoiling and intercropping. The authors accordingly made the conclusion that the combined use of subsoiling and intercropping under rain-fed drought stress conditions can be a useful method to increase soybean yield production.

Hati et al. (2006) evaluated the effects of chemical fertilization (NPK) and organic manure (farmyard manure, FYM) on the physical properties of soil, root growth, water use efficiency, and soybean yield in cropping with mustard. FYM at 10 Mg and NPK at the essential rate for soybean production in 3 years increased the carbon content of the surface soil (0–15 cm) from the initial value of 4.4–6.2 g/kg and, related to the control, increased soybean grain yield and water use efficiency by 103% and 76%, respectively. The plots fertilized with FYM and NPK had a higher mean weight diameter (0.50 mm) and a greater percentage of water stable aggregates (55%) than chemically fertilized plots (0.44% and 49%) and control plots (0.41% and 45.4%).

The NPK and FYM treatment also resulted in a higher saturated hydraulic conductivity (13.32×10^{-6} m/s) in the 0–7.5 cm related to the NPK (10.53×10^{-6} m/s) and control plots (8.61×10^{-6} m/s). The lowest bulk density in the 0–7.5 cm was resulted by FYM and NPK treatment (1.18 Mg/m^3), compared with the highest bulk density, which was resulted by the control treatment (1.30 Mg/m^3). However, the effects of FYM and NPK treatment on the saturated hydraulic conductivity and bulk density at the subsurface (22.5–30 cm) were not significant. The highest root length density (RLD) up to the 30 cm depth was related to treatment FYM and NPK, which was 31.9% and 70.5% higher than the NPK and control plots, respectively. There was a significant and negative correlation ($r = -0.88**$) between RLD and the soil resistance to penetrometer (soil strength).

The emission of N_2O, N_2, and CO_2 from nitrate-fertilized soil plots, as affected by compaction stress, soil moisture, and rewetting, was investigated by Ruser et al. (2006) in the lab. The undisturbed soil samples were fertilized with $^{15}NO_3$ (150 kg/ha). The soil cores were collected from different parts of a potato field with different bulk densities including the ridges (1.03 g cm^3), the interrow (1.24 g cm^3), and the area compacted by tractor (1.64 g cm^3), and their moisture rate was adjusted to 40–98% according to the rate of water-filled pore space (WFPS). A high rate of denitrification (N_2O emission) resulted in WFPS >70% by the compaction stress.

N_2 was produced at the highest level of moisture (WFPS >90%), which was significantly smaller than the emission of N_2O most of the time. The compaction stress did not affect the emission of CO_2 under different moisture levels, just in the case of 98% WFPS, which the emission of CO_2 (respiration of soil) significantly decreased. Rewetting the dry soil resulted in the highest level of N_2O emission from all the treatments. The effect of rewetting was intensified with the higher rate of water. The authors accordingly indicated that the emission of N_2O from the soil can be affected by the increased availability of carbon and the use of O_2 for microbial activities as a function of soil drying and rewetting.

Place et al. (2008) evaluated the root growth of different soybean and weed genotypes including sicklepod (*Senna obtusifolia* L.) and Palmer amaranth (*Amaranthus palmeri* S. Wats) under compaction stress to obtain N from the deeper soil depth. Using soil columns the compaction stress was created to the depth of 8 cm. According to the results the two weed genotypes were able to grow into the compacted soil more effectively than the four soybean genotypes. However, root growth into the compacted soil was not correlated with the growth of the aerial part.

Although the growth of weed genotypes, including the roots and the aerial part, was sustained even when the roots were not able to grow into the stressed soil, this was not the case for the soybean genotypes, even in a highly fertile soil. The weeds were also able to acquire a high rate of $^{15}N-NO_3$ from the patches underneath the compacted soil. The authors accordingly made the conclusion that the weeds had a competitive growth over soybeans, explaining their growth dynamic in a compacted soil.

The response of soybeans to corn stover grazing was investigated by Clark et al. (2004) in a 3-year experiment using two 19-ha agricultural fields with a variable range of soil textures. The experiment was a split plot using a corn–soybean rotation, under chisel-plowed and grazed treatments for a certain time. The grazed plots were then disked or remained nontilled. Different soil and plant properties including soil bulk density, moisture content, aggregate stability, soil resistance, surface strength, the rate of cover residue, and the population of soybean plants and their yield were determined for each tillage treatment.

Although soil bulk density was not affected, soil resistance increased up to a 10 cm depth in plots grazed in October and November. Grazing did not affect the population of the subsequent soybean plants; however, increased soil resistance decreased soybean yield ($r^2 = 0.36$). If during the grazing, the soil degree was $0°C$ or less the soybean yield increased ($r^2 = 0.72$). The authors accordingly indicated that grazing during the frozen period of the year or disking of soil before planting soybeans resulted in the minimal adverse effects of grazing on the soil properties and soybean yield.

The use of NT is a suitable method for the erodible fields; however, the subsequent yield reduction may be a concern for the farmers. Accordingly, Vetsch et al. (2007) evaluated the effects of rotational full width tillage related to long-term NT and zone tillage on the yield and economical significance of corn and soybeans. The effects of rotational tillage on the yield of corn and soybeans were the highest. If cropping corn and soybeans is favorable, although rotational tillage may result in the highest corn and soybean yield, it may not produce the highest yield rate, economically.

The effects of soil strength on the establishment of soybean crops were investigated by Tola et al. (2008). The experimental treatments, including the cultivated field and medium and high soil strength, were established, and the edible soybean was used as the test crop. The results showed that there were not differences related to plant population, height, yield, and yield components among different treatments. The authors accordingly indicated that more research was essential to evaluate the effects of nontilled practices on the growth and yield of soybeans under different soil strengths.

The effects of different soybean germplasm on their root growth and architecture were investigated by Manavalan et al. (2010). Root properties are able to affect root behavior under different conditions including stress. Harvesting the roots from the soil is not easy, and as a result, alternative methods may be used for the proper harvesting of roots from the soil. Accordingly, the authors suggested two different methods for properly screening the root properties, including the cone and the tube method under controlled conditions using turface and sand.

For the cone method, 34 soybean genotypes were evaluated for the length of their tap root and biomass 12 days following planting in a growth chamber in eight replicates. The results obtained by the cone method were validated by the tube method for eight soybean genotypes 21 days following planting. A coefficient determination of 0.72 showed that there was a suitable correlation between the two methods. Accordingly, the cone method may be used as a suitable method for assessing the root properties and architecture of different soybean genotypes.

Stipesevic et al. (2009) investigated the effects of different tillage methods including: (1) CT using moldboard plowing, (2) diskharrowing (DH), and (3) NT on the zinc (Zn) uptake of wheat and soybeans in a 4-year period. The concentration of Zn in both plants decreased with reducing the rate of tillage in the order of CT > DH > NT, which might have been due to a decreased root growth as a result of less root disturbance under DH and NT treatments. Related to soybeans, the less Zn uptake by winter wheat than the optimum level resulted in a higher P:Zn rate with reducing the level of tillage. Accordingly, under reduced tillage, a higher rate of Zn fertilization is essential, although for the exact recommendations, more research must be conducted.

Bavin et al. (2009) investigated the effects of reduced tillage and cover cropping on the emission of greenhouse gases including CO_2, N_2O, and CH_4 in a corn/soybean rotation. During the 2-year experimental period (2004–2005), cover cropping increased the rate of CO_2 production at a rate of 222.7 g/cm^2 higher than the reduced tillage treatment, which was due to the decomposition of organic residue. The emission rate of N_2O was similar in both treatments in 2004 and was 100.1 mg N/m^2 higher in conventional tillage in 2005 following N fertilization. The two most important parameters controlling N_2O emission were N fertilization and fertilizer type in both treatments. The emission of CH_4 in both treatments was negligible and not at a detectable level. Hence the loss of N_2O is an important component of the greenhouse gas budget.

Farmaha et al. (2012) investigated the effects of the P and K rate and placement (broadcast and 15-cm placement) under nontill and strip-till practices using a 3-year experiment on the rate of soybean seed oil and protein content, the growth of the aerial part and its P and K uptake, and soil moisture. The tillage practices and the chemical fertilization (P at 0, 12, 24, and 36 kg/ha yr and K at 0, 42, 84, and 168 kg/ha yr) were selected as the main plots and subplots, respectively. Strip till and fertilization resulted in the higher seed oil and protein content than nontill practice.

Just the nontill practice increased the seed protein content when fertilized with P, while the strip-till increased the seed protein content when P fertilization was not used, indicating that strip till is more efficient to make P available to the crop than nontill treatment. The effect of P and K on the rate of seed oil and protein concentration was not significant. Leaf area index was higher for strip-till than the other treatments by the V2 stage, indicating the importance of early fertilization. The other advantage of strip-till over nontill was the higher soil moisture during the reproductive stage.

Rhizobium and Compaction Stress

With respect to the importance of soil microbial communities for the health and sustainability of agriculture, Gil et al. (2011) investigated the effects of tillage practices on the activities of soil microbes using a long-term experiment started in 1992. Two tillage practices, non and reduced tillage, and two crop rotations, soybean monoculture and corn–soybean, were used as the experimental treatments on the basis of a split-plot experiment. The structure of total microbial community was assessed using the phospholipids fatty acid (PLFA) method.

The electrophoresis method of 18S rRNA was used to evaluate the effects of agricultural practices on the soil fungal communities. The minimum level of soil PLFA content was related to the reduced tillage and soybean monoculture, indicating the adverse effects of reduced tillage when combined with soybean monoculture on the community of soil microbes. The PLFA method indicated that soil microbial communities were affected by crop rotation. The fungal communities, as indicated by the electrophoresis method, were a function of tillage practices and crop rotation.

The effects of different tillage methods, including reduced tillage (RT) and conventional tillage (CT), on the process of N fixation by soybeans and *B. japonicum* were investigated by Kihara et al. (2011) under the humid conditions of Western Kenya using the ^{15}N dilution method. Crop residue (CR) was also used as a treatment under the soybean–maize rotation and intercropping. The N content was determined in soybean leaves. The rate of N biological fixation was in the range of 41–65% of the total N content, with a higher rate of 55.6% under RT ($p < .05$) than CT (46.6%).

Total rate of fixed N under (RT + CT) was higher related to the other treatments including intercropping by at least 55% and rotation by 34%. The annual rate of N balances removed by soybean and maize grain were better under NT (−9 to −32 kg N/ha) than CT (−40 to −60 kg N/ha). Using p resulted in a higher nodule weight (3–16 times) ($p < .05$) than the control treatment. Soybean residue decreases the accumulation of N in the soil. The authors accordingly indicated that the higher rate of biological N fixation is resulted by RT + CR than CT.

Agricultural practices such as tillage, fertilization, and liming can affect soil properties including microbial communities, organic matter, pH, and nitrification rates. Using three different experimental sites, including one at the Rothamsted Research in the United Kingdom and two in the United States, the soil microbes, which are best correlated with agricultural practices and the factors that may regulate their abundance, were investigated. Using the method of 16S rRNA sequencing, the activity and abundance of bacteria and archaea were determined. The bacteria, which were most correlated with the cropping practices, were the ones from the nitrogen cycle, including the ammonium oxidizing archaea, mostly Ca Nitrososphaera with a positive correlation, and the N fixing bacteria, *Bradyrhizobium*, with a negative correlation (Zhalnina et al., 2013).

The time following the agricultural practices decreased the number of Ca Nitrososphaera and increased the number of *Bradyrhizobium*. Archaeal abundance was positively correlated with soil pH and NH_3 concentration and *Bradyrhizobium* abundance was negatively correlated with such factors. The authors indicated that with respect to the nature of the experiments, including a wide range of edaphoclimatic conditions for a long time, the high correlations of the two investigated microbes make them suitable representatives for evaluating the effects of agricultural practices on the activity and abundance of soil microbes.

The effects of different tillage practices on the yield and nodulation of soybeans were evaluated by Jug et al. (2005). To avoid the negative effects of moldboard plowing on the soil properties and the environment, the effects of reduced and nontill methods were also investigated in the experiment. In the years 2002 and 2003, reduced tillage significantly decreased plant density, the weight of 1000 grains, and grain yield. However, reduced tillage positively affected the nodulation by rhizobium, as the highest number and weight of nodules per plant were obtained under reduced tillage.

Campos et al. (2001) and Campos and Gnatta (2006) indicated that the nontill method is a suitable tillage practice, positively affecting the growth and activity of *Bradyrhizobium*. In their experiments, they conducted a 5-year research work (1994–1999) to evaluate the efficiency of *Bradyrhizobium* as determined by biological N fixation under nontill practice. Treatments included noninoculated control, 200 kg of N fertilizer, and the commercial strains of *Bradyrhizobium* and were tested under the nontill method. N fertilization did not increase nodulation nor soybean grain yield. There was not a significant difference between the noninoculated treatment and the inoculated treatments. They accordingly indicated that such a lack of response to inoculation may be due to the established population of efficient *Bradyrhizobium*, with a high number, and to the presence of favorable soil conditions for the process of biological N fixation including moisture and heat, which are present under nontilled practice.

In some parts of Canada and the United States, the soybean is grown as a prevalent annual crop under the nontill method, due to its environmental benefits. Most of the time the soybean is rotated with maize; however, because of increasing maize yields, the use of reduced and conventional tillage has become favored by the farmers. Accordingly, the rate of maize residue has increased, adversely affecting the nontill method by affecting the soil N, moisture and heat, soybean nodulation, and soybean germination, growth, and yield. Accordingly, efficient agricultural practices must be implemented to increase the efficiency of maize residue for soybean production under nontill methods (Vanhie et al., 2015).

Methods of Alleviating Compaction Stress
Biological Methods

Using field and greenhouse experiments, Miransari et al. (2006, 2007, 2008, 2009a,b) and Miransari (2013) hypothesized, tested, and proved that with respect to the effects of arbuscular mycorrhizal (AM) fungi on plant growth and yield production, it is possible to alleviate the adverse effects of compaction stress. AM fungi are the soil fungi, establishing a symbiotic association with their nonspecific host plant and increasing its growth and yield production by the enhanced uptake of water and nutrients under different conditions including stress (Compant et al., 2010; Miransari, 2010; Bothe, 2012).

Under greenhouse conditions, corn and wheat plants were subjected to the compaction stress, and their growth and their nutrient uptake were determined in a 2-year research work. The objectives were to (1) evaluate the effects of compaction stress on the growth of corn and wheat and (2) test the hypothesis that the incaution of corn and wheat seeds with different species of AM fungi might alleviate the stress under unsterilized and sterilized conditions. Three different fungal species, including the Iranian species of *Glomus mosseae* and *Glomus etunicatum* and the Canadian species of *G. mosseae*, received from Glomales in vitro Collection, Canada, were used for the experiments.

AM fungi increased corn root fresh and dry weights by 94% and 100%, respectively, under the compaction stress. Species of mycorrhizal fungi with different origins were able to alleviate the stress; however, their efficiency was a function of the stress level and the interaction with the other soil microbes. Mycorrhizal fungi significantly increased wheat growth (the root and the aerial part), and wheat grain yield under different levels of stress in both sterilized and unsterilized soils. However, the growth of the root was more enhanced by AM fungi than the growth of the aerial part, resulting in a higher root/aerial rate. Mycorrhizal fungi were able to increase wheat growth under the stress; however, its effectiveness was a function of stress level: as the stress increased, the efficiency of the fungi in alleviating the stress decreased.

The greenhouse experiments were conducted according to the following details. The surface soil of the research field of the Soil and Water Research Institute in Karaj, Iran was collected, air dried, and sieved. A part of it was sterilized using an autoclave at 121°C for an hour and placed in pots with a height and diameter of 20 cm. The pot soils were compacted using a 2-kg weight with an 18-cm diameter released from a 20-cm height 4, 12, and 20 times to create the favorable levels of A stress.

Corn and wheat seeds were planted in the pots and inoculated with different species of mycorrhizal fungi. Soil bulk density was determined three times and in six replicates during the growing period using a 100 cm^3 cylinder. The corresponding values ranged from 1.18 to 1.54 g cm^3. Using a penetrometer, soil resistance was determined three times during the growing period at the determined soil moisture. Different plant properties were determined during the growing period, harvest, and following the harvest, including the uptake of nutrients by corn and wheat (Miransari et al., 2007, 2008, 2009a,b; Miransari, 2013).

The fungi spores are able to germinate in the absence or presence of the host plant, although for proceeding with the further stages of the symbiosis, the presence of the host plant is essential. The germination of the fungal spores results in the production of an extensive hyphal network, which is able to grow into the soil around the host plant roots, including the mycorhizosphere, and significantly increase the uptake of water and nutrients by the host plant. The hyphal network is able to develop two different organelles, the vesicle and the arbuscule, which can make the fungi act more efficiently (Rillig et al., 2003; Pozo and Azcón-Aguilar, 2007).

The vesicle is the storage organelle, with a high number of vacuole helping the fungi under different conditions including stress. For example, under the salinity stress the high number of vacuole are able to absorb a high concentration of salt ions and hence make the host plant alleviate the stress. Arbuscule is the branched-like structure, which is used as the water and nutrient interface between the fungi and the host plant (Bonfante and Genre, 2008; Talaat and Shawky, 2014).

Under stresses such as compaction stress the enhanced uptake of P by the fungi can increase root growth. Interestingly, Pupin et al. (2009) found that the level of urease decreases under the stress, but the production of phosphatase increases. They also indicated that under the compaction stress the population and activities of soil fungi increases, but the population and activities of soil bacteria including the soil nitrifying bacteria decreases. Mycorrhizal fungi are also able to increase the host plant tolerance under the stress by producing the antioxidant enzymes (Kohler et al., 2008).

Under field conditions a 2-year field experiment was conducted using different passing (1 m/s) of a John Deere tractor (with the weight of 3986 kg and the tire width of 40 cm) at a certain soil moisture to establish the stress. The corn seeds were then planted and inoculated with different species of AM fungi. In the first year, three levels of compaction stress and in the second year, four levels of the stress were created in the field in four replicates. Different plant and soil properties to the depth of 30 cm were evaluated during corn growth and at harvest. The impedance of the soil was determined using a penetrometer three times for each plot and three times during the season. Soil density was also determined using a $100\,cm^3$ cylinder.

Each plot consisted of four rows with 60 cm spacing, planted with corn seeds (variety 704) at 20 cm spacing in plots of 4×2.4 m. The species of AM fungi, including Iranian and Canadian *G. mosseae* (received from Glomales in vitro Collection, Canada) and *G. etunicatum*, were used to inoculate the corn seeds at planting. The fungal inoculums were produced using sorghum plants, and their number was determined using the method of Most Probable Number. The field was irrigated when essential and was also fertilized (hand broadcast) when the seeds germinated with urea (260 kg/ha), potassium sulfate (160 kg/ha), and triple superphosphate (40 kg/ha), according to the soil analyses (Miransari et al., 2006).

The following parameters determined plant height, plant leaf weight, and plant nutrient uptake including N, P, K, Fe, Zn, Mn, and Cu using the standard methods (Miransari et al., 2009a,b) as well as corn grain yield at harvest. Data were analyzed using Statistical Analysis System (SAS). It was indicated that tractor passing resulted in compaction stress. Although the fungi were able to alleviate the stress at different levels of the stress, it was the most effective at the moderate level of compaction stress by increasing nutrient uptake and corn yield.

It has been indicated that under the compaction stress the fungal hyphae are able to grow into the finest root micropores, where even the host plant root hairs are not able to grow and absorb water and nutrients. The fungi are also able to produce a glycoprotein called glomalin, which can improve the soil structure (for example, under the compaction stress) and hence increase the growth of the host plant (Hammer and Rillig, 2011; Singh et al., 2013).

It has been indicated that glomalin production can be the response of mycorrhizal fungi to the environmental stresses. Hammer and Rillig (2011) showed that under salinity stress, *Glomus intraradices* is able to produce glomalin. Kohler et al. (2010) investigated the single and combined effects of mycorrhizal fungi, *G. mosseae*, and plant growth-promoting rhizobateria on the production of glomalin under different levels of salinity stress in the rhizosphere of *Lactuca sativa*. With increasing the level of salinity the aggregate stability of soil decreased (29% less than the control treatment). Although the highest level of salinity stress decreased the rate of glomalin in the soil, the inoculated treatments produced a higher rate of glomalin related to the control treatment.

Agronomical Methods

In some parts of India, high rates of rainfall, drought, low infiltration rates, and subsequent water ponding, during different growth stages, are the main causes decreasing the production of soybean yield (<1 t/ha) (Mohanty et al., 2007). Tillage practices, especially deep tillage (subsoiling using chisel plow), may alleviate the water stress in this region by increasing the rate of infiltration. Accordingly, the effects of different tillage practices, including conventional (S1), conventional and subsoiling in alternative years with chisel plow (S2), and conventional and yearly subsoiling (S3) as the main plots and three nutrient treatments including 0% NPK (N0), 100% NPK (N1), and 100% NPK + farmyard manure at 4 t/ha (N2) as the subplots were evaluated on

(1) the rate of infiltration, (2) water use efficiency, (3) root length and density mass, and (4) the yield of rain-fed soybeans.

The minimum rate of infiltration was resulted by S3 at 22%, 28%, and 20% for the 17.5, 24.5, and 31.5 cm depth, respectively, compared with S1, and the decreased values related to S2 were 17%, 19%, and 13%, respectively. The rate of bulk density following the 15 days of tillage practice significantly decreased in the subsurface (15–30 cm) in S3 equal to 1.39, followed by S2 at 1.41 and S3 at 1.58 mg/m^3. A higher root length density (RLD) and root mass density (RMD) of soybeans resulted at the 0–15 cm depth, using the yearly subsoiling. At the depth of 15–30 cm, S3 resulted in a significantly higher RLD (1.04 cm/cm^3) related to S2 (0.92 cm/cm^3) and S1 (0.65 cm/cm^3).

The basic rate of infiltration by the yearly subsoiling (5.65 cm/h) was greater than the conventional tillage (1.84 cm/h). Water storage characteristics (0–90 cm) followed a similar trend. The higher rate of infiltration and water storage resulted in greater soybean grain yield and water use efficiency for soybeans under subsoiling related to the conventional tillage. Water use efficiency under S3 (17 kg/ha cm) was significantly higher than S2 (16 kg/ha cm) and S1 (14 kg/ha cm). The higher rate of soybean grain production by subsoiling was at 20% compared with conventional tillage, although not different in S3 and S2.

Under N2, a greater rate of RLD, RMD, water use efficiency, and grain yield resulted related to N1 and N0. The rate of soybean grain in N2 (1517 kg/ha) was significantly higher than N1 (1392 kg/ha) and N0 (898 kg/ha) treatments, which had not been treated with manure. Accordingly, the authors indicated that subsoiling and treating soil with NPK nutrients (30 kg N, 26 kg P and 25 kg K) and organic matter such as manure can be an effective method for the alleviation of compaction stress on the grain yield of soybeans.

Reviewing the research on the compaction stress, in a 15-year period, Hamza and Anderson (2005) have suggested the following practical techniques for preventing, avoiding, and delaying the compaction stress: (1) decreased pressure on the soil using lighter agricultural machineries, or increasing the contact area of the wheels, (2) reduced number of machinery passes and the grazing intensity, (3) using the machinery as well as grazing at the appropriate moisture, (4) increasing the rate of organic matter in the field by using crop residue or manure, (5) using controlled traffic or limiting the use of agricultural machinery to the certain areas of the field, (6) lifting the hard pan by subsoiling, (7) providing the plant with the appropriate rate of nutrients to make the plant tolerate adverse conditions such as stress, and (8) using crop rotations with deeply rooted crop plants.

In a review by Nawaz et al. (2013), it was indicated that the level of compaction stress is determined in the soil by (1) soil moisture, texture, and structure; (2) soil bulk density, porosity, and strength; (3) under wet conditions, the modification of soil properties by the compaction stress can alter the mobility cycle of nutrients and results in the emission of greenhouse gases; (4) a high level of compaction stress alters root morphology and physiology, decreases plant height, and delays and decreases the germination rate; (5) the adverse effects of compaction stress on the soil biodiversity are by reducing the microbial population and their enzymatic activities, as well as the population of soil fauna; and (6) recent progress related to the stress may be used to model the effects of stress.

Williams and Weil (2004) indicated that using cover crops, which are deeply rooted, can alleviate the compaction stress by creating channels in the soil. Such channels are able to make the water accessed by soybean roots more easily and also the soybean roots will be able to grow in the channels. The other important effect of cover crops is by the production of residue and mulch, which are able to keep the soil moisture, affect the soil structure, and be used as a source of organic matter for soybeans and soil microbes.

Conclusion and Future Perspectives

The compaction stress is among the most important stresses affecting plant growth and yield production, worldwide, which is mostly due to the use of agricultural machinery, especially at high soil moisture. Although using agricultural machinery in the field is unavoidable in most cases, it is essential to find the optimum rate of using such equipment with respect to the plant and soil properties, agricultural practices, climate, etc. Accordingly, the optimum rate of crop production must be indicated with respect to the present conditions, while resulting in the sustainability of agriculture. The presence of a balance among different parameters, which are essential for crop production including soybeans, is essential for the optimum production of crop yield, while economically and environmentally recommendable. It may be possible to define and conduct the most optimum properties and conditions, which eventually results in the highest efficiency for yield production. Similar to the other crop plants, the soybean and its symbiotic rhizobium, *B. japonicum*, are not tolerant species under the compaction stress. Although it is not easy to find one straight method for the alleviation of compaction stress, a combination of methods, techniques, and strategies may be more effective for alleviating or delaying the stress, including (1) the use of biological methods such as using the soil microbes, including mycorrhizal fungi; (2) using different biochemicals (plant and microbial products) for the pretreatment of seeds and rhizobium before inoculation; (3) producing strong soybean species, which are able to produce a strong network of roots, especially under stress; (4) subsoiling; (5) using organic matter, including manure, plant residue, etc.; (6) using cover cropping with a strong root network; (7) using the appropriate tillage method, such as nontill or zero tillage practices; (8) controlling traffic of machinery and grazing at the appropriate soil moisture; (9) properly rotating crop plants, which have the ability to grow their roots deeply into the soil; and (10) intercropping soybeans with other crop or cover plants.

References

Ball, B.C., 2013. Soil structure and greenhouse gas emissions: a synthesis of 20 years of experimentation. European Journal of Soil Science 64, 357–373.

Bavin, T.K., Griffis, T.J., Baker, J.M., Venterea, R.T., 2009. Impact of reduced tillage and cover cropping on the greenhouse gas budget of a maize/soybean rotation ecosystem. Agriculture, Ecosystems & Environment 134, 234–242.

Beutler, A., Centurion, J., Silva, A., Centurion, M., Leonel, C., Freddi, O., 2008. Soil compaction by machine traffic and least limiting water range related to soybean yield. Pesquisa Agropecuária Brasileira 43, 1591–1600.

Bonfante, P., Genre, A., 2008. Plants and arbuscular mycorrhizal fungi: an evolutionary-developmental perspective. Trends in Plant Science 13, 492–498.

Bothe, H., 2012. Arbuscular mycorrhiza and salt tolerance of plants. Symbiosis 58, 7–16.

Botta, G., Jorajuria, D., Balbuena, R., Rosatto, H., 2004. Mechanical and cropping behavior of direct drilled soil under different traffic intensities: effect on soybean (*Glycine max* L.) yields. Soil and Tillage Research 78, 53–58.

Campos, B.C., Hungria, M., Tedesco, V., 2001. Biological nitrogen fixation efficiency by strains of *Bradyrhizobium* in soybean under no-tillage. Revista Brasileira de Ciência do Solo 25, 583–592.

Campos, B., Gnatta, V., 2006. Inoculant and foliar fertilizer in soybean under no-tillage cultivated in areas with established *Bradyrhizobium* populations. Revista Brasileira de Ciência do Solo 30, 69–76.

Carter, M., 2005. Long-term tillage effects on cool-season soybean in rotation with barley, soil properties and carbon and nitrogen storage for fine sandy loams in the humid climate of Atlantic Canada. Soil and Tillage Research 81, 109–120.

Chen, G., Weil, R., 2011. Root growth and yield of maize as affected by soil compaction and cover crops. Soil and Tillage Research 117, 17–27.

Clark, J.T., Russell, J.R., Karlen, D.L., Singleton, P.L., Busby, W.D., Peterson, B.C., 2004. Soil surface property and soybean yield response to corn stover grazing. Agronomy Journal 96, 1364–1371.

Compant, S., Van Der Heijden, M.G., Sessitsch, A., 2010. Climate change effects on beneficial plant–microorganism interactions. FEMS Microbiology Ecology 73, 197–214.

Defossez, P., Richard, G., 2002. Models of soil compaction due to traffic and their evaluation. Soil and Tillage Research 67, 41–64.

Flowers, M., Lal, R., 1998. Axle load and tillage effects on soil physical properties and soybean grain yield on a mollic ochraqualf in northwest Ohio. Soil and Tillage Research 48, 21–35.

Farmaha, B.S., Fernández, F.G., Nafziger, E.D., 2012. Soybean seed composition, aboveground growth, and nutrient accumulation with phosphorus and potassium fertilization in no-till and strip-till. Agronomy Journal 104, 1006–1015.

Frac, M., Siczek, A., Lipiec, J., 2009. The Effect of Mulching and Soil Compaction on Fungi Composition and Microbial Communities in the Rhizosphere of Soybean. EGU General Assembly Vienna, Austria, p. 2596.

Gil, S.V., Meriles, J., Conforto, C., Basanta, M., Radl, V., Hagn, A., Schloter, M., March, G.J., 2011. Response of soil microbial communities to different management practices in surface soils of a soybean agroecosystem in Argentina. European Journal of Soil Biology 47, 55–60.

Ghosh, P., Mohanty, M., Bandyopadhyay, K., Painuli, D., Misra, A., 2006. Growth, competition, yield advantage and economics in soybean/pigeonpea intercropping system in semiarid tropics of India: I. Effect of subsoiling. Field Crops Research 96, 80–89.

Gupta, S., Sharma, P., Defranchi, S., 1989. Compaction effects on soil structure. Advances in Agronomy 42, 311–338.

Hammer, E.C., Rillig, M.C., 2011. The influence of different stresses on glomalin levels in an arbuscular mycorrhizal fungus—salinity increases glomalin content. PLoS One 6, e28426.

Hamza, M.A., Anderson, W.K., 2005. Soil compaction in cropping systems a review of the nature, causes and possible solutions. Soil and Tillage Research 82, 121–145.

Hati, K., Mandal, K., Misra, A., Ghosh, P., Bandyopadhyay, K., 2006. Effect of inorganic fertilizer and farmyard manure on soil physical properties, root distribution, and water-use efficiency of soybean in Vertisols of central India. Bioresource Technology 97, 2182–2188.

Horn, R., Doma, H., Sowiska-Jurkiewicz, A., Van Ouwerkerk, C., 1995. Soil compaction processes and their effects on the structure of arable soils and the environment. Soil and Tillage Research 35, 23–36.

Ishaq, M., Hassan, A., Saeed, M., Ibrahim, M., Lal, R., 2001. Subsoil compaction effects on crops in Punjab, Pakistan: I. Soil physical properties and crop yield. Soil and Tillage Research 59, 57–65.

Izumi, Y., Yoshida, T., Iijima, M., 2009. Effects of subsoiling to the nontilled field of wheat-soybean rotation on the root system development, water uptake, and yield. Plant Production Science 12, 327–335.

Jensen, E., Peoples, M., Boddey, R., Gresshoff, P., Hauggaard-Nielsen, H., Alves, B., Morrison, M., 2012. Legumes for mitigation of climate change and the provision of feedstock for biofuels and biorefineries. A review. Agronomy for Sustainable Development 32, 329–364.

Jug, D., Blažinkov, M., Redžepović, S., Jug, I., Stipešević, B., 2005. Effects of different soil tillage systems on nodulation and yield of soybean. Poljoprivreda 11, 38–43.

Kahnt, G., Hijazi, L.A., Rao, M., 1986. Effect of homogeneous and heterogeneous soil compaction on shoot and root growth of field bean and soybean. Journal of Agronomy and Crop Science 157, 105–113.

Kihara, J., Martius, C., Bationo, A., Vlek, P., 2011. Effects of tillage and crop residue application on soybean nitrogen fixation in a tropical ferralsol. Agriculture 1, 22–37.

Kohler, J., Hernández, J.A., Caravaca, F., Roldán, A., 2008. Plant-growth-promoting rhizobacteria and arbuscular mycorrhizal fungi modify alleviation biochemical mechanisms in water-stressed plants. Functional Plant Biology 35, 141–151.

Kohler, J., Caravaca, F., Roldán, A., 2010. An AM fungus and a PGPR intensify the adverse effects of salinity on the stability of rhizosphere soil aggregates of *Lactuca sativa*. Soil Biology and Biochemistry 42, 429–434.

Long, S., 2001. Genes and signals in the *Rhizobium*-legume symbiosis. Plant Physiology 125, 69–72.

Manavalan, L., Guttikonda, S., Nguyen, V., Shannon, J., Nguyen, H., 2010. Evaluation of diverse soybean germplasm for root growth and architecture. Plant and Soil 330, 503–514.

Mathew, R., Feng, Y., Githinji, L., Ankumah, R., Balkcom, K., 2012. Impact of no-tillage and conventional tillage systems on soil microbial communities. Applied and Environmental Soil Science 2012.

Mazzoncini, M., Di Bene, C., Coli, A., Antichi, D., Petri, M., Bonari, E., 2008. Rainfed wheat and soybean productivity in a long-term tillage experiment in central Italy. Agronomy Journal 100, 1418–1429.

Mattioni, N., Schuch, L., Villela, F., Mertz, L., Peske, S., 2012. Soybean seed size and quality as a function of soil compaction. Seed Science and Technology 40, 333–343.

Micucci, F., Taboada, M., 2006. Soil physical properties and soybean (*Glycine max*, Merrill) root abundance in conventionally-and zero-tilled soils in the humid Pampas of Argentina. Soil and Tillage Research 86, 152–162.

Miransari, M., Bahrami, H.A., Rejali, F., Malakouti, M.J., 2006. Evaluating the effects of arbuscular mycorrhizae on corn (*Zea mays* L.) yield and nutrient uptake in compacted soils. Soil and Water Journal 1, 106–122 (In Persian, Abstract in English).

Miransari, M., Bahrami, H.A., Rejali, F., Malakouti, M.J., Torabi, H., 2007. Using arbuscular mycorrhiza to reduce the stressful effects of soil compaction on corn (*Zea mays* L.) growth. Soil Biology and Biochemistry 39, 2014–2026.

Miransari, M., Bahrami, H.A., Rejali, F., Malakouti, M.J., 2008. Using arbuscular mycorrhiza to alleviate the stress of soil compaction on wheat (*Triticum aestivum* L.) growth. Soil Biology and Biochemistry 40, 1197–1206.

Miransari, M., Bahrami, H.A., Rejali, F., Malakouti, M.J., 2009a. Effects of soil compaction and arbuscular mycorrhiza on corn (*Zea mays* L.) nutrient uptake. Soil and Tillage Research 103, 282–290.

Miransari, M., Bahrami, H.A., Rejali, F., Malakouti, M.J., 2009b. Effects of arbuscular mycorrhiza, soil sterilization, and soil compaction on wheat (*Triticum aestivum* L.) nutrients uptake. Soil and Tillage Research 104, 48–55.

Miransari, M., 2010. Contribution of arbuscular mycorrhizal symbiosis to plant growth under different types of soil stress. Plant Biology 12, 563–569.

Miransari, M., 2013. Corn (*Zea mays* L.) growth as affected by soil compaction and arbuscular mycorrhizal fungi. Journal of Plant Nutrition 36, 1853–1867.

Mohanty, M., Bandyopadhyay, K., Painuli, D., Ghosh, P., Misra, A., Hati, K., 2007. Water transmission characteristics of a Vertisol and water use efficiency of rainfed soybean (*Glycine max* (L.) Merr.) under subsoiling and manuring. Soil and Tillage Research 93, 420–428.

Nawaz, M.F., Bourrie, G., Trolard, F., 2013. Soil compaction impact and modelling. A review. Agronomy for Sustainable Development 33, 291–309.

Or, D., Ghezzehei, T., 2002. Modeling post-tillage soil structural dynamics: a review. Soil and Tillage Research 64, 41–59.

Place, G., Bowman, D., Burton, M., Rufty, T., 2008. Root penetration through a high bulk density soil layer: differential response of a crop and weed species. Plant and Soil 307, 179–190.

Pozo, M.J., Azcón-Aguilar, C., 2007. Unraveling mycorrhiza-induced resistance. Current Opinion in Plant Biology 10, 393–398.

Pupin, B., da Silva Freddi, O., Nahas, E., 2009. Microbial alterations of the soil influenced by induced compaction. Revista Brasileira de Ciencia do Solo 33, 1207–1213.

Ramos, J., Imhoff, S., Pilatti, M., Vegetti, A., 2010. Morphological characteristics of soybean root apexes as indicators of soil compaction. Scientia Agricola 67, 707–712.

Rillig, M.C., Treseder, K.K., Allen, M.F., 2003. Global change and mycorrhizal fungi. In: Mycorrhizal Ecology. Springer Berlin Heidelberg, pp. 135–160.

Ruser, R., Flessa, H., Russow, R., Schmidt, G., Buegger, F., Munch, J., 2006. Emission of N_2O, N_2 and CO_2 from soil fertilized with nitrate: effect of compaction, soil moisture and rewetting. Soil Biology and Biochemistry 38, 263–274.

Salvagiotti, F., Cassman, K.G., Specht, J.E., Walters, D.T., Weiss, A., Dobermann, A., 2008. Nitrogen uptake, fixation and response to fertilizer N in soybeans: a review. Field Crops Research 108, 1–13.

Singh, P.K., Singh, M., Tripathi, B.N., 2013. Glomalin: an arbuscular mycorrhizal fungal soil protein. Protoplasma 250, 663–669.

Soane, B., Van Ouwerkerk, C., 1995. Implications of soil compaction in crop production for the quality of the environment. Soil and Tillage Research 35, 5–22.

Stipesevic, B., Jug, D., Jug, I., Tolimir, M., Cvijovic, M., 2009. Winter wheat and soybean zinc uptake in different soil tillage systems. Cereal Research Communications 37, 305–310.

Talaat, N.B., Shawky, B.T., 2014. Protective effects of arbuscular mycorrhizal fungi on wheat (*Triticum aestivum* L.) plants exposed to salinity. Environmental and Experimental Botany 98, 20–31.

Tola, E., Kataoka, T., Burce, M., Okamoto, H., 2008. Soybean growth at different soil compactions for the adoption of zero tillage cultivation system. In: ISHS Acta Horticulturae 824: International Symposium on Application of Precision Agriculture for Fruits and Vegetables.

Toliver, D., Larson, J., Roberts, R., English, B., De La Torre Ugarte, D., West, T., 2012. Effects of no-till on yields as influenced by crop and environmental factors. Agronomy Journal 104, 530–541.

Vanhie, M., Deen, W., Lauzon, J.D., Hooker, D.C., 2015. Effect of increasing levels of maize (*Zea mays* L.) residue on no-till soybean (*Glycine max* Merr.) in northern production regions: a review. Soil and Tillage Research 150, 201–210.

Vetsch, J.A., Randall, G.W., Lamb, J.A., 2007. Corn and soybean production as affected by tillage systems. Agronomy Journal 99, 952–959.

West, T., Post, W., 2002. Soil organic carbon sequestration rates by tillage and crop rotation. Soil Science Society of America Journal 66, 1930–1946.

Wilhelm, W., Wortmann, C., 2004. Tillage and rotation interactions for corn and soybean grain yield as affected by precipitation and air temperature. Agronomy Journal 96, 425–432.

Williams, S., Weil, R., 2004. Crop cover root channels may alleviate soil compaction effects on soybean crop. Soil Science Society of America Journal 68, 1403–1409.

Xavier, G.R., Correia, M.E.F., de Aquino, A.M., Zilli, J.É., Rumjanek, N.G., 2010. The structural and functional biodiversity of soil: an interdisciplinary vision for conservation agriculture in Brazil. In: Soil Biology and Agriculture in the Tropics. Springer Berlin Heidelberg, pp. 65–80.

Zhalnina, K., de Quadros, P.D., Gano, K.A., Davis-Richardson, A., Fagen, J.R., Brown, C.T., et al., 2013. Ca. Nitrososphaera and *Bradyrhizobium* are inversely correlated and related to agricultural practices in long-term field experiments. Frontiers in Microbiology 4.

Soybeans, Stress, and Nutrients 12

M. Miransari
AbtinBerkeh Scientific Ltd. Company, Isfahan, Iran

Introduction

Crop plants including soybeans (*Glycine max* L.) require nutrients for their growth and yield production. Nutrients, including nitrogen (N), phosphorous (P), potassium (K), sulfur (S), calcium (Ca), magnesium (Mg), iron (Fe), zinc (Zn), manganese (Mn), and copper (Cu), handle different morphological and physiological functioning in plants (Marschner, 1995). Under nutrient-deficient conditions, plants use some mechanisms to alleviate the stress and hence increase the uptake of nutrients. For example, plant roots produce organic products, which are able to enhance the solubility and hence the availability of nutrients in the rhizosphere. Soil microbes can also importantly affect the uptake of nutrients by crop plants using different mechanisms, including the symbiotic and nonsymbiotic association with their host plants and the production of organic products (Alam, 1999; Munns, 2002; Essa, 2002).

The soybean is a legume plant, which is able to establish a symbiotic association with its specific rhizobium, *Bradyrhizobium japonicum*, and acquire most of its essential N for growth and yield production, and the remaining part is from chemical or organic fertilization. However, it must be noted that although chemical fertilization is a quick method of providing a plant with its essential N, it is not recommended, economically and environmentally, especially if used at extra amounts, because it is subjected to leaching. The major constraints for the production of soybeans are unfavorable environmental stresses such as salinity, drought, pH, nutrient deficiency, etc. (Schulze, 2004; Hungria et al., 2005; Silvente et al., 2012; Miransari, 2016).

Lassaletta et al. (2014) investigated the annual N trade of food and feed among different nations of the world (Fig. 12.1), assuming 16% rate of N in proteins from 1961 to 2010 (Table 12.1). They indicated that the trade rate of N in such a period increased eight times (from 3 to 24 TgN), which is one-third of the total soybean production worldwide. The authors also determined the transfer of N among different parts of world using 12 regions in the reference years of 1986 and 2009 (Table 12.2). There has been a significant increase in the exchange rate of N among the world regions, which is mainly due to the increased rate of population and the higher use of N proteins. However, interestingly, only a few nations, including the United States, Brazil, and Argentina are feeding the other parts of the world. With respect to their results the authors indicated the new insights related to the dependency of N food among different regions of the world and the increasing significance of food and feed trade with respect to the global N cycle.

Environmental Stresses in Soybean Production. http://dx.doi.org/10.1016/B978-0-12-801535-3.00012-7

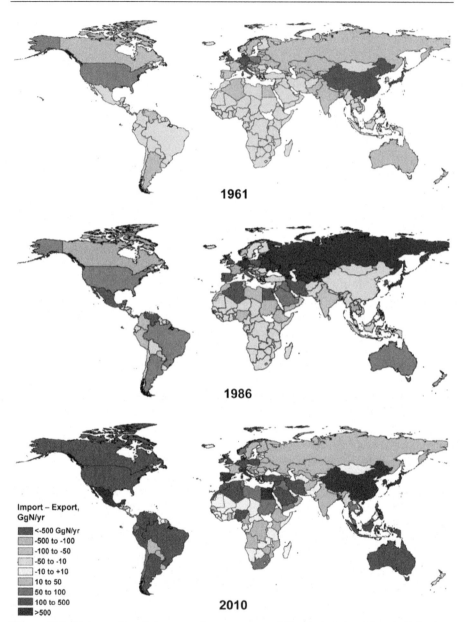

Figure 12.1 The rate of net N import and export among different nations of the world for the years 1961, 1986, and 2010 (Lassaletta et al., 2014).
With kind permission from Springer, License number 3800910438047.

Table 12.1 Net Rate of N Trade Among Different Nations From 1961 to 2010

1961			1986			2010		
Crop	GgN	%	Crop	GgN	%	Crop	GgN	%
Soybeans	249	17	Soybeans	1647	36	Soybeans	5816	39
Wheat	782	53	Wheat	1719	37	Soybean cake	4589	31
Barley	128	9	Barley	346	8	Wheat	2817	19
Maize	217	15	Maize	895	19	Maize	1630	11
Rice	105	7						
Total trade of N	1481			4607			14,852	

Modified from Lassaletta, L., Billen, G., Grizzetti, B., Garnier, J., Leach, A., Galloway, J., 2014. Food and feed trade as a driver in the global nitrogen cycle: 50-year trends. Biogeochemistry 118, 225–241.

Table 12.2 **Net N import in the 12 Regions, Evaluated in 1986 and 2009**

	N Import (GgN)	
Region	**1986**	**2009**
Africa	116	338
China	−308	2895
Maghreb and Middle East	535	1656
North America	−3433	−4850
South and West Central America	332	1330
India	−47	−188
Europe	2170	2337
Japan and South Korea	1486	2296
Former Soviet Union	555	−622
Oceania	−461	−458
South American soy nations	−1002	−5263
Southeast Asia	57	528

The positive and the negative values are the indicators of the net imports and exports, respectively (Lassaletta et al., 2014)

Stress can decrease the uptake of nutrients by crop plants, because under stress the growth of different plant parts, including the roots and the aerial part, is adversely affected and hence less nutrients and water can be absorbed by plants and as a consequence plant growth decreases. It is accordingly important to find methods, which can alleviate the adverse effects of stresses on plant growth and yield production by increasing the rate of water and nutrient uptake. The use of soil microbes such as arbuscular mycorrhizal (AM) fungi and plant growth-promoting rhizobacteria (PGPR) are among such methods (Miransari et al., 2008; Miransari, 2014; Li et al., 2015).

The use of nutrient-efficient crop plants is among the useful methods, which can increase the rate of crop yield for the increasing world population. Such an ability of nutrient-efficient crop plants is affected by the following mechanisms: (1) the uptake of nutrients and their subsequent utilization, (2) the response of plants to nutrient deficiency under stress, and (3) the methods of breeding used to enhance plant nutrient uptake and efficiency (Fageria et al., 2008).

Increasing plant uptake and utilization of nutrients can be done using the following mechanisms: (1) improving the morphology and physiology of plant roots, (2) a better distribution of nutrients in crop plants, and (3) a balance between the source and the sink (Fageria et al., 2008). Abiotic stresses including salinity, drought, cold, and nutrient deficiency can significantly decrease the rate of yield production worldwide (Atkinson and Urwin, 2012; Li et al., 2015).

Under stress the ability of crop plants to absorb and utilize nutrients is significantly affected, and due to a reduced nutrient uptake, plant growth and yield production decreases. Accordingly, for the survival of crop plants under stress, suitable morphological and physiological processes must be used by the crop plant (Alam, 1999; Pessarakli, 1999). It is possible to enhance the nutrient efficiency of crop plants under stress by developing genotypes, which are more efficient in the uptake and utilization of nutrients. This is due to the improvement in acquisition, translocation, utilization, and remobilization by plants and their enhanced tolerance under environmental stresses (Raun and Johnson, 1999; Good et al., 2004).

In this chapter, some of the most recent details, advancements, and findings related to the nutrients essential for soybean growth and yield production and how their availability can increase under different conditions including stress are presented, reviewed, and analyzed.

Plant Biotechnology and Nutrient Uptake

One important aspect related to the use of biotechnology is improved crop nutrition. The improvement in the use of molecular and genetic techniques is an important step toward the production of crop foods with the suitable rate of nutrients, including micronutrients. The improved levels of nutrients in crops, which can be fulfilled using biotechnological techniques, are essential for human health (McGloughlin, 2010).

Biofortifying the edible parts of crop plants is important for the improved nutrition of humans, especially in parts of the world, which are facing nutrient deficiency. The soybean is an important source of minerals, proteins, and nutrients. It is accordingly important to determine the genetic mechanisms, which control the seed composition of crop plants including soybeans, for the improvement of seed nutrient quality. Ramamurthy et al. (2014) investigated the genetic properties, which regulate soybean seed composition, using three different soybean genotypes produced by crossing with the Williams genotype.

The authors detected 40 quantitative trait loci (QTLs) for 18 traits, including seed weight and the traits that control the concentration of different macro- and micronutrients, the rate of N:S, and the traits controlling the concentrations of methionine and cysteine. Different genes, including the ones that regulate the translocation of nutrients and the metabolism of N and amino acids, were recognized. There were QTLs with no genes, indicating that finding more details related to such genetic properties may indicate the new genes, which affect the quality and quantity of nutrients in soybean seeds.

The results indicated that the weight of seeds and their mineral concentrations were not correlated, suggesting that it is possible to improve the seed quality of nutrients without altering seed weight. The inverse correlation between the rates of N:S and most of the nutrients showed it is possible to use such a rate for the determination of seed nutrient concentration and hence for the genetic modification of soybean plants (Ramamurthy et al., 2014).

Genetic modification for the production of tolerant and efficient crop plants under nutrient-deficient conditions is a suitable method to enhance plant growth and yield production under such conditions. The use of suitable agronomic practices can also effectively enhance plant nutrient uptake under different conditions including stress. Such agronomic practices include (1) fertilization and liming, (2) the use of soil amendments such as manure, crop residue, mulch, etc., and (3) the control of weeds and pests.

The production of more tolerant and efficient crop plants under nutrient-deficient conditions, by genetic techniques, affect plant morphology (for example, root growth and plant physiology). For example, modifying plant genes, which affect root morphology and root hair production, can affect the uptake of different nutrients, especially P, by plants (Richardson, 2001).

Among the most important effects of transgenic crop plants are their effects on the growth and functioning of soil microbes, which are able to affect the soil environment by the following mechanisms: (1) mineralization of organic matter, (2) immobilization of soil nutrients, (3) the reactions of oxidation and reduction, (4) solubilization of nutrients, and (5) biological N fixation (Motavalli et al., 2004).

The authors reviewed the effects of transgenic crop plants on the microbial activities and on the behavior of soil nutrients. The properties of soil and climate, which also can affect such activities and behavior, were also presented. The related genetic traits, which were used for the production of transgenic crop plants with improved plant nutrition, were (1) enhanced crop plant Fe tolerance in alkaline soils, (2) improved assimilation of soil N, and (3) increased uptake of soil organic and inorganic P.

The following are among the most important effects of transgenic crop plants on the properties of soil microbes: (1) changes in the activities of soil microbes resulting from the altered rate and composition of root products, (2) alteration of the microbial community due to the changes in the cultivation practices, essential for transgenic crop plants such as tillage, and use of chemicals such as pesticide and fertilization, and (3) altered functioning of soil microbes due to the transfer of genes from the transgenic crop plants (Kent and Triplett, 2002; Liu et al., 2005; Hartmann et al., 2009).

Among the most important parameters, which affect plant response to the nutrients, is the properties of plant roots and its interactions with the rhizosphere. Shen et al. (2012) presented the mechanisms, which can affect root uptake of nutrients, including (1) modifying root growth (root size and growth, root architecture, and root distribution), (2) synchronizing root growth and nutrient supply, and (3) regulating the processes of the rhizosphere. The regulation of rhizosphere processes can be done by the following: (1) using different crop genotypes, (2) using nutrients at the root environment, (3) production of organic anions and phosphatases, (4) acidification of the rhizosphere, and (5) interactions in the rhizosphere. Accordingly, if the interactions of the root and rhizosphere are properly managed using the appropriate methods, it is possible to enhance the efficiency of nutrient uptake and crop production (Bertin et al., 2003; Zaidi et al., 2003).

Soil Salinity and Plant Nutrient Uptake

Under salinity stress the excess concentration of ions such as Na^+ and Cl^- can adversely affect plant growth and yield production. Accordingly, water deficiency and the toxicity of ions can negatively affect plant growth. Salinity decreases the growth and permeability of roots and as a result the uptake of nutrients and water by plant reduces. Under salinity stress the availability of macro- and micronutrients to the plant decreases (Sairam and Tyagi, 2004).

The rate of agricultural fields, which are salty or subjected to the salinity stress, is at 10^6 ha. Legumes are among the sensitive or moderately tolerant crop plants under salinity stress (Lauchli, 1984). The tolerance of legumes under salinity stress is a function of crop genotype, rhizobium strains, the properties of soil, and climatic conditions. The mechanisms used by legumes under salinity stress to tolerate the stress include the exclusion of salt from the plant leaf and/or the localization of salt in specific tissues, for example, in the cellular vacuole. Under high salinity stress, rhizobia are not able to get dormant and hence must have the ability of tolerating the stress (Velagaleti et al., 1990; Munns, 2002; Lu et al., 2009; Miransari, 2013; Miransari et al., 2013).

Higher salt concentration in older leaves, compared with young leaves, results in the senescence of the older leaves. Salt stress also adversely affects plant growth as a result of cytoplasm malfunctioning, leakage of membranes, and loss of water and hence the related water potential. The ability of soybean plants to allocate salt to the cellular vacuole is an important parameter determining soybean tolerance under salinity stress. The role of plant hormones, including abscisic acid (ABA), is also important under stress, which results in controlling the stomata activity (Miransari, 2013; Miransari et al., 2013).

A balanced nutrient presence under salinity stress can make the crop plant tolerate the stress more efficiently. This can be achieved by using an optimum rate of chemical, organic, and biological fertilization. The use of tolerant crop plants and microbial species are also of great significance under stress, as they can handle the stress more efficiently. A balanced and suitable fertilization increases the growth of crop plants under salinity stress and makes the plant survive the stress. However, the important point is the determination of the suitable type and rate of fertilization under salinity stress (Miransari, 2016).

There is a wide range of differences in the responses of crop plants under salt stress. The successful production of crop plants under salinity stress is a function of water and nutrient uptake. Using different methods, it is possible to alleviate the stress of salinity and hence increase the growth and yield of crop plants. For example, the use of tolerant crop plants and treating the soil by leaching can make the crop plant to grow under salt stress and produce yield (Tuteja, 2007; Miransari, 2016).

Under stress the early stages of nodulation and N fixation, including the exchange of signal molecules, specifically genistein, are adversely affected and hence the rate of biologically fixed N decreases. Accordingly, Miransari and Smith (2007) hypostasized, tested, and proved that it is possible to alleviate the stress of salinity under field and greenhouse conditions using genistein. *Bradyrhizobium japonicum* inoculums were pretreated with genistein at μM concentrations and were used to inoculate soybean seeds in the field and in the greenhouse. The results indicated that genistein was able to significantly alleviate the stress and increase soybean nodulation, growth, and yield (Miransari and Smith, 2007).

Although the presence of all nutrients is essential for plant growth and yield production, with respect to the role of potassium in plants, it is the most important nutrient affecting plant growth and yield production under drought and salinity stress. Potassium is able to control plant behavior under water and salt stress by affecting the activities of the stomata, different plant enzymes, and decreasing the rate of sodium to potassium (Pettigrew, 2008; Munns and Tester, 2008; Daei et al., 2009).

Soil Acidity and Plant Nutrient Uptake

Soil acidity is another problem adversely affecting plant growth and yield production. The most important limiting factors under the stress of acidity are the high concentrations of Al, Fe, Mn, and H with toxic effects, which are not favorable to plant growth. Among the chemical methods, liming is the most practical one, alleviating the stress of acidity on plant growth. The other important factor negatively affecting plant growth and yield production under acidic conditions is the deficiency or unavailability of different macro- and micronutrients, including N, P, K, Ca, Mg, Fe, and Zn. It is also possible to use acidity-tolerant crop plants and soil microbes, which are able to alleviate the stress of acidity on plant growth (Caires et al., 2008).

For example, under acidic conditions the following details significantly decrease plant growth and yield production: (1) lower levels of soil fertility; (2) the presence of elements at unfavorable concentrations (Fe, Mn, H, and Fe); (3) deficiency and unavailability of some nutrients including different macronutrients, Fe and Zn; (4) unfavorable properties of soil due to the presence of soil compaction, hard pan, leaching, and volatilization of nutrients; (5) less activities of soil microbes; (6) less efficiency of chemical fertilization (N < 50%, P < 20%, K < 40–70% and micronutrients <5–10%); (7) using an unsuitable rate of soil amendments such as lime; (8) abiotic stresses; (9) using nontolerant genotypes without a suitable nutrient use efficiency; and (10) biotic stresses (Raun and Johnson, 1999; Fageria and Baligar, 2005; Fageria and Moreira, 2011).

A practical method for the alleviation of acidity stress is liming, which increases the efficiency of nutrient uptake by crop plants. Liming can improve the properties of the soil and hence increases the uptake of nutrients by the following processes: (1) decreasing the toxic effects of elements such as H, Al, and Mn, (2) increasing the rate of cation exchange capacity (CEC), (3) enhancing the biological activities of soil microbes, (4) increasing the availability of P, Ca, and Mg, (5) improving the structure of the soil, (6) stimulating the process of nitrification, (7) enhancing biological N fixation, and (8) decreasing the availability of K and micronutrients (Foy, 1984; Glasener and Palm, 1995; Caires et al., 2008).

A set of different mechanisms are used by tolerant crop plants under acidity stress including (1) the selective permeability or polymerization of the plasma membrane, (2) the production of the pH barrier (mucilage or chelating ligands) at the interface of root and soil, (3) the internal chelation by organic acids (such as carboxylic and citric), (4) the metal binding by proteins and enzymes, and (5) Al and Mn compartmentation in the vacuole (Passioura, 2002; Garvin and Carver, 2003).

Plants also use the following mechanisms to tolerate the acidity stress and acquire their essential nutrients: (1) enhancing the activities of soil microbes by their root products, (2) the increased microbial activities result in: (a) the higher rate of organic matter mineralization, (b) the release of nutrients for plant use, (c) the higher rate of symbiotic association with the soil microbes, (d) higher nutrients availability, and (3) increasing the pH of the rhizosphere to alleviate the toxic effects of H, Al, and Mn (Lambers et al., 2008a, 2008b; Horst et al., 2010).

Because different plant activities and biochemical processes are controlled by nutrients, under nutrient-deficient conditions, crop plants are more susceptible to biotic stresses. Accordingly, it is essential to provide plants with the optimum level of nutrients so that they can produce the high rate of yield and be resistant under biotic and abiotic stresses. The deficiency and unavailability of micronutrients make crop plants more susceptible under biotic stresses, because, for example, Cu, B, and Mn are used by plant cells for the production of lignin and phenol, and Si increases plant tolerance to pathogenic invasion (Orcutt and Nilsen, 2000; Atkinson and Urwin, 2012).

Soybeans, Macronutrients, and Stress

Stress significantly decreases the uptake of macronutrients by crop plants, including soybeans. The adverse effects of stress are on plant morphology and physiology and on the chemical properties of soil nutrients. Under stress, plant growth and hence the uptake of nutrients decrease. Alleviation of stress can increase plant growth and yield production by enhancing the uptake of nutrients and water by plants. The soybean and its symbiotic *B. japonicum* are not tolerant under stress, and their growth and activities decrease. It is mainly due to the decreased uptake of nutrients and soybean growth under stress. Using the tolerant species of soybean and *B. japonicum* strains under stress can increase the efficiency of both symbionts and hence their growth and activities (Miransari, 2010, 2013; Miransari et al., 2013).

Using the cultivation method of intercropping, Li et al. (2001) investigated the nutrient uptake of maize, wheat, and soybeans using two different field experiments. In the first experiment, two levels of P fertilization (0 and 53 kg P/ha), three methods of cropping (just maize, just wheat, and wheat/maize intercropping), and two plant densities for maize and wheat were used. In the second place a similar experiment was conducted; however, P was used at 33 kg/ha without the treatment of plant density. Intercropping significantly increased the yield and nutrient uptake of maize, wheat, and soybeans compared with just wheat, maize, or soybeans, except for K uptake by maize. Intercropping wheat with maize resulted in a 40–70% wheat yield increase and with soybeans resulted in a 28–30% increase.

In a similar manner, Zhang and Li (2003) reviewed the effects of intercropping with the sole cultivation of different crop plants. They found that the intercropping of wheat with soybeans and maize significantly increased the yield of crop plants, compared with just using one of them. The intercropping cultivation resulted in an improved uptake of different nutrients including N, P, and Fe.

Salvagiotti et al. (2008) investigated the effects of N chemical fertilization on the process of biological N fixation by analyzing field data sets (combinations of site treatment year) of 637 from 1966 to 2006. Data indicated that each one kg increase in the accumulation of N in the aerial plant part resulted in a 0.013 Mg increase in soybean seed yield. The least and the highest rate of N uptake for the slope of the regression equations were equal to 0.0064 and 0.0188 Mg grain kg^{-1} of N, respectively.

According to the analysis, an average of 50–60% of soybean N requirement was supplied by the process of biological N fixation and in most cases the use of N chemical fertilization to provide a plant with its essential N was a must. With increasing the rate of soybean yield, the difference between the rate of crop N uptake and the N from the process of biological N fixation increased. It was more likely that the response of soybean yield to N fertilization increased with increasing the rate of crop yield (>4.5 Mg/ha).

The use of chemical N fertilization resulted in a negative exponential association with N_2 fixation when the fertilization was used on the soil surface or was incorporated into the soil surface. The authors accordingly suggested that the deep use of N fertilization with a little rate of release or the late use of N fertilization during the reproductive growth stage may be suitable methods, which result in higher soybean response to N fertilization under the conditions of high soybean yielding.

The authors indicated that more research is essential to determine how the contribution of biological N fixation to the production of soybean seed yield may increase using the new methods of inoculum production, using more precise and efficient cultivation practices, and most importantly, using the complete and precise measurements of biological N fixation and the efficiency uptake of chemical N fertilization by plant and soil N during the growing season. With respect to such details, it is possible to present a more applicable recommendation for the use of biological N fixation and chemical N fertilization under the conditions of high-yielding soybeans (Salvagiotti et al., 2008).

Among the important parameters affecting the availability of soil nutrients is the age of the soil. In young soil the availability of N is limited, decreasing plant productivity and growth; however, in old soil the concentration of P significantly decreases, which is due to the process of leaching and the loss of the surface soil. Two important traits affecting N and P uptake by plants are the mycorrhizal symbiosis and the cluster growth of roots. Mycorrhizal fungi are able to increase the uptake of P significantly under different conditions including the P limited conditions. The cluster growth of roots is prevalent in conditions such as compaction stress, which is mainly related to the properties of soil (Lynch and Brown, 2008; Wang et al., 2010a).

Soybeans, Nitrogen, and Stress

Nitrogen is among the most important macronutrients significantly affecting plant growth and yield production. Nitrogen is essential for different plant components and functioning, including chlorophyll, different enzymes, proteins, cellular components (especially cell walls), and the other plant components. N is especially essential for the vegetative growth of crop plants. The efficiency of chemical N fertilization is less than 50%; however, it is a fast method of providing crop plants with their essential N. Different processes result in the loss of N from the soil including leaching, denitrification, volatilization, and immobilization. The role of plant roots in the uptake of N is important for plant growth and yield production (Fageria and Baligar, 2005). N deficiency affects root growth by affecting the number of lateral roots. N fertilization increases root length, root length density, rooting depth, and root surface area (Miller and Cramer, 2005).

Although the optimum rate of N is essential for crop growth, higher rates of N can adversely affect plant growth and yield production by increasing the period of crop vegetative growth, resulting in crop lodging by decreasing the period of reproductive growth. It is hence important to provide crop plants with a balanced rate of N fertilization, which is of economic and environmental significance (Miransari and Mackenzie, 2015).

Schipanski et al. (2010) investigated the effects of different parameters, including the level of soil N on the process of N fixation by soybeans. Using 13 different fields, the authors investigated how the rate of N fixation is affected by soil organic matter and soil N. Different soil N fractions were determined, including the labile and the recalcitrant N sources. Accordingly, soybean dependency on the process of biological N fixation ranged from 36% to 82%, and the rate of total N fixed in the soybean biomass ranged from 40 to 224 kg N/ha. There was a negative and significant correlation between the soil N fractions and the rate of N fixed by the process of biological N fixation.

Javaid and Mahmood (2010) investigated the growth, nodulation, and yield of soybeans as affected by *B. japonicum* inoculation and a commercial microbial fertilizer, under field conditions using *Trifolium alexandrinum* green manure or farmyard manure at 20 tons/ha each. With the green manure treatment the use of *B. japonicum* significantly increased the number and the weight of nodules as well as the number and the weight of pods and the weight of the aerial part by 65%, 55%, and 17%, respectively. Using the treatment of farmyard manure, *B. japonicum* significantly increased the weight of nodules and the aerial part as well as the number of pods by 45% and 47%, respectively. Although there were some antagonistic effects between *B. japonicum* and the commercial microbial fertilization, in the treatment of farmyard manure, the soybean plants, coinoculated with *B. japonicum* and the commercial microbial fertilization, resulted in the highest weight of the aerial part and the number and the weight of pods, compared with the other treatments. The authors accordingly indicated that the combined use of *B. japonicum* and the commercial microbial fertilization with farmyard manure can significantly increase the nodulation, growth, and yield of soybean plants under field conditions.

Cook and Trlica (2016) investigated the effects of tillage and fertilization on crop yield and properties of soil in a 45-year period. A set of 21 continuous corn and 24 corn–soybean products were tested during the experimental period on a soil without suitable drainage. The treatments of four tillage practices, including nontillage (NT), alternate tillage (AT), chisel tillage (ChT), and moldboard plow (MP), and a set of five fertilization treatments, including a control, just N, NPK broadcast, N + NPK starter, and NPK + NPK starter were used during the experiment.

The use of NPK fertilization resulted in a similar rate of yield production under tilled and NT practices with an average yield of 8.73 t/ha for continuous corn and 11.93 t/ha and 3.70 t/ha for the rotation of corn and soybeans. If the rate of soil N, P, and pH were similar but soil N was less, the essential rate for optimum crop production, NT resulted in a lower rate of crop yield than tilled practices. The use of NT increased soil organic matter to 27.6 g/kg; however, the average of the corresponding values for the other treatments was equal to 20.5 g/kg, with the highest rate of increase in the top 5 cm of soil.

Although NT resulted in the stratification of P and K, it did not increase crop yield. The results indicated that the use of NPK fertilization resulted in the production of yield under NT similar to tilled practices even under the unsuitable drainage conditions. The authors accordingly indicated that the farmers can produce a high rate of yield under NT if the suitable rate of NPK is used. Such a practice is also of economic and environmental significance as it decreases the rate of nutrient losses and improves the properties of soil.

Stress decreases soybean growth and yield production by negatively affecting soybeans and *B. japonicum* growth and activity and hence the process of biological N fixation. The decreased rate of N fixation under stress reduces soybean growth and yield production. However, it has been indicated that it is possible to alleviate the adverse effects of different stresses such as suboptimal root zone temperature, acidity, and salinity on the process of biological N fixation using the signal molecule genistein (Miransari and Smith, 2007, 2008, 2009). The authors hypostasized that stress adversely affects the initial stages of biological N fixation, including the exchange of the signal molecules including genistein. They accordingly proved under greenhouse and field conditions that if the inoculums of *B. japonicum* are preincubated with genistein, it is possible to alleviate the stress on soybean growth and yield production.

Under stresses such as salinity, the concentration of N protein compounds including amino acids and proteins increases in plant. However, it is possible to alleviate the adverse effects of stress on N behavior in plants by maintaining the important molecules in plant osmoregulation, altering the storage form of N in plants and alleviating the oxidative stress by scavenging the produced free radicals in plant and cellular detoxification (Mansour, 2000).

Different components of crop plant seeds have been modified to increase their use by humans. For example, in soybeans: (1) the balance of amino acids have been modified and (2) the rate of lysine and tryptophan, oleic acid, flavonoids, and the availability of P have increased by increasing the rate of phytase (Kinney and Knowlton, 1998; Hirschi, 2009).

Soybeans, Phosphorous, and Stress

Phosphorous is the other important nutrient that is essential for soybean growth and yield production; soybeans can absorb P using chemical, organic, and biological fertilization. Similar to N, chemical fertilization can also provide a plant with its essential P, but it is not economically and environmentally recommendable, as P products are not highly soluble and accumulate in the environment. Using organic matter, which is a good source of nutrients for the use of plants and microbes, is a suitable method for providing crop plants with their essential nutrients. However, high amounts of organic matter must be used to provide plants with their essential nutrients (Chiera et al., 2002; Six et al., 2004).

The use of soil microbes is also a good method of providing plants with their essential P, as the soil microbes, specifically PGPR (including *Bacillus*), are able to increase the solubility of P by the production of different enzymes such as phosphatase. Research has indicated the positive effects of P solubilizing bacteria on the growth of different crop plants, including soybeans, under different conditions, including stress, by enhancing the solubility and hence the availability of soil P (Pal, 1998; Vassilev et al., 2006; Afzal and Bano, 2008; Li et al., 2008; Paul and Lade, 2014; Miransari, 2014).

Similar to N, phosphorous (P) is also essential for plant growth and yield production, and with respect to its chemical properties, it is deficient in the soil for the use of crop plants most of the time. The P compounds in the soil are not highly soluble and are subjected to desolubilization. It is the most important reason for the increased rate of P chemical fertilization in the soil. P is essential for plant morphology and physiology, and its deficiency can significantly decrease plant growth, especially root growth and yield. Although P produces adenosine triphosphate (ATP) in plants, affecting different plant activities during different growth stages, the most important role of P in plants is at the reproductive stage, because P is an important component of sugar, DNA, RNA, phospholipids, and proteins (Blevins, 1994).

The soybean is produced under different climatic conditions, including tropical, subtropical, and temperate areas, which are continuously subjected to P deficiency. Accordingly, P deficiency is a major constraint to the production of soybeans worldwide. The other important point is the decreased resources of P and the increased expenses related to the production of chemical P fertilization. Crop plants, which are efficient in the utilization of nutrients, can produce a higher rate of yield using less nutrients, compared with inefficient crop plants (Fageria et al., 2008).

Accordingly, the development and use of P efficient soybean genotypes is a suitable method for the increased production of soybeans under P limited conditions. Such details can be useful for the researchers to find the mechanisms that affect P efficiency in soybeans and the subsequent genetic strategies, which can improve such abilities, especially the ones that affect root morphology and physiology (Wang et al., 2010a). The authors have been able to produce P efficient genotypes, which are able to grow more effectively under P deficient conditions (acidic conditions), with the P concentration of less than 20 mg/kg, producing a yield comparable to the yield of standard genotypes, which grow under P sufficient conditions.

The following mechanisms can be used by P efficient genotypes under P limited conditions: (1) changes in root architecture and morphology, (2) the activation of P transporters, (3) alteration of symbiosis, (4) higher activity of internal phosphatase, and (5) higher production of phosphatases and organic acids by soybean roots into the rhizosphere (Gahoonia and Nielsen, 2004; Wang et al., 2010a).

Accordingly, finding and altering the traits, which can affect soybeans under P deficient conditions is an important step in the production of P efficient genotypes. With respect to the above mentioned details the following traits are among the most important ones affecting soybean response under P deficient conditions: (1) root architecture and morphology, (2) exudates of roots, and (3) the process of symbiosis (Wang et al., 2004, 2009; Liu et al., 2008).

The deficiency of P is among the most important limiting factors affecting the growth and production of soybeans. Development of P efficient genotypes, which can effectively use soil P and P chemical fertilization, is among the useful methods for the production of soybeans under P deficient conditions. Root biology is also an important field, which can be used for the improved use of P by soybeans. This is an effort, which can be done by researchers from different plant- and soil-related areas. Such kinds of investigations also result in the enhanced uptake of P by plants and increases the possibility of producing soybeans under P deficient conditions, resulting in a more sustainable production of soybeans (Ma et al., 2001; Liang et al., 2010; Wang et al., 2011).

Ramesh et al. (2011) investigated the effects of different *Bacillus* isolates on the enzymatic activities of the rhizosphere and soybean P nutrition. The results indicated the significant effects of *Bacillus* inoculation on the rhizosphere enzymatic activities, including the activity of fluorescein diacetate, phytase, and phosphatase and hence the P nutrition of soybeans compared with the uninoculated control. The most efficient isolates were *Bacillus amyloliquefacians* and *Bacillus cereus*.

Inoculation with *Bacillus* decreased the rate of phytic acid-P, as a rate of total P in the seeds of soybeans, related to the control treatment, resulting in a better digestibility of soybean seeds and hence their increased feed efficiency. *Bacillus* inoculation resulted in a significant and positive correlation between phytase and alkaline, acid phosphatases activity on the available P content of soil and soybean seeds. The authors accordingly indicated how such enzymes can affect the acquisition and mobilization of P in soybeans (Ramesh et al., 2011).

The availability of P is usually high for the use of soybeans because it reacts with Ca, Fe, and Al, resulting in the production of insoluble phosphate products. However, with respect to the increasing price of P fertilizer, using P fertilization by farmers must be done with more care. With respect to the limitation of P, increasing soybean yield in the parts of the world that are not of high yield production is an important field of research and practice.

Using the biological methods, specifically using the soil microbes, which are able to enhance the availability of P products in the soil (P solubilizing bacteria), is among the most useful strategies and techniques for the sustainable production of crop plants. Accordingly, Afzal et al. (2010) investigated the effect of *Bradyrhizobium* and P solubilizing bacteria (*Pseudomonas*) on the yield of soybeans in the presence of P chemical fertilization using pot and field experiments in a 2-year experiment. Soybean seeds were treated with the inoculums of N fixing *Bradyrhizobium* and P solubilizing bacteria.

Their results indicated that the combined use of N fixing bacteria and P solubilizing bacteria in the presence of chemical P fertilization increased the rate of soybean yield in the pots and in the field by 38% and 12%, respectively, compared with the single use of P chemical fertilization. *Bradyrhizobium japonicum* resulted in the production of IAA (74.64 µg/mL) and gibberellic acid (261.2 µg/mL). Similarly, the *Pseudomonas* strain produced indole acetic acid (IAA) and gibberellic acid at 8.034 µg/mL and 1766 µg/ mL, respectively. The use of coinoculation and P chemical fertilization resulted in a 46% and 33% higher survival by *B. japonicum* and *Pseudomonas* compared with the single use of bacteria.

With respect to the deficiency of P worldwide, as previously mentioned, the development of P efficient crop plants is of great significance. Accordingly, such a definition is for crop plants, which are able to produce yield under the certain rate of P in the soil. The efficiency of crop plants can be defined by the efficiency of P acquisition and the efficiency of P utilization. The efficiency of P acquisition is the crop's ability of absorbing P from the soil, and the efficiency of P utilization is the crop's ability of converting the absorbed P to plant biomass (Hinsinger, 2001; Wang et al., 2010b). Accordingly, it is possible to enhance P efficiency in crop plants by improving the efficiency of P uptake and P utilization. Such factors are determined by the properties of crop species and the environment. When the availability of P is not high in the soil, P absorption efficiency is important, affecting P uptake by the plant; however, when the availability of P is high in the soil, P utilization efficiency by the plant is the important process affecting P behavior in the plant. Hence it is important that the crop plant be modified for both processes under different environmental conditions. However, because for a higher P uptake, more P chemical fertilization must be used, which is of economic and environmental concern, modifying crop plants for P utilization efficiency is more practical and applicable (Wang et al., 2010b).

The advancements in the field of molecular biology have resulted in the recognition of more genes and QTLs, which can affect plant P uptake and utilization, although most research has been related to *Arabidopsis thaliana*. The presence of P transporters of high affinity in a crop results in the higher uptake of P under P limited conditions. The P transporters of less affinity make the absorbed P translocate to different parts of the plant under the high concentration of P. Although there is not much known about the modification of P uptake efficiency and P utilization efficiency, the presence of the new biotechnological techniques can make the researchers modify the crop plants, which are of high efficiency in the uptake and utilization of P, and hence less P chemical fertilization will be used. However, more knowledge is essential for the recognition of mechanisms, affecting P uptake and utilization by crop plants. The combined efforts of plant physiologists and plant biotechnologists are also essential for the production of crop plants with higher P efficiency. Also the important point is the rate of carbon use by crop plants when the P efficiency increases using biotechnological techniques (Wang et al., 2010b).

Although the soybean is able to establish a symbiosis with its specific rhizobium, *B. japonicum*, it is also able to establish a symbiosis with the soil mycorrhizal fungi and acquire water and nutrients. Such symbiotic association with both rhizobium and AM fungi is a tripartite association and can significantly enhance soybean growth and yield production. In a tripartite symbiotic association between soybeans, rhizobium, and mycorrhizal fungi, the bacteria provide the soybean host plant with most of its essential N, and the fungi are also able to supply the host plant with other nutrients, especially phosphorous. The interactions between the bacteria and the fungi can also importantly affect their growth and activities and the growth and yield of soybeans.

Wang et al. (2011) investigated the tripartite association between soybeans, *B. japonicum*, and AM fungi as affected by root architecture and the availability of N and P. Using two different soybean genotypes with contrasting root architecture a field experiment was conducted to investigate how such an association (root nodulation and mycorrhizal colonization) can be affected by such a tripartite association. The authors also conducted a pot experiment to find the effects of the tripartite association on the uptake of N and P.

The tripartite association was significantly affected by the root architecture and at the lower concentration of P the genotype with the deep root resulted in a greater mycorrhizal colonization, compared with the surface root. However, at the higher concentration of P, a greater nodulation resulted from the deep root soybean. Depending on the availability of N and P, the bacteria and the fungi synergistically and positively affected soybean growth.

The tripartite association significantly improved soybean growth and N and P uptake under the lower concentration of N and or P; however, under the suitable concentrations of N and P, soybean growth was not affected by the tripartite association. Root architecture importantly affected the response of soybeans to the coinoculation with rhizobium and mycorrhizal fungi, as the genotype with the deep root was more responsive than the genotype with the surface root. The authors accordingly indicated that such findings can be used for the proper inoculation of soybeans with *B. japonicum* and mycorrhizal fungi, especially under N and P deficient conditions.

Soybeans, Potassium, and Stress

Potassium is also an important plant nutrient affecting different plant growth and functioning, including root growth, photosynthesis, water uptake, the behavior of assimilates, and enzymatic activities. However, under K deficient conditions the number and the size of plant leaves decrease, and hence the rate of photosynthesis and plant leaf area as well as the production of crop yields decreases (Blevins, 1994). Different methods may be used to enhance the K efficiency of plants, including the use of higher K, more efficient methods of using K, and producing transgenic plants, which are able to use K more efficiently. There are soil microbes, which are able to increase the availability of K in the soil by the production of different enzymes. However, K can also be supplied to the plant using chemical and organic fertilization (Zhang et al., 2011).

Under stress the production of reactive oxygen species can adversely affect cellular activities and plant growth. However, it has been indicated that potassium is able to alleviate the stress by reducing the production of reactive oxygen species, which is due to the decreased activity of NAD(P)H oxidases, resulting in the maintenance of the photosynthesis process by transporting electrons. Under potassium-deficient conditions the fixation rate of CO_2 and the rate of photosynthate utilization by plants decreases significantly (Cakmak, 2005).

Zheng et al. (2008) investigated if it is possible to alleviate salinity stress on the growth of wheat plants under greenhouse conditions using potassium nitrate (KNO_3). Plant growth, the rate of K/Na, and the activity of antioxidant enzymes were evaluated during the experiment. Using two different wheat genotypes (salt sensitive and salt tolerant) under greenhouse conditions, the seedlings were grown using Hoagland nutrient solution with 6 mM KNO_3 and using no salt. The seedlings were then subjected to the treatments of salinity and KNO_3. When the plants were almost 22 cm high, 30 d after germination, they were harvested and different morphological and physiological parameters were determined.

Although under stress the growth of seedlings decreased, the combined use of KNO_3 and NaCl alleviated the adverse effects of stress on the growth and physiology of wheat genotypes by increasing plant growth including root growth, increasing the activities of antioxidant enzymes, decreasing electrolyte leakage, the contents of soluble sugar, and malondialdehyde. A higher rate of potassium was accumulated in the wheat-tolerant genotype, compared with the higher rate of Na accumulated in the salt-sensitive genotype. The rate of antioxidant enzymes and sugar content was more stable in the salt-tolerant genotype. The authors accordingly indicated that KNO_3 can more effectively alleviate the stress of salinity in the wheat-tolerant genotype compared with the sensitive genotype.

Under salinity stress the deficiency of potassium and the toxicity of Na, decreases plant growth and yield production (Maathuis and Amtmann, 1999). It is accordingly important for plants to absorb a suitable rate of K^+ under salinity stress, which results in the increased concentration of Na^+ in plants. The important trait for tolerant crop plants under salinity stress is their higher affinity for K^+ compared with sodium, which should be used for the production of tolerant genotypes (Zhang et al., 2010).

Soybeans, Sulfur, and Stress

Sulfur is also essential for plant growth and yield production. The important functioning of S in plants include (1) its presence in the structure of the two amino acids, methionine and cysteine, which are essential for the production of proteins, (2) activation of different enzymes, (3) production of nodules in legumes, (4) production of chlorophyll, (5) production of the N fixation enzyme, (6) production of oil in oilseeds, (7) enhancing plant tolerance under drought stress, (8) controlling pathogens, (9) production of vitamins and hormones, (10) for the oxidation-reduction reactions, (11) alleviation of heavy metals stress in plants, and (12) alleviation of salinity stress (Tabatabai, 1984; Hell, 1997; Scherer, 2001; Fageria, 2008).

Sulfur is not essential at high amounts for soybean growth and just 20 kg/ha can be enough for soybean growth and yield production. Such amounts of S may be supplied using chemical fertilization or by using the soil bacteria, *Thiobacillus* (with sulfur substrates), which is able to oxidize the soil S and hence make it available for the use of crop plants (Miransari and Smith, 2007; Zhao et al., 2008). Using a pot experiment, Zhao et al. (2008) investigated the effects of sulfur on soybean root and leaf growth and on the soil microbial activities in 2004 and 2005. Three different sulfur treatments were used including the control, 30 and 60 mg/kg.

The results indicated that sulfur significantly increased the number and the weight of side roots, plant chlorophyll content, and soybean yield. The results also indicated the number of soil microbes, including bacteria, fungi, and actinomycetes, and the activity of different enzymes, including urease, catalase, polyphenoloxidase, and phosphatase. The authors accordingly indicated that it is possible to increase soybean yield by using elemental sulfur, which also enhances the activities of soil microbes.

Soybean, Calcium, Magnesium, and Stress

Calcium and magnesium are usually available at essential rates in the soils, especially under arid and semiarid climatic conditions; however, in humid areas and due to the high rate of leaching, crop plants may face Ca and Mg deficiency. Under such conditions the use of products such as calcite is recommended (Burton et al., 2000; White and Broadley, 2005, 2009).

Although plants are less subjected to calcium deficiency, even under acidic conditions, under highly leached conditions, crop plants may face Ca deficiency. However, the important role of Ca in the soil is to detoxify the negative effects of elements such as aluminum, manganese, and heavy metals, which are not favorable to plant growth, especially at high concentrations. Calcium is able to improve root growth and enhances the efficiency of water utilization by plants. Calcium at 12 Mg lime per hectare improved the root growth of soybeans compared with the control treatment (Fageria and Moreira, 2011). Magnesium is also essential for plant growth and yield production. The most important role of Mg in plants is its presence in the structure of chlorophyll. Mg is also present in the other plant physiological processes.

Soybeans, Micronutrients, and Stress

Similar to macronutrients, micronutrients are also essential for plant growth and yield production, however, at much lower amounts. Different functions in plants are done by micronutrients such as the oxidation–reduction processes and the activity of enzymes. The deficiency of micronutrients is prevalent because of (1) a high demand for micronutrients, especially under high-yielding agricultural fields, (2) production of crop plants on soils with a low rate of micronutrients, (3) less use of organic fertilization, including manure, crop residues, and compost, (4) using chemical fertilization with a low rate of macronutrients, (5) different climatic factors, which may decrease the availability of macronutrients in the soil, and (6) the use of lime under acidic conditions (Fageria et al., 2002).

Due to the high rate of yield production, micronutrient deficiency is prevalent worldwide. The other important aspect is the deficiency of micronutrients in crop plants, which is usually less than the essential rate for human health. Such issues may be corrected using the following methods: (1) the use of soil and foliar fertilization, (2) the use of organic amendments, which increase the uptake and hence the content of micronutrients in the edible parts of plants, and (3) the proper use of agronomic practices (Khoshgoftarmanesh et al., 2010). Accordingly, the authors have reviewed how it is possible to enhance the rate of micronutrients in crop plants using the suitable agronomic methods, and how such contents may be affected by genetic practices in different crop plant genotypes.

The important point about the use of grains is their nutrient concentrations. If the rate of nutrients is not at the essential level for human use, it must be increased using the appropriate methods. Among such methods, biofortification is of great significance, which is providing the staple foods with the suitable rate of micronutrients. Plants are able to provide humans with almost all the essential nutrients and vitamins, with the exception of vitamins D and B_{12}. However, most crop foods including wheat, cassava, rice, and maize do not have the essential level of micronutrients for human use. Most of the time, such nutrients are found at uneven rates in different parts of the plant. Biofortification is a combination of genetic, agronomic, and breeding practices, which increase the levels of micronutrients in the edible parts of crop plants.

Fertilization is among the most prevalent methods of biofortification; however, factors such as the mobility of minerals in plants, the properties of soil, the method of fertilization, and the plant tissue of high mineral concentration can affect the efficiency of fertilization and hence biofortification (Zhu et al., 2007). Accordingly, such a method can be used for particular places. For example, because of the mobility of Zn in the soil, the use of $ZnSO_4$ can significantly increase crop yield and nutrient contents in crop plants (White and Broadley, 2005).

However, the mobility of Fe in the soil is not high and is subjected to binding to soil and changing into Fe(III), and as a result, using biofortification for Fe has not been beneficial (Grusak and DellaPenna, 1999). High concentrations of micronutrients can also adversely affect plant growth and microbial activities. The other disadvantage with fertilization is that they are not economically and environmentally recommended. Accordingly, such a method can be used for just some specific crop plants.

With respect to the presence of a high rate of genetic differences in crop plant genotypes, significantly affecting the availability of nutrients, traditional breeding has also been used to modify the content of nutrients in plants. When there are not genetic differences among different genotypes, the use of transgenic methods may be really useful. A high number of crop plants have been genetically modified for the rate of micro- and macronutrients to be beneficial for human use. Although the use of transgenic techniques is beneficial to plant biologists, due to some difficulties the commercial use of such methods is not easy (Powell, 2007). Two different methods have been used for the improvement of crop nutrient contents: (1) increased efficiency and concentration of nutrients in the edible parts and (2) enhancing bioavailability of nutrients in crop plants. The two most important micronutrients, which are genetically modified in crop plants, are Fe and Zn (Palmgren et al., 2008).

Soybeans, Iron, Zinc, Manganese, Copper, and Stress

The other nutrients, specifically micronutrients including Fe, Zn, Mn, and Cu, can be absorbed by plants using chemical, organic, and biological fertilization. With respect to the chemical properties of micronutrients, they are not highly soluble in the soil, especially in the arid and semiarid areas, which is due to the presence of Ca and Mg in the soil, resulting in the precipitation of micronutrients (Welch and Graham, 2004).

However, in humid areas and due to the higher solubility of micronutrients in the soil, crop plants may not face the deficiency of micronutrients, and sometimes the high concentration of micronutrients in the soil can be toxic to plants and microbes. Soil microbes can also enhance the availability of micronutrients by producing different products such as siderophores, which can chelate micronutrients and hence make them more absorbable by crop plants and the other soil microbes (Chen et al., 2011; Paul, 2014).

Iron is one of the most important micronutrients essential for a high number of activities in crop plants; however, due to its chemical properties, it is not highly available and hence it is widely deficient in different parts of the world (Welch et al., 1991; Marschner, 1995). The soybean is among the crop plants that is continuously subjected to Fe deficiency, adversely affecting plant growth and yield production. However, at high concentrations, such as high levels of moisture or acidic conditions, Fe can have negative effects on plant growth.

Zinc as an important micronutrient has different functions in plants, and its deficiency decreases plant growth and yield production in different parts of the world. The importance of micronutrients has also been indicated for human health (Welch, 2008). Manganese deficiency is prevalent under the high rate of moisture (not favorable drainage).

Iron deficiency is the most important nutrient deficiency in the world, affecting about three billion people. The mechanism used by grasses to absorb iron is different from the other plants. The abundant iron in the soil is Fe(III), which is absorbed by plants and is converted to Fe(II). The activity of different chelating proteins increases the rate of Fe(II) in different parts of the plant. Usually with increasing the rate of Fe in genetically modified plants the rate of Zn also increases, indicating that there is a cross talk between the related pathway. This shows that under such conditions the production of nicotianamine in plants increases, resulting in the enhanced mobility of Fe and Zn. The modification of nicotianamine in plants doubles the level of Fe and Zn (Ishimaru et al., 2006).

The other method used for the biofortification of nutrients in crop plants is their increased bioavailability by modifying the related proteins. For example, the expression of ferritin, which is an iron storage protein, can increase the rate of seed iron by three or four times. It is also possible to increase the availability of nutrients in crop plants by catabolizing the antinutrients. For example, phytate is an antinutrient, which can chelate nutrients and decrease their bioavailability. A useful method is the combined use of expressed Fe storage proteins and the use of phytase (which analyses phytate) produced by a fungi. This has been done in maize and rice, resulting in the maximum bioavailability of iron (Drakakaki et al., 2005). The deficiency of copper (Cu) is prevalent in different parts of the world, such as soils with a high rate of lime and organic matter, which can absorb Cu and decrease its availability in the soil. At high levels, Cu can adversely affect root growth more than the growth of the aerial part (Marschner, 1995).

Conclusion and Future Perspectives

The important effects of macro- and micronutrients, including N, P, K, S, Ca, Mg, Fe, Zn, Mn, and Cu on the growth and yield production of different crop plants, including soybeans, under different conditions, including stress, have been presented and analyzed. Nitrogen, as the most important macronutrient, can significantly affect plant growth and yield production. However, at extra amounts and due to the effects of N on crop vegetative growth, it decreases crop growth and yield production due to lodging and hence the decreased time of the reproductive stage. The soybean acquires most of its essential N by the process of biological N fixation with its symbiotic *B. japonicum*, but a part of N must also be supplied by using N chemical fertilization. Under stress the growth and activity of both soybeans and *B. japonicum* is adversely affected, as the plant and the bacteria are not tolerant under stress. Different methods have been used with success to alleviate the adverse effects of stress on soybean nitrogen fixation, growth, and yield, including the use of tolerant plant genotypes and rhizobium strains and the use of the signal molecule, genistein, which is produced by soybean roots in the initial stages of biological N fixation. Under stress the production of genistein decreases by soybean roots; however it has been indicated that if the *B. japonicum* inoculums are pretreated with genistein, it is possible to alleviate the adverse effects of stress on the process of biological N fixation and hence soybean growth and yield production. The other nutrients, including P, K, Ca, Mg, Fe, Zn, Mn, and Cu, are also essential for soybean growth and yield production, and their deficiency can significantly decrease soybean N fixation, growth, and yield production under different conditions including stress. The related details have been presented in this chapter.

References

Afzal, A., Bano, A., 2008. Rhizobium and phosphate solubilizing bacteria improve the yield and phosphorus uptake in wheat (*Triticum aestivum*). International Journal of Agriculture Biology 10, 85–88.

Afzal, A., Bano, A., Fatima, M., 2010. Higher soybean yield by inoculation with N-fixing and P-solubilizing bacteria. Agronomy for Sustainable Development 30, 487–495.

Alam, S., 1999. Nutrient Uptake by Plants under Stress Conditions. In: Pessarakli, M. (Ed.), Handbook of Plant and Crops Stress. Marcel Dekker, New York, pp. 285–313.

Atkinson, N., Urwin, P., 2012. The interaction of plant biotic and abiotic stresses: from genes to the field. Journal of Experimental Botany 63, 3523–3543.

Bertin, C., Yang, X., Weston, L., 2003. The role of root exudates and allelochemicals in the rhizosphere. Plant and Soil 256, 67–83.

Blevins, D., 1994. Uptake, Translocation, and Function of Essential Mineral Elements in Crop Plants. In: Peterson, G. (Ed.), Physiology and Determination of Crop Yield. ASA, CSSA, and SSSA, Madison, WI, pp. 259–275.

Burton, M., Lauer, M., McDonald, M., 2000. Calcium effects on soybean seed production, elemental concentration, and seed quality. Crop Science 40, 476–482.

Caires, E., Garbuio, F., Churka, S., Barth, G., Correa, J., 2008. Effects of soil acidity amelioration by surface liming on no-till corn, soybean, and wheat root growth and yield. European Journal of Agronomy 28, 57–64.

Cakmak, I., 2005. The role of potassium in alleviating detrimental effects of abiotic stresses in plants. Journal of Plant Nutrition and Soil Science 168, 521–530.

Chen, Y., Liu, S., Li, H., Li, X.F., Song, C.Y., Cruse, R.M., Zhang, X., 2011. Effects of conservation tillage on corn and soybean yield in the humid continental climate region of Northeast China. Soil and Tillage Research 115, 56–61.

Chiera, J., Thomas, J., Rufty, T., 2002. Leaf initiation and development in soybean under phosphorus stress. Journal of Experimental Botany 53, 473–481.

Cook, R., Trlica, A., 2016. Tillage and fertilizer effects on crop yield and soil properties over 45 years in Southern Illinois. Agronomy Journal 108, 415–426.

Daei, G., Ardekani, M.R., Rejali, F., Teimuri, S., Miransari, M., 2009. Alleviation of salinity stress on wheat yield, yield components, and nutrient uptake using arbuscular mycorrhizal fungi under field conditions. Journal of Plant Physiology 166, 617–625.

Drakakaki, G., Marcel, S., Glahn, R., Lund, E., Pariagh, S., et al., 2005. Endosperm-specific coexpression of recombinant soybean ferritin and *Aspergillus phytase* in maize results in significant increases in the levels of bioavailable iron. Plant Molecular Biology 59, 869–880.

Essa, T., 2002. Effect of salinity stress on growth and nutrient composition of three soybean (*Glycine max* L. Merrill) cultivars. Journal of Agronomy and Crop Science 188, 86–93.

Fageria, N., Baligar, V., Clark, R., 2002. Micronutrients in crop production. Advances in Agronomy 77, 185–268.

Fageria, N., Baligar, V., 2005. Enhancing nitrogen use efficiency in crop plants. Advances in Agronomy 88, 97–185.

Fageria, N. (Ed.), 2008. The Use of Nutrients in Crop Plants. CRC Press. ISBN: 9781420075106.

Fageria, N., Baligar, V., Li, Y., 2008. The role of nutrient efficient plants in improving crop yields in the twenty first century. Journal of Plant Nutrition 31, 1121–1157.

Fageria, N., Moreira, A., 2011. The role of mineral nutrition on root growth of crop plants. Advances in Agronomy 110, 251–331.

Foy, C., 1984. Physiological effects of hydrogen, aluminum and manganese toxicities in acid soils. In: Adams, F. (Ed.), Soil Acidity and Liming, second ed. SSSA, ASA and CSSA, Madison Wisconsin, pp. 57–97.

Gahoonia, T., Nielsen, N., 2004. Root traits as tools for creating phosphorus efficient crop varieties. Plant and Soil 260, 47–57.

Garvin, D., Carver, B., 2003. Role of the genotype in tolerance to acidity and aluminum toxicity. In: Rengel, Z. (Ed.), Handbook of Soil Acidity. Marcel Dekker, New York, pp. 387–406.

Glasener, K., Palm, C., 1995. Ammonia volatilization from tropical legume mulches and green manures on unlimited and limed soils. Plant and Soil 177, 33–41.

Good, A., Shrawat, A., Muench, D., 2004. Can less yield more? Is reducing nutrient input into the environment compatible with maintaining crop production? Trends in Plant Science 9, 597–605.

Grusak, M., DellaPenna, D., 1999. Improving the nutrient composition of plants to enhance human nutrition and health. Annual Reviews of Plant Physiology Plant Molecular Biology 50, 133–161.

Hartmann, A., Schmid, M., Van Tuinen, D., Berg, G., 2009. Plant-driven selection of microbes. Plant and Soil 321, 235–257.

Hell, R., 1997. Molecular physiology of plant sulfur metabolism. Planta 202, 138–148.

Hinsinger, P., 2001. Bioavailability of soil inorganic P in the rhizosphere as affected by root-induced chemical changes: a review. Plant and Soil 237, 173–195.

Hirschi, K., 2009. Nutrient biofortification of food crops. Annual Review of Nutrition 29, 401–421.

Horst, W., Wang, Y., Eticha, D., 2010. The role of the root apoplast in aluminium-induced inhibition of root elongation and in aluminium resistance of plants: a review. Annals of Botany 106, 185–197.

Hungria, M., Franchini, J., Campo, R., Graham, P., 2005. The importance of nitrogen fixation to soybean cropping in South America. In: Nitrogen Fixation in Agriculture, Forestry, Ecology, and the Environment. Springer Netherlands, pp. 25–42.

Ishimaru, Y., Suzuki, M., Tsukamoto, T., Suzuki, K., Nakazono, M., et al., 2006. Rice plants take up iron as an Fe^{3+}-phytosiderophore and as Fe^{2+}. Plant Journal 45, 335–346.

Javaid, A., Mahmood, N., 2010. Growth, nodulation and yield response of soybean to biofertilizers and organic manures. Pakistan Journal of Botany 42, 863–871.

Kent, A., Triplett, E., 2002. Microbial communities and their interactions in soil and rhizosphere ecosystems. Annual Reviews in Microbiology 56, 211–236.

Khoshgoftarmanesh, A., Schulin, R., Chaney, R., Daneshbakhsh, B., Afyuni, M., 2010. Micronutrient-efficient genotypes for crop yield and nutritional quality in sustainable agriculture. A review. Agronomy for Sustainable Development 30, 83–107.

Kinney, A.J., Knowlton, S., 1998. Designer oils: the high oleic acid soybean. In: Roller, S., Harlanders, S. (Eds.), Genetic Modification in the Food Industry: A Strategy for Food Quality Improvement. Blackie Acad. Prof. Publ./Thomson Sci., London, pp. 193–213.

Lambers, H., Chapin III, F., Pons, T., 2008a. Mineral nutrition. In: Lambers, H., Chapin III, F., Pons, T. (Eds.), Plant Physiological Ecology. Springer New York, pp. 255–320.

Lambers, H., Raven, J., Shaver, G., Smith, S., 2008b. Plant nutrient-acquisition strategies change with soil age. Trends in Ecology & Evolution 23, 95–103.

Lassaletta, L., Billen, G., Grizzetti, B., Garnier, J., Leach, A., Galloway, J., 2014. Food and feed trade as a driver in the global nitrogen cycle: 50-year trends. Biogeochemistry 118, 225–241.

Lauchli, A., 1984. Salt exclusion: an adaptation of legumes for crops and pastures under saline conditions. In: Staples, R., Toeniessen, G. (Eds.), Salinity Tolerance in Plants. Strategies for Crop Improvement. John Wiley and Sons, New York, pp. 171–187.

Li, L., Sun, J., Zhang, F., Li, X., Yang, S., Rengel, Z., 2001. Wheat/maize or wheat/soybean strip intercropping: I. Yield advantage and interspecific interactions on nutrients. Field Crops Research 71, 123–137.

Li, J., Wang, E., Chen, W., Chen, W., 2008. Genetic diversity and potential for promotion of plant growth detected in nodule endophytic bacteria of soybean grown in Heilongjiang province of China. Soil Biology and Biochemistry 40, 238–246.

Li, X., Zeng, R., Liao, H., 2015. Improving crop nutrient efficiency through root architecture modifications. Journal of Integrative Plant Biology 58, 193–202.

Liang, Q., Cheng, X., Mei, M., Yan, X., Liao, H., 2010. QTL analysis of root traits as related to phosphorus efficiency in soybean. Annals of Botany 106, 223–234.

Liu, B., Zeng, Q., Yan, F., Xu, H., Xu, C., 2005. Effects of transgenic plants on soil microorganisms. Plant and Soil 271, 1–13.

Liu, L., Liao, H., Wang, X., Yan, X., 2008. Regulation effect of soil P availability on mycorrhizal infection in relation to root architecture and P efficiency of *Glycine max*. Chinese Journal of Applied Ecology 19, 564–568.

Lu, K., Cao, B., Feng, X., He, Y., Jiang, D., 2009. Photosynthetic response of salt-tolerant and sensitive soybean varieties. Photosynthetica 47, 381–387.

Lynch, J., Brown, K., 2008. Root strategies for phosphorus acquisition. In: White, P., Hammond, J. (Eds.), The Ecophysiology of Plant-Phosphorus Interactions. Springer Netherlands. ISBN: 978-1-4020-8434-8, pp. 83–116.

Ma, Z., Bielenberg, D., Brown, K., Lynch, J., 2001. Regulation of root hair density by phosphorus availability in *Arabidopsis thaliana*. Plant, Cell & Environment 24, 459–467.

Mansour, M., 2000. Nitrogen containing compounds and adaptation of plants to salinity stress. Biologia Plantarum 43, 491–500.

Marschner, H., 1995. Mineral Nutrition of Higher Plants, second ed. Academic Press, New York.

Maathuis, F., Amtmann, A., 1999. K^+ nutrition and Na^+ toxicity: the basis of cellular K^+/Na^+ ratios. Annals of Botany 84, 123–133.

McGloughlin, M., 2010. Modifying agricultural crops for improved nutrition. New Biotechnology 27, 494–504.

Miller, A., Cramer, M., 2005. Root nitrogen acquisition and assimilation. In: Miller, A., Cramer, M. (Eds.), Root Physiology: From Gene to Function. Springer Netherlands. ISBN: 978-1-4020-4098-6, pp. 1–36.

Miransari, M., Smith, D., 2007. Overcoming the stressful effects of salinity and acidity on soybean nodulation and yields using signal molecule genistein under field conditions. Journal of Plant Nutrition 30, 1967–1992.

Miransari, M., Smith, D., 2008. Using signal molecule genistein to alleviate the stress of suboptimal root zone temperature on soybean-*Bradyrhizobium* symbiosis under different soil textures. Journal of Plant Interactions 3, 287–295.

Miransari, M., Smith, D., 2009. Alleviating salt stress on soybean (*Glycine max* (L.) Merr.)–*Bradyrhizobium japonicum* symbiosis, using signal molecule genistein. European Journal of Soil Biology 45, 146–152.

Miransari, M., Bahrami, H., Rejali, F., Malakouti, M., 2008. Using arbuscular mycorrhiza to alleviate the stress of soil compaction on wheat (*Triticum aestivum* L.) growth. Soil Biology and Biochemistry 40, 1197–1206.

Miransari, M., 2010. Contribution of arbuscular mycorrhizal symbiosis to plant growth under different types of soil stress. Plant Biology 12, 563–569.

Miransari, M., 2013. Soil microbes and the availability of soil nutrients. Acta Physiologiae Plantarum 35, 3075–3084.

Miransari, M., Riahi, H., Eftekhar, F., Minaie, A., Smith, D., 2013. Improving soybean (*Glycine max* L.) N_2 fixation under stress. Journal of Plant Growth Regulation 32, 909–921.

Miransari, M., 2014. Plant growth promoting rhizobacteria. Journal of Plant Nutrition 37, 2227–2235.

Miransari, M., Mackenzie, A., 2015. Development of soil N testing for wheat production using soil residual Mineral N. Journal of Plant Nutrition 38, 1995–2005.

Miransari, M. (Ed.), 2016. Abiotic and Biotic Stresses in Soybean Production. Academic Press. ISBN: 978-0-12-801730-2.

Motavalli, P., Kremer, R., Fang, M., Means, N., 2004. Impact of genetically modified crops and their management on soil microbially mediated plant nutrient transformations. Journal of Environmental Quality 33, 816–824.

Munns, R., 2002. Comparative physiology of salt and water stress. Plant, Cell & Environment 25, 239–250.

Munns, R., Tester, M., 2008. Mechanisms of salinity tolerance. Annual Review of Plant Biology 59, 651–681.

Orcutt, D., Nilsen, E., 2000. In: The Physiology of Plants under Stress: Soil and Biotic Factors, vol. 2. John Wiley & Sons. ISBN: 978-0-471-17008-2.

Pal, S., 1998. Interactions of an acid tolerant strain of phosphate solubilizing bacteria with a few acid tolerant crops. Plant and Soil 198, 169–177.

Palmgren, M., Clemens, S., Williams, L., Kramer, U., Borg, S., et al., 2008. Zinc biofortification of cereals: problems and solutions. Trends in Plant Science 13, 464–473.

Passioura, J., 2002. Review: environmental biology and crop improvement. Functional Plant Biology 29, 537–546.

Paul, E., 2014. Soil Microbiology, Ecology and Biochemistry. Academic press. ISBN: 9780124159556.

Paul, D., Lade, H., 2014. Plant-growth-promoting rhizobacteria to improve crop growth in saline soils: a review. Agronomy for Sustainable Development 34, 737–752.

Pessarakli, M. (Ed.), 1999. Handbook of Plant and Crops Stress. Marcel Dekker, New York.

Pettigrew, W., 2008. Potassium influences on yield and quality production for maize, wheat, soybean and cotton. Physiologia Plantarum 133, 670–681.

Powell, K., 2007. Functional foods from biotech—an unappetizing prospect? Nature Biotechnology 25, 525–531.

Ramesh, A., Sharma, S., Joshi, O., Khan, I., 2011. Phytase, phosphatase activity and P-nutrition of soybean as influenced by inoculation of *Bacillus*. Indian Journal of Microbiology 51, 94–99.

Ramamurthy, R., Jedlicka, J., Graef, G., Waters, B., 2014. Identification of new QTLs for seed mineral, cysteine, and methionine concentrations in soybean [*Glycine max* (L.) Merr.]. Molecular Breeding 34, 431–445.

Raun, W., Johnson, G., 1999. Improving nitrogen use efficiency for cereal production. Agronomy Journal 91, 357–363.

Richardson, A., 2001. Prospects for using soil microorganisms to improve the acquisition of phosphorus by plants. Australian Journal of Plant Physiology 28, 897–906.

Salvagiotti, F., Cassman, K., Specht, J., Walters, D., Weiss, A., Dobermann, A., 2008. Nitrogen uptake, fixation and response to fertilizer N in soybeans: a review. Field Crops Research 108, 1–13.

Sairam, R., Tyagi, A., 2004. Physiology and molecular biology of salinity stress tolerance in plants. Current Science 86, 407–421.

Scherer, H., 2001. Sulphur in crop production—invited paper. European Journal of Agronomy 14, 81–111.

Schipanski, M., Drinkwater, L., Russelle, M., 2010. Understanding the variability in soybean nitrogen fixation across agroecosystems. Plant and Soil 329, 379–397.

Schulze, J., 2004. How are nitrogen fixation rates regulated in legumes? Journal of Plant Nutrition and Soil Science 167, 125–137.

Shen, J., Li, C., Mi, G., Li, L., Yuan, L., Jiang, R., Zhang, F., 2012. Maximizing root/rhizosphere efficiency to improve crop productivity and nutrient use efficiency in intensive agriculture of China. Journal of Experimental Botany 64, 1181–1192.

Silvente, S., Sobolev, A., Lara, M., 2012. Metabolite adjustments in drought tolerant and sensitive soybean genotypes in response to water stress. PLoS One 7, e38554.

Six, J., Bossuyt, H., Degryze, S., Denef, K., 2004. A history of research on the link between (micro) aggregates, soil biota, and soil organic matter dynamics. Soil and Tillage Research 79, 7–31.

Tabatabai, M., 1984. Importance of sulphur in crop production. Biogeochemistry 1, 45–62.

Tuteja, N., 2007. Mechanisms of high salinity tolerance in plants. Methods in Enzymology 428, 419–438 (Chapter 24).

Vassilev, N., Vassileva, M., Nikolaeva, I., 2006. Simultaneous P-solubilizing and biocontrol activity of microorganisms: potentials and future trends. Applied Microbiology and Biotechnology 71, 137–144.

Velagaleti, R., Marsh, S., Kramer, D., Fleischman, D., Corbin, J., 1990. Genotypic differences and nitrogen fixation among soybean (*Glycine max* L.) cultivars grown under salt-stress. Tropical Agriculture 67, 169–177.

Wang, L., Liao, H., Yan, X., Zhuang, B., Dong, Y., 2004. Genetic variability for root hair traits as related to phosphorus status in soybean. Plant and Soil 261, 77–84.

Wang, X., Wang, Y., Tian, J., Lim, B., Yan, X., Liao, H., 2009. Overexpressing AtPAP15 enhances phosphorus efficiency in soybean. Plant Physiology 151, 233–240.

Wang, X., Yan, X., Liao, H., 2010a. Genetic improvement for phosphorus efficiency in soybean: a radical approach. Annals of Botany 106, 215–222.

Wang, X., Shen, J., Liao, H., 2010b. Acquisition or utilization, which is more critical for enhancing phosphorus efficiency in modern crops? Plant Science 179, 302–306.

Wang, X., Pan, Q., Chen, F., Yan, X., Liao, H., 2011. Effects of co-inoculation with arbuscular mycorrhizal fungi and rhizobia on soybean growth as related to root architecture and availability of N and P. Mycorrhiza 21, 173–181.

Welch, R., Allaway, W., House, W., Kubota, J., 1991. Geographic distribution of trace element problems. In: Mortvedt, J., Cox, F., Shuman, L., Welch, R. (Eds.), Micronutrients in Agriculture, second ed. Soil Science Society of America, Madison,WI, pp. 31–57.

Welch, R., Graham, R., 2004. Breeding for micronutrients in staple food crops from a human nutrition perspective. Journal of Experimental Botany 55, 353–364.

Welch, R., 2008. Linkages between trace elements in food crops and human health. In: Alloway, B. (Ed.), Micronutrient Deficiencies in Global Crop Production. Springer, New York, pp. 287–317.

White, J., Broadley, M., 2005. Biofortifying crops with essential mineral elements. Trends in Plant Science 10, 586–593.

White, P., Broadley, M., 2009. Biofortification of crops with seven mineral elements often lacking in human diets–iron, zinc, copper, calcium, magnesium, selenium and iodine. New Phytologist 182, 49–84.

Zaidi, A., Khan, M., Amil, M., 2003. Interactive effect of rhizotrophic microorganisms on yield and nutrient uptake of chickpea (*Cicer arietinum* L.). European Journal of Agronomy 19, 15–21.

Zhang, F., Li, L., 2003. Using competitive and facilitative interactions in intercropping systems enhances crop productivity and nutrient-use efficiency. Plant and Soil 248, 305–312.

Zhang, J., Flowers, T., Wang, S., 2010. Mechanisms of sodium uptake by roots of higher plants. Plant and Soil 326, 45–60.

Zheng, Y., Jia, A., Ning, T., Xu, J., Li, Z., Jiang, G., 2008. Potassium nitrate application alleviates sodium chloride stress in winter wheat cultivars differing in salt tolerance. Journal of Plant Physiology 165, 1455–1465.

Zhang, Y., Wang, E., Li, M., Li, Q.Q., Zhang, Y., et al., 2011. Effects of rhizobial inoculation, cropping systems and growth stages on endophytic bacterial community of soybean roots. Plant and Soil 347, 147–161.

Zhao, Y., Xiao, X., Bi, D., Hu, F., 2008. Effects of sulfur fertilization on soybean root and leaf traits, and soil microbial activity. Journal of Plant Nutrition 31, 473–483.

Zhu, C., Naqvi, S., Gomez-Galera, S., Pelacho, A., Capell, T., Christou, P., 2007. Transgenic strategies for the nutritional enhancement of plants. Trends in Plant Science 12, 548–555.

Index

Printed in the United States
By Bookmasters